Technology and the Wealth of Nations

Technology and the Wealth of Nations

Edited by
Nathan Rosenberg,
Ralph Landau,
and David C. Mowery

Stanford University Press
Stanford, California

Stanford University Press
Stanford, California

Copyright © 1992 by the Board of Trustees of the
Leland Stanford Junior University

Printed in the United States of America

CIP data appear at the end of the book

Preface

This volume incorporates the outcome of several years' work in the Technology and Economic Growth Program of Stanford's Center for Economic Policy Research. The papers included here originated as materials presented at a conference held at Stanford University in September 1989. Subsequently, the papers were substantially modified and, where appropriate, brought up to date by the authors. In addition, the editors have prepared an introduction that weaves together the main themes of the separate papers and suggests some appropriate conclusions.

In brief, the volume emphasizes that it is the factors affecting the commercialization of technology, rather than the development of the technology itself, that constitute the missing link in discussions of science and technology policy, as well as in macroeconomic policy discussions directed toward issues of international competitiveness. In order for technological forces to play a more effective role in shaping economic performance, it is necessary to weave the micro- and macroeconomic factors together. We are indebted to the Technology and Economic Growth Program, to the Pine Tree Trust in New York, and to the Sloan Foundation, for the funding that supported this research.

N.R.
R.L.
D.C.M.

Contents

Part III International Contrasts in Technology Commercialization

Part IV The International Environment and the Commercialization Process

Part V Managing the Commercialization of Technology

Contributors

SERGIO BARABASCHI—Deputy General Manager, Ansaldo S.p.A. and president of Ansaldo Research.

B. DOUGLAS BERNHEIM—John L. Weinberg Professor of Economics and Business Policy, Princeton University.

MICHAEL BOSKIN—Chairman, President's Council of Economic Advisers; Professor of Economics, on leave, Stanford University.

STEVEN DURLAUF—Associate Professor, Department of Economics, Stanford University.

RALPH GOMORY—President of the Alfred P. Sloan Foundation, member of the President's Council of Advisers on Science and Technology, and former Senior Vice President of IBM.

BRUCE GUILE—Director, Program Office, U.S. National Academy of Engineering.

WILLIAM G. HOWARD, JR.—Senior Fellow, National Academy of Engineering. Former vice-president for R&D, Motorola.

PETER HUBER—Senior Fellow, Manhattan Institute; lawyer and engineer.

KEN-ICHI IMAI— Director of Research, Stanford Japan Center, Kyoto.

RALPH LANDAU—Consulting Professor of Economics and of Chemical Engineering, Stanford University, and Research Fellow at The Kennedy School, Harvard University; co-director of programs on technology, economy, and policy at both universities.

LAWRENCE LAU—Professor of Economics, Stanford University.

FRANCO MALERBA—Professor, Universitá Commerciale 'Luigi Bocconi', Istituto di Economia Politica, Milan.

RONALD McKINNON—Eberle Professor of Economics, Department of Economics, Stanford University.

DAVID MOWERY—Associate Professor, William A. Haas School of Business, University of California–Berkeley.

RICHARD NELSON—Henry R. Luce Professor of International Political Economy, School of International and Public Affairs, Columbia University.

DAVID ROBINSON—Economist, International Monetary Fund.

NATHAN ROSENBERG—Fairleigh S. Dickinson, Jr. Professor of Public Policy, Department of Economics, Stanford University; Director, Program on Technology and Economic Growth, Stanford Center for Economic Policy Research.

KIYONORI SAKAKIBARA—Professor in the Department of Commerce and the Institute of Business Research, Hitotsubashi University, Tokyo.

JOHN SHOVEN—Charles R. Schwab Professor of Economics, Stanford University; Director, Center for Economic Policy Research,

DAVID TEECE—Mitsubishi Bank Professor, William A. Haas School of Business, University of California–Berkeley.

D. ELEANOR WESTNEY—Associate Professor of Management, Corporate Strategy and International Management Group, Sloan School of Management, M.I.T..

Technology and the Wealth of Nations

1

Introduction

NATHAN ROSENBERG, RALPH LANDAU, AND DAVID C. MOWERY

Productivity growth is indispensable to growth in national income and wealth. Research carried out over the last 30 years demonstrates that technological change is an important contributor to productivity growth, and therefore to growth in the income and wealth of nations. The last decade of the twentieth century presents great challenges and opportunities to policy makers and managers concerned with improving technological and economic performance. The Cold War that consumed much of the wealth of nations has come to an end. It contributed to inequities among and within nations, and reduced the rate of improvement of the standard of living for many. The postwar period also saw a great increase in global population and growing pressure on environmental and resource constraints. The time is ripe for a deeper study of ways to increase the wealth of nations, one that tries to overcome the long-standing analytic separation of the microeconomic and macroeconomic levels of study and policy. Technological change is an essential ingredient in the processes of growth examined in this volume. Investment in technological improvement and in the complementary assets and activities needed to support innovation is a positive sum strategy for improving living standards.

The papers in this volume were first presented at a conference on "The Commercialization of New Technologies" organized by the Technology and Economic Growth Program of Stanford University's Center for Economic Policy Research. The conference was organized in recognition of the fact that U.S. economic performance in the global economy of the 1990s will rest more than ever on the ability of U.S. managers, workers, and firms to reap the commercial returns to new technologies, both as rapid developers of these technologies and as effective adopters of advanced technologies. More rapid productivity growth, on which a resumption of growth in U.S. earnings and household incomes depends, requires that new technologies be applied more effectively within this economy.

The importance of effective management and performance in the commercialization of new technologies reflects the changing environment within which the United States finds itself in the final decade of this century. The postwar decades of undisputed U.S. scientific and technological hegemony have been replaced by a situation in which U.S. firms in particular are challenged by foreign competitors in some fields, and struggling to regain their former positions in other fields. Although the U.S. scientific research establishment arguably has lost little if any of its postwar pre-eminence, the same cannot be said with respect to the performance of U.S. firms as developers and adopters of new technologies. Technological and scientific knowledge move across national boundaries more rapidly, and non-U.S. firms have improved their abilities to develop, commercialize, and adopt new technologies. As a result, the "lead time" that U.S. firms have enjoyed during much of the postwar period in their innovative efforts has shrunk to the vanishing point. No longer can U.S. firms rest comfortably on a technological "gap" between foreign firms and themselves—the windows of opportunity for acquiring competitive advantage through introducing and commercializing new technologies have shrunk considerably.

Dealing with this new environment requires that economists in particular improve their understanding of the underlying dynamics of technology commercialization. No longer can theorists prescribe to policy makers on the assumptions that all firms are in possession of, or have equal access to, all technologies, that technological knowledge is costlessly transferred and absorbed, or that the macroeconomic environment of inflation, risk, and capital costs is unimportant. Moreover, the challenges posed to U.S. policy and practice by the experience of foreign firms and their public policies must be given greater weight in formulating new approaches to improving economic performance.

Economic policy and performance alike will be linked more and more closely to technology-related issues in the future, and the factors affecting commercialization are now recognized as critical to this linkage. The papers in this volume highlight the role in technology commercialization of macroeconomic influences, the legal environment, and domestic and international interfirm "networks," to name a few influences. Moreover, as the papers make clear, the analytic frameworks employed by both macroeconomic and microeconomic theory must be revised in important ways.

The papers in this volume are initial contributions to three important tasks. First, this volume draws on new developments in theoretical and empirical analysis to begin the task of integrating the macroeconomic and microeconomic dimensions of technological innovation and commercialization. Second, the papers extend and enrich the macroeconomic analysis of growth, capital formation and international economic interactions to highlight the influences on technology commercialization of macroeconomic variables. Indeed, one unifying theme in this volume is the importance of the macroeconomic environment in the technological performance of this nation's firms. Third, the papers extend and enrich the microeconomic analysis of technological innovation and commercialization, and provide guidance for managers seeking to improve performance in both of these areas. The papers cover five broad topic areas:

1. the conceptual framework for the analysis of innovation and commercialization;
2. the influence on technology commercialization of factors beyond those within research and development;
3. the structure and performance of foreign economies in technology commercialization;
4. the influence of the international environment and the process of "internationalization" on U.S. performance in technology development and commercialization; and
5. the implications of these new analytic perspectives for the management of technology commercialization.

Although many of the papers devote some attention to public policy issues and recommendations, the authors were asked to focus primarily on the implications of their analysis for the commercialization process. These papers raise challenging issues for economic theorists, empirical analysts, and policy makers, but they are contributions to a broadening debate, rather than definitive resolutions of the issues. Among other things, this research should contribute to a scholarly dialogue between students of macroeconomic growth theory and scholars focusing on the microeconomic dynamics of technological change. As these papers show, new developments in theory and empirical analysis have created exciting opportunities for this dialogue to enrich our understanding of the technology commercialization process in ways that may inform policy as well.

1 Conceptual Overview

The paper by Boskin and Lau is an important contribution to a new analysis of economic growth that goes beyond the classic treatment in Solow 1956.[1] Solow's model held that the rate of growth of a perfectly competitive economy is independent of the rate of savings or investment, that is, growth is unaffected by the share of total output that is reinvested. Empirical studies by Solow and others, including Abramovitz and Denison, of long-run growth in the U.S. economy corroborated this model in concluding that a key influence on growth in per capita output was not captured by conventional measures of labor or capital. Boskin and Lau's analysis of postwar growth in five industrial economies, by contrast, concludes that technological change and capital formation are strongly complementary. The "capital-augmenting" character of technological change means that nearly 70% of growth in total U.S. output during 1964–85 is due to the combined effects of technological change and capital formation. The Boskin-Lau analysis suggests that the economic payoff to technological innovation is enhanced by high rates of capital formation; it is not independent of growth in capital inputs.

Nelson's analysis of the nature of technological knowledge also extends the conventional economic view of technology. As Nelson points out, the conceptualization of technology in much economic theorizing as a "book of blueprints," readily accessible to all comers and entirely codified, ignores considerable evidence to the contrary. Indeed, much of the extensive public and private institutional infrastructure for the creation and utilization of technology in advanced industrial economies is based largely on the fact that the transfer and utilization of technological knowledge is costly and complex. Nonetheless, Nelson argues, different types of scientific and technological knowledge vary in their ease of codification and transferability, as well as in their inherently proprietary or public character. As a result, different institutional structures will produce different blends of public and proprietary technological and scientific knowledge, and these bodies of knowledge will vary in the ease with which other individuals or firms can exploit them.

Durlauf's paper makes important contributions to the analysis of international differences in growth behavior, but also provides some valuable links between the microeconomic effects of economic institutions

---

[1]Solow, R. M. 1956. "A Contribution to the Theory of Economic Growth, *Quarterly Journal of Economics*, vol. 70, no. 1.

and policy and the long-run growth path of the overall economy. Although the concept of "spillovers" among industries and technologies has long been an important part of microeconomic analysis of technological change, Durlauf notes several important potential implications of such interaction effects at the macroeconomic level. High investment in one industry can induce high investment in other industries, supporting higher growth rates and additional investment. Multiple equilibria, in the form of alternative growth paths, influenced by different domestic regimes of institutions and policy, become possible. Government policies toward investment and production affect both the short-run business cycle and long-run growth.

Institutions and policy thus affect the long-run performance of the economy, and do so in part through their effects on investment behavior. Durlauf's hypothesis that multiple equilibria are possible receives some support from the demonstration in his paper that the rates of growth in total output in six industrial economies have not converged during the 1950–85 period.

The paper by Landau and Rosenberg on successful technology commercialization in the international chemicals industry complements the analyses of Durlauf, Nelson, and Boskin and Lau in several important respects. Landau and Rosenberg show the important role of institutional and resource constraints in shaping the evolution of entire industrial sectors within different industrial economies along contrasting trajectories. The historical analysis of the chemicals industry in Landau and Rosenberg, in its suggestion of a strong path-dependent element to the growth of this sector, provides important microeconomic corroboration of Durlauf's analysis.

Landau and Rosenberg's discussion of the growth in capital requirements of the chemicals industry, and of the complex interaction between the scale of production establishments and technological evolution, is also consistent with the analysis in the Boskin-Lau paper of the interaction between capital formation and technological change in long-run economic growth. Finally, the paper's discussion of the growth of a "science of manufacturing" in the chemicals industry (chemical engineering) within an academic institution (M.I.T.), highlights the relationships among the U.S. institutions that perform research and train scientists and engineers in the public and private sectors. The historical evolution of chemical engineering research and practice underlines the complexity of the relationship between public and private knowledge, know-how, and practice that receives extensive attention in Nelson's paper in this part.

2 "Non-R&D" Influences on Technology Commercialization

The papers in part I of this volume emphasize the need to integrate the technology development activities of firms with their other functions, pointing out the importance for commercial success of inputs in addition to those lumped under the broad rubric of "research and development." The three papers in this part expand on this theme, which also receives attention in the discussion elsewhere of the influence of the investment climate on technology commercialization (see the paper by McKinnon and Robinson below). The influences on technology commercialization discussed by the papers in this part include elements of the firm's external environment as well as elements that may be internal to the firm.

The paper by Boskin and Lau suggests that capital formation, and international differences in capital formation, could affect the contrasting technological and economic performance of different national economies. For example, the ability of Japanese firms to rapidly adopt new technologies, to succeed at "cyclical" innovation (see the paper by Gomory below), and to develop and exploit dynamic learning networks, all are supported by rates of investment in physical capital that exceed those of the U.S. economy. Bernheim and Shoven analyze the cost of capital in the United States and Japan, in order to evaluate the validity of one widely cited reason for higher rates of capital formation in the Japanese economy. Despite the enormous quantity of research on the relative cost of capital in these economies, there remains little clarity or consistency in the definition of the cost of capital in the United States and Japan. Bernheim and Shoven argue that risk premia, tax systems, and interest rates all affect the cost of capital and must be taken into account in measuring this quantity.

Their analysis confirms the conclusions of other studies in suggesting that the cost of capital in Japan is lower than in the United States, and further suggests that the gap between the cost of capital in these two economies widened during the 1980s. Several other conclusions in this paper underline the links among economic policy, the cost of capital, and rates of capital formation, and point to deficiencies in U.S. performance in all three areas. Bernheim and Shoven find that Japanese capital markets display lower risk premia, thereby favoring relatively risky investments, such as those associated with the commercialization of new technologies. They also conclude that the U.S. tax system tends to discriminate against high-risk investments and favors debt over equity finance.

The paper by Teece also directs attention to the importance of factors beyond those typically considered in discussions of the management of technology commercialization. Teece focuses on the "complementary assets" that firms must accumulate to successfully commercialize new technologies. His paper confronts a perennial riddle in the analysis of technological innovation that is becoming uncomfortably familiar to many U.S. firms: why do the inventors or first developers of new technologies so often fail to reap the lion's share of the profits from these technologies? As Teece shows, reaping the commercial returns from a new technology under "normal" circumstances (i.e., those of the real world, in which patent or other intellectual property protection is imperfect at best) requires that the innovator gain access to other key "complementary" assets, such as manufacturing expertise, marketing and distribution channels, etc. The ease and cost of gaining access to these complementary assets will critically affect the share of the profits from innovation that accrue to the pioneer. Teece's analysis is particularly skeptical of the potential for the so-called "hollow" firm that attempts to specialize in the research and design of new technologies, contracting with other firms for production and all other related activities. His discussion thus suggests that there may be limits to the "network firm" described by Imai below.

The legal environment is an important and little-studied influence on technology commercialization. Huber's paper describes the effects on the incentives and risks faced by the would-be commercializer of new technologies in the United States that have resulted from the extension and expansion of the legal doctrine of strict liability in product liability cases. The author notes that the U.S. judiciary is virtually unique among the justice systems of the industrial world in this interpretation of product liability, which (along with other changes in tort law and decisions) has increased the risks faced by the individual or firm developing and commercializing a new technology. The innovative performance of the British economy, whose product liability system is lauded by Huber, suggests that reform in product liability law may be a necessary, but is not a sufficient, condition for improving performance in technology commercialization. Nevertheless, the issues raised by Huber have been echoed in more recent studies of innovation in such technologies as oral contraceptives, and may create significant obstacles to improved economic performance.

In his important studies of the historical development of the modern corporation in the United States and other industrial economies, Alfred Chandler (1977, 1990) argued that competitive advantage was achieved when manufacturing firms expanded the range of activities coordinated by managers, rather than by the visible hand of the market.[2] Administrative coordination of such activities as industrial research and manufacturing, or manufacturing and product distribution, allowed the emergent firms of the late nineteenth century United States to exploit economies of scale and scope. The papers in this part, by contrast, demonstrate that the cost of capital, the interaction between intrafirm innovative activities and other functions within the firm, and between these activities and the external environment, are also crucial to technology commercialization.

3   International Comparisons of the Macroeconomic and Microeconomic Environment of Technology Commercialization

Comparison of the recent performance of the U.S. and foreign industrial economies suggests that there exist important institutional and policy-related differences between these economies that affect technological innovation and commercialization. The papers in this part consider several aspects of the contrasting structure and environment of technology commercialization in other industrial economies.

The structure and performance of the Japanese economy, especially in the area of technology commercialization, contrast quite sharply with those of the U.S. economy in recent years, and raise important issues for students of innovation and commercialization. Imai's paper examines the organization of the innovation process within and (a particularly important dimension of Japanese structure and performance) among Japanese industrial firms, focusing on the characteristics of innovation that have enabled Japanese firms to adapt to enormous changes in their competitive and technological environments. Imai emphasizes the role of dynamic learning within networks that span multiple functions within the Japanese firm, while linking different firms, for example, users and suppliers of new technologies. These interactions are facilitated by high levels of investment in both physical and human capital, since they often are centered on the development and installation of new capital goods technologies. Imai also emphasizes the dynamic character of the innovation and commercialization processes in the Japanese economy in his discussion of the response of Japanese firms to the demands of "systemic" innovation, their growing investments in R&D, and the increased investment by Japanese firms in basic research.

---

[2]See Chandler 1977, *The Visible Hand*, and Chandler 1990, *Scale and Scope* (Cambridge, MA: Harvard University Press.)

The organization of the technology innovation and commercialization processes within Europe is also undergoing considerable change, partly in response to intensified competition from the United States and Japan. Malerba analyzes the changing structure of private and public institutions for the management and support of technology commercialization in the European semiconductor and computer industries. Consistent with other authors in this volume, Malerba distinguishes among different types of technology within this sector. "Basic" systems and components are relatively standardized, characterized by high R&D costs, economies of scale, and entry barriers. "System applications," which includes customized computer hardware and software as well as application-specific integrated circuits (ASICs), display lower entry barriers and capital-intensity.

Malerba's paper discusses the significant changes in the organization of innovation within this sector in Europe, as firms in both the "basic" and "applications" segments have pursued a wide array of intra-European (and, to a lesser extent, extra-European) strategic alliances to tap external sources of technology and user requirements. Many of these developments in the computer and semiconductor industries follow the hypotheses and analysis of Barabaschi's paper (see below). Malerba also discusses the changing structure of public policy towards technology innovation and commercialization. He describes the rise of EC-sponsored and other regional programs to a place of importance that overshadows the national programs of many member states, the increased emphasis within these programs on cooperative R&D, and the gradual shift within national programs away from the former "national champion" strategies. Malerba's paper complements that of Imai in describing a regional "innovation system" that is undergoing considerable change, much of which has been inspired by the example of Japan. Interestingly, the growing integration of national European R&D systems has been complemented by increased financial integration, which has been facilitated in part by the European Monetary System (EMS) of stable exchange rates. McKinnon and Robinson argue that greater foreign exchange rate stability may be associated with greater domestic financial stability, thereby supporting more rapid technology development and commercialization.

4 The International Environment and the Internationalization Process

One of the most significant changes in the structure of the postwar U.S. economy is the dramatic increase in this economy's openness to international financial and economic influences during the past twenty years.

The share of GNP accounted for by either imports or exports has more than doubled during this period, and international capital and investment flows have bound together the financial and manufacturing operations of the industrial economies much more tightly. The papers in this part of the volume examine several aspects of the interaction between the economies of the United States and other industrial nations.

McKinnon and Robinson focus on the contribution of U.S. international monetary policy to increased domestic financial instability and lower long-term investment. The authors argue that the U.S. policy of continual but uncertain dollar devaluation in the floating exchange-rate era of the 1970s and 1980s was associated with higher domestic inflation, and with higher and less stable domestic interest rates in the United States. Higher nominal interest rates have raised the cost of capital and increased the amount of borrowing needed for financing investment in relatively long-lived assets. The short time horizons of U.S. managers that many observers criticize are in part a rational response to this increasingly uncertain, unstable, and costly financial environment that, according to McKinnon and Robinson, results from U.S. international monetary policy. The authors agree with Bernheim and Shoven in concluding that risk premia in U.S. capital markets have increased and the attractiveness of debt finance has increased relative to that of equity as a result of these developments.

Interestingly, despite the greater integration among global financial markets during the last 25 years, McKinnon and Robinson demonstrate that the floating rate era, by introducing higher risk premia into financial decisions, has widened the gaps in domestic interest rates among industrial economies. The McKinnon-Robinson analysis complements those of other authors in this volume who emphasize the interaction of financial and real phenomena in the investment and technology commercialization behavior of managers within advanced economies.

The past two decades have witnessed significant growth in the international investment, sourcing, and R&D activities of global firms from the United States and other industrial economies. The paper by Westney and Sakakibara discusses the challenges to Japanese firms' R&D management posed by the rapidly expanding international manufacturing and, increasingly, R&D activities of these firms. Along with the paper by Imai, this discussion underscores the challenges created by the success of innovative Japanese firms in such industries as automobiles and consumer electronics. Both Imai and Westney and Sakakibara emphasize the importance of long-term employment and employee rotation within the Japanese firm as a means of familiarizing employees with the technological foundations of firm products and processes. As the R&D networks utilized by the Japanese firm become increasingly international in scope and incorporate non-Japanese employees and firms, however, these long-term personnel employment and rotation policies may become less feasible. The authors also note that the strongly hierarchical authority relationships that characterize many Japanese R&D facilities may create problems in dealing with non-Japanese or "nonpermanent" employees.

The paper by Mowery analyzes another dimension of the internationalization of the activities of U.S. firms, focusing on the growth of international collaborative ventures in technology commercialization. Mowery argues that the relative decline of the technological capabilities of U.S. firms, along with the increasing costs and risks of new product development, increasing nontariff trade barriers, and (in some industries) the importance of "strategic" technology and trade policies of the U.S. and foreign governments, have all contributed to increased international collaboration. Although international joint ventures long have characterized the foreign activities of U.S. firms in industries like petroleum or mining, these "new" international ventures include different activities and inevitably result in higher levels of technology transfer to and from U.S. firms.

In many respects, the growth of international collaborative ventures represents an effort by many U.S. firms to acquire critical "complementary assets," as Teece's paper in this volume denotes them, from foreign sources in order to facilitate the commercialization of new technologies. This paper also complements the analysis of others in emphasizing the differences among industries in the structure and activities included in international collaborative ventures. Although these undertakings do result in more rapid international technology transfer, there is little evidence that this flow consists overwhelmingly of technology "exports" from U.S. to foreign firms. Nor does the comparison of eight U.S. manufacturing industries provide strong support for the argument that U.S. antitrust policy forces U.S. firms to collaborate with foreign, rather than domestic, enterprises.

5 Improving Management of Technology Commercialization

The final group of papers draws on the extensive practical experience of their authors in high-level management of corporate technology development and commercialization in the United States and Europe. Their analysis of the management challenges of technology commercialization in today's environment emphasizes some important weaknesses in U.S.

performance. All three papers also stress the different approaches that are needed to manage development and commercialization of technologies that differ in markets, maturity, dependence on science, and capital-intensity. These differences merit greater attention from scholars.

Gomory focuses on two different processes for the commercialization of new technologies. "Ladder" innovation processes, in Gomory's view, involve major contributions from science, and must be distinguished from the far more common (and commercially important) "cyclical" process of technology development, which draws on numerous small, incremental improvements in products and processes. U.S. firms have demonstrated considerable weakness in their management of the "cyclical" innovation process, with unfortunate consequences for their competitive performance.

Closely related to the Gomory analysis is that of Howard and Guile. Like Gomory and Nelson, they distinguish among different innovation processes, developing the categories of "technology-driven," "production or design driven," and "market-driven." Each of these processes has different requirements for management techniques and organization—but all require senior management that is familiar with the underlying technology, as opposed to one professing an ability to manage "by the numbers." The authors note that the application of quantitative techniques of analysis to innovation investment decisions, although defensible as a general approach, is sensitive to the quality of the underlying data on which decisions are based. Moreover, application of these quantitative techniques must be carefully adjusted to the specific circumstances of a given technology development project, rather than applied according to some rigid template. The authors also offer some important insights into the use by U.S. firms of research consortia that complement the observations of Barabaschi (see below), especially in their emphasis on the complementarity of in-house and cooperative R&D. Barabaschi's paper examines the growing need for managers to exploit the external technological environment of their firms. This need, which is also discussed in Mowery's paper in Part IV, has been heightened by several developments. These factors include the rapidly rising costs of in-house R&D, the steadily falling costs of the information technologies needed to monitor information produced outside the firm, the need for successful commercializers of new technologies to master a broader array of industrial technologies, and the increasing importance of "basic" or (in Nelson's terms) "public" knowledge from such institutions as universities. The paper provides important insights from the viewpoint of a European manager with considerable experience in the design and management of

R&D networks, and raises a number of issues that are pursued in Imai's discussion of the "network firm" in Japan.

The analyses of the innovation process by Gomory and Howard and Guile complement those of Boskin and Lau and Nelson in several important respects. "Cyclical" and "production or design driven" innovation processes tend to be far more capital-intensive, relying as they do on simultaneous management of several generations of new technology, than the "ladder" process, which involves the reduction to practice of a scientific research breakthrough. A portion of U.S. firms' weaknesses in cyclical innovation therefore may reflect the relatively high cost of capital for U.S. firms, illustrating an important complementarity between innovative performance and capital formation.

The characteristics of the knowledge underpinning different types of innovation will differ along lines noted by Nelson. The "ladder" or "technology-driven" processes will draw more heavily on relatively "public," basic scientific knowledge. "Cyclical" or "production or design driven" innovation processes, however, will rely on firm-specific, noncodified know-how and expertise. The "market-driven" innovation process may exploit technological knowledge that is specific to various firms, but is freely shared because of the desire of firms to exploit complementarities among otherwise proprietary technologies (see the paper by Imai).

6 Conclusions and Policy Implications

Taken as a group, the papers gathered in this volume demonstrate the need for an approach to the analysis of technological innovation that is broader in its inclusion of elements "outside" of R&D and that takes into account the influence of the macroeconomic and microeconomic environment on the innovative performance of nations and firms. For too long, elegant neoclassical models of economic growth have enforced a conceptual division between the processes of capital formation and those of innovation. The economic record of the past four decades, as well as the recent empirical work of Boskin and Lau and others that is presented in this volume and elsewhere, suggesting that technology tends to be capital-augmenting, should contribute to an end to this conceptual division.

This integrated view of the interaction between innovation and capital formation also suggests that public policies supporting investment may strengthen national innovative and economic performance. To an unprecedented extent, technological progress is embodied in physical capital—examples include information technologies, biotechnology, and new materials. Human capital formation, another area in which U.S. performance recently has been wanting, is also critical to the realization of the commercial and economic possibilities of new technologies. Unfortunately, current U.S. policy provides less support for investment in human capital or (with the dubious exception of real estate) physical capital embodying technological advances.

Our policy conclusions thus may be succinctly summarized as follows: many of the policies that are most important in affecting the innovative performance of nations lie well outside of the conventional boundaries of "science and technology policy," and include macroeconomic, international financial, regulatory, and tax policies. Indeed, the most important policy levers extend well beyond the categories lumped together as "industrial policy" in the era during which this concept had some political cachet. Without serious attention to the effects of these broader areas of policy on innovative performance, "targeted" technology initiatives, "strategic" industry policies, and the like will have modest positive effects at best, and may instead prove harmful to overall economic performance.

Part I

Conceptual Overview

2

Capital, Technology, and Economic Growth

MICHAEL J. BOSKIN AND LAWRENCE J. LAU

Enhanced capital, labor and technical progress (or equivalently, total factor productivity) are the three principal sources of the economic growth of nations. The rate of growth of labor is generally constrained by the rate of growth of population. For industrialized countries, the rate of growth of the labor force is seldom higher than two percent per annum, even with international migration. Consequently, the rate of growth of capital (physical and human) and technical progress have been found to account for a significant proportion of economic growth by a long line of distinguished economists: Abramovitz (1956), Denison (1962, 1967, 1979, 1985), Griliches (Griliches and Jorgenson 1966, Jorgenson and Griliches 1967, 1972), Jorgenson (see previous references; also Jorgenson, Gollop, and Fraumeni 1987), Kendrick (1961, 1973), Kuznets (1965, 1966, 1971, 1973), and Solow (1957), to name only a few. For example, Jorgenson, Gollop, and Fraumeni (1987) find that between 1948 and 1979, capital formation accounted for 46 percent of the economic growth of the United States, growth of labor input accounted for 31 percent,

The authors are, respectively, Chairman, Council of Economic Advisers (on leave from Stanford University), and Professor of Economics, Stanford University. The authors wish to thank the Ford Foundation, The John M. Olin Foundation, the Pine Tree Charitable Trust, and the Technology and Economic Growth Program of the Center for Economic Policy Research, Stanford University, for financial support. They are grateful to Moses Abramovitz, Paul David, Zvi Griliches, Bert Hickman, Dean Jamison, Lawrence Klein, and Ralph Landau for helpful discussions and to Denise Hare and Jong-Il Kim for able research assistance. The views expressed herein are the authors' and do not necessarily reflect those of the institutions with which they are affiliated.

and technical progress accounted for 24 percent (Jorgenson, Gollop, and Fraumeni 1987, p. 21).

The importance of the contributions of capital and technical progress to the growth of output can be readily understood with the help of some simple arithmetic. Starting with an aggregate production function:

$$(2.1) \quad Y_t = F(K_t, L_t, t)$$

where Y_t, K_t, and L_t are the quantities of aggregate real output, physical capital and labor respectively at time t, and t is an index of chronological time, the rate of growth of output can be expressed in the familiar equation of growth accounting by taking natural logarithms of both sides of equation (2.1) and differentiating it totally with respect to t:

$$(2.2) \quad \frac{d\ln Y_t}{dt} = \frac{\partial \ln F}{\partial \ln K}(K_t, L_t, t)\frac{d\ln K_t}{dt} + \frac{\partial \ln F}{\partial \ln L}(K_t, L_t, t)\frac{d\ln L_t}{dt}$$
$$+ \frac{\partial \ln F}{\partial t}(K_t, L_t, t),$$

where $\frac{d\ln Y_t}{dt}$, $\frac{d\ln K_t}{dt}$ and $\frac{d\ln L_t}{dt}$ are the instantaneous proportional rates of change of the quantities of real output, capital and labor respectively at time t; $\frac{\partial \ln F}{\partial \ln K}$ and $\frac{\partial \ln F}{\partial \ln L}$ are the elasticities of real output with respect to capital and labor respectively at time t; and $\frac{\partial \ln F}{\partial t}$ is the instantaneous rate of growth of output holding the inputs constant, or equivalently, the rate of technical progress.

The three terms on the right-hand side of equation (2.2) may be identified as the contribution of capital, labor, and technical progress respectively to the growth in output.[1]

The production elasticity of output with respect to *measured* labor input can typically be estimated as approximately 0.6 for industrialized countries[2] Thus, given the rate of growth of measured labor force, which is typically no higher than two percent per annum in industrialized countries, the maximum rate of growth that can be accounted for by the growth in labor input is on the order of 1.2 percent. Any growth in output in excess of 1.2 percent per annum in an industrialized country is attributable to the growth in the capital input and to technical progress. For an industrialized country that grows at 3 percent per annum, approximately 60 percent of the growth in output may be attributed to physical capital and technical progress. In the short and intermediate runs, physical capital is even more important for another reason—it is the only input that can be readily var-

---

[1] In almost all such formulations, technical progress is taken to be exogenous.

[2] See, for example, Boskin and Lau 1990 for estimates of the production elasticity of labor for industrialized countries.

ied. Human capital and technical progress can be influenced only in the longer run.

Most aggregate production function and growth-accounting studies (see, e.g., Abramovitz 1956; Griliches and Jorgenson 1966, 1967, 1972; Kendrick 1961, 1973; Kuznets 1971, 1973; Solow 1957) have been conducted under one or more of the traditionally maintained hypotheses of constant returns to scale in capital and labor,[3] neutrality of technical progress, and profit maximization with competitive output and input markets.[4] The validity (or lack thereof) of each of these hypotheses affects the measurement of technical progress and the decomposition of economic growth into its sources. The objective of the present study is to present alternative estimates of technical progress as well as the contributions of each of the sources of growth—capital, labor, and technical progress—without maintaining these assumptions.

In Section 1, the conventional methodology of the measurement of technical progress and growth accounting is reviewed, with the possible effects of the various traditional assumptions on the estimate of technical progress identified. In Section 2, the results of the growth accounting exercises of the various authors for the United States over different time periods are surveyed and summarized. In Section 3, a new approach for studying the relationship between output, inputs and technical progress, first employed by Boskin and Lau (1990), based on the direct econometric estimation of an aggregate meta-production function, is introduced.[5] This new approach does not require the traditionally maintained assumptions. Section 4 contains a very brief discussion of the data and the statistical model used. In Section 5, the major new findings of Boskin and Lau (1990) are summarized. In Section 6, alternative measurements of technical progress and growth accounting based on the estimated aggregate meta-production function estimated in Boskin and Lau 1990 are

---

[3] An exception is Denison (1962, 1967, 1979, and 1985), who assumes that there are three inputs—capital, labor and land—and that the degree of returns to scale is 1.1, that is, if all three inputs are increased by one percent, real output is increased by 1.1 percent.

[4] As increasing returns to scale is assumed by Denison (1962, 1967, 1979, 1985), it is inconsistent to assume profit maximization with respect to all inputs with competitive output and input markets. By using the factor cost shares as weights in calculating the contribution of each of the factors, Denison implicitly assumes cost minimization, under which the ratio of the factor cost shares may be equated to the ratio of the production elasticities.

[5] The term "meta-production function" is due to Hayami and Ruttan (1970, 1985). See also Lau and Yotopoulos 1989 and Boskin and Lau 1990.

presented and compared with those of the conventional approach. Some concluding remarks are made in Section 7.

1 The Conventional Methodology of Growth Accounting

The point of departure of most studies of productivity and growth accounting is the aggregate production function in equation (2.1) and the growth-accounting identity in equation (2.2). The purpose of growth accounting is to determine from the empirical data how much of the change in real output between say $t = 0$ and $t = T$ can be attributed to changes in the inputs, capital and labor, and technology, respectively. The first term on the right-hand side of equation (2.2) represents the instantaneous contribution of the growth in capital input to the growth in real output. Note that the contribution of capital is the product of the production elasticity of capital and the rate of growth of capital. If the rate of growth of capital is low, then the contribution of capital may be low even with a high production elasticity of capital, and vice versa. Similarly, the second term represents the instantaneous contribution of the growth in labor input and the third term represents the instantaneous contribution of technical progress. Together, the three terms add up to the instantaneous rate of growth of aggregate real output at time t.

Equation (2.2) is the fundamental equation of growth accounting. Its function is twofold. First, it can be used as the basis of growth accounting—a decomposition of the growth in real output into the growth in its three separate sources—capital, labor and technology, each represented by one of the three separate terms. Second, equation (2.2) may be rewritten as:

$$(2.3) \quad \frac{\partial \ln F}{\partial t}(K_t, L_t, t) =$$
$$\frac{d \ln Y_t}{dt} - \left[\frac{\partial \ln F}{\partial \ln K}(K_t, L_t, t) \frac{d \ln K_t}{dt} + \frac{\partial \ln F}{\partial \ln L}(K_t, L_t, t) \frac{d \ln L_t}{dt} \right],$$

which can be used to obtain an estimate of technical progress (or equivalently, the growth in total factor productivity), given the values of the different quantities on its right-hand side. However, in general, not every quantity on the right-hand side of equation (2.2) or (2.3) can be directly observed. In fact, only the rates of growth of aggregate real output, capital, and labor are directly measurable as data. The elasticities of output with respect to capital and labor must be separately estimated, often requiring additional assumptions. Moreover, one should note that the instantaneous rate of technical progress, $\frac{\partial \ln F}{dt}(K_t, L_t, t)$, depends on K_t and L_t as well as t, except in the case of neutral technical progress.[6] One

---

[6]More specifically, except in the case that the production function takes the form:

may be tempted to sum up the instantaneous rate of technical progress in each period to arrive at an estimate of the technical progress over T periods as:

$$(2.4) \quad \int_0^T \frac{\partial \ln F}{\partial t}(K_t, L_t, t)dt$$

However, equation (2.4) can be rigorously justified only if either (1) technical progress ($= \partial \ln F/dt$) is (multiplicatively) neutral, that is,

$$(2.5) \quad \ln F(K_t, L_t, t) = \ln A(t) + \ln f(K_t, L_t)$$

so that $\partial \ln F/\partial t$, the instantaneous rate of technical progress, is actually independent of K_t and L_t; or (2) capital and labor remain constant over time, so that:

$$(2.6) \quad \int_0^T \frac{\partial \ln F}{dt}(K_t, L_t, t)dt = \int_0^T \frac{\partial \ln F}{\partial t}(K_0, L_0, t)dt$$
$$= \ln F(K_0, L_0, T) - \ln F(K_0, L_0, 0),$$

which is precisely the logarithmic growth in real output holding inputs constant. However, if technical progress is not neutral, then to the extent that K_t and L_t change over time, the rate of technical progress over many periods cannot be simply cumulated from one period to the next.

Thus, in general, when equation (2.3) is used to measure technical progress over time, three basic hypotheses are maintained: constant returns to scale, neutrality of technical progress, and profit maximization with competitive output and factor markets. Under profit maximization with competitive markets, the elasticity of output with respect to labor is equal to the share of labor cost in the value of total output. One can thus use the labor share, which is directly observable, as an estimate for the production elasticity of labor. Constant returns to scale in production implies that the sum of the production elasticities of capital and labor is exactly equal to unity, so that the elasticity of capital can be readily estimated as one minus the elasticity of labor when the latter is known. The constant returns to scale assumption is crucial to the estimation of the production elasticity of capital because while, in principle, it can be independently estimated by the share of capital cost in the value of total output under the assumption of profit maximization, in practice, data on capital cost are seldom available or reliable. Finally, neutrality of technical progress implies that $\frac{\partial \ln F}{\partial t}$ is independent of capital and labor. This assumption is used to justify the cumulation of successive

---

$Y_t = A(t)f(K_t, L_t)$. Neutrality of technical progress actually implies a somewhat weaker condition: $Y_t = F(f(K_t, L_t), t)$, where $f(.)$ is a real-valued function of capital and labor.

estimates of technical progress over time even as K_t and L_t change their values.

However, even equations (2.2) and (2.3) cannot be directly implemented because data are not available in continuous time. For example, data on aggregate real output and inputs are typically available as the total flow between say time t and time $t+1$. Technical progress between time t and time $t+1$ is given by the integral of equation (2.3) between time t and time $t+1$:[7]

$$(2.7) \qquad \int_t^{t+1} \frac{\partial \ln F}{\partial t}(K_t, L_t, t) dt = \int_t^{t+1} \frac{d \ln Y_t}{dt} dt$$

$$- \int_t^{t+1} \frac{\partial \ln F}{\partial \ln K}(K_t, L_t, t) \frac{d \ln K_t}{dt} dt$$

$$- \int_t^{t+1} \frac{\partial \ln F}{\partial \ln L}(K_t, L_t, t) \frac{d \ln L_t}{dt} dt$$

However, since data are not available in continuous time, a discrete approximation of the integrals in equation (2.7) is usually made. One such approximation is the Törnqvist index, given by:[8]

$$(2.8) \qquad \int_t^{t+1} \frac{\partial \ln F}{\partial t}(K_t, L_t, t) dt \simeq (\ln Y_{t+1} - \ln Y_t)$$

$$- \frac{(S_{K(t+1)} + S_{Kt})}{2}(\ln K_{t+1} - \ln K_t)$$

$$- \frac{(S_{L(t+1)} + S_{Lt})}{2}(\ln L_{t+1} - \ln L_t)$$

where S_{Kt} and S_{Lt} are the values of the production elasticities of capital and labor respectively at time t. Other approximations are, of course, possible.

A leading alternative to using equation (2.2) or (2.3) (or the derivative equations (2.7) or (2.8)) to measure technical progress or to account for growth is the direct econometric estimation of the aggregate production function, $Y_t = F(K_t, L_t, t)$, from data on aggregate real output and capital and labor inputs. Such direct estimation of the production function does not, in principle, require any assumption beyond that of the functional form. In particular, it does not require the traditionally maintained assumptions of constant returns to scale, neutrality of technical progress, and profit maximization with competitive output and input markets. In practice, however, the hypothesis of constant returns to scale is often maintained, especially when data of a single country are analyzed, because it reduces the number of independent parameters to be estimated and thereby mitigates the

---

[7]This is sometimes referred to as the Divisia index.

[8]For a discussion of the Törnqvist index, see, for example, Jorgenson, Gollop, and Fraumeni 1987.

possible multi-collinearity among the data on capital and labor inputs and time.[9]

It is tempting, once an aggregate production function is estimated, to use the estimated production function to derive estimates of the production elasticities of capital and labor, and then to use these estimated elasticities to implement equation (2.2) or (2.3). However, such a procedure generally results in an estimate of technical progress that is statistically defective. An example will make this point clear. Suppose the aggregate production function is assumed to have the Cobb-Douglas form, and the following equation is estimated:

$$(2.9) \quad \ln Y_t = \ln A_0 + \alpha \ln K_t + \beta \ln L_t$$

then $\hat{\alpha}$ and $\hat{\beta}$ are the estimated production elasticities of capital and labor respectively. Substituting these estimated elasticities into, say, equation (2.8), the right-hand side may be immediately recognized as simply the first difference of the residuals from the estimated equation (2.9), which, under the usual assumptions, have zero expected values! (If ordinary least-squares is used to estimate equation (2.9), the residuals will actually sum to zero over the sample.) Thus, any estimate of technical progress obtained this way, whether for a single period or over the whole sample period, is meaningless.[10]

The appropriate way to obtain an estimate of technical progress if an aggregate production function is to be directly estimated is to include a term (or terms)[11] for the time trend in the aggregate production function to capture the effect of technical progress. Once this is done, however, equation (2.2) or (2.3) becomes redundant, because all the information on technical progress is contained in the estimated aggregate production function. In general, if the aggregate production function is specified correctly (with or without the time trend term(s)), the residuals cannot be interpreted as estimates of technical progress, and if the aggregate production function is specified incorrectly (say, by omitting the time trend term(s) when they should be included), the estimated parameters,

---

[9]However, the problem of multi-collinearity may also be partially mitigated by pooling time-series data across countries. See the discussions in Lau and Yotopoulos 1989 and Boskin and Lau 1990.

[10]One may further note that if in fact the time trend term belongs in the aggregate production function in equation (2.9), not including it in the estimation will in general lead to biased estimates of the production elasticities, unless the capital and labor input series are orthogonal to the time trend, a very unlikely situation.

[11]One may wish to include more than one term in order to allow technical progress to be nonlinear (nonexponential) over time.

and hence any functions of these parameters, are biased and unlikely to be useful.

Finally, what are some of the pitfalls of maintaining the traditional assumptions of constant returns to scale, neutrality of technical progress, and profit maximization with competitive output and input markets in the measurement of technical progress and growth accounting? First, it should be recognized that for an economy in which aggregate real output and inputs are all growing over time, it is in general difficult to identify separately the effects of returns to scale and technical progress— either one can be used as a substitute explanation for the other. Thus, to the extent that there are increasing returns to scale, maintaining the hypothesis of constant returns to scale results in an overestimate of technical progress; and to the extent that there are decreasing returns to scale, maintaining the hypothesis results in an under-estimate. This is most apparent from an examination of equation (2.3). If there are increasing returns to scale, the assumption of constant returns to scale implies that the sum of the production elasticities is underestimated and hence at least one but possibly both of the production elasticities are underestimated, implying that the rate of technical progress will be overestimated. A further implication is that the contributions of the capital and labor inputs to economic growth will also be underestimated. The reverse is true if there are decreasing returns to scale.

Second, as pointed out earlier, if technical progress is non-neutral, then the rate of technical progress at time t will vary depending on the quantities of capital and labor inputs at time t. Moreover, technical progress over many periods cannot be expressed simply as a cumulative sum of the technical progress that has occurred over the individual periods, nor can it be expressed simply as an average. In Figure 2.1 we present a particularly simple example of non-neutral technical progress for the one-output and one-input case. The production function shifts from F_0 at $t = 0$, to F_1 at $t = 1$, and remains stationary at $t = 2$. However, the shift is non-neutral—it is positive for the lower range of values of X, the input, but is zero for the higher range of values of X. Let Y_0, Y_1, and Y_2, and X_0, X_1, and X_2 be the measured quantities of outputs and inputs at $t = 0$, 1, and 2 respectively. It is easy to see that if the quantity of input is held constant at X_2, there is no technical progress whereas if the quantity of input is held constant at X_0, there is significant technical progress (equal to T_0/Y_0). It is also clear that technical progress over the entire period cannot be estimated by simply adding up the technical progress that has occurred in each of the individual periods.

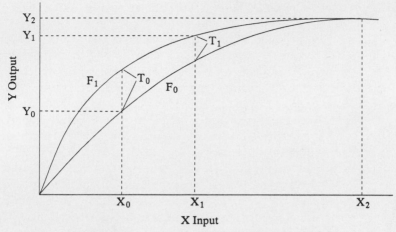

FIGURE 2.1 Non-Neutral Technical Progress

Third, the hypothesis of profit maximization (or cost minimization) with competitive output and input markets implies the equality of the production elasticities of capital and labor with the respective factor shares. To the extent that there are constraints to instantaneous adjustments of inputs to their desired levels and/or monopolistic or monopsonistic influences in the output and input markets, the production elasticities may deviate from the factor shares, and an estimate of technical progress obtained from equation (2.3) or (2.8), using data on factor shares, will be subject to biases. In addition, the estimated contributions of capital and labor inputs to economic growth will also be subject to biases. For example, if the output market is monopolistic, then the factor shares are likely to underestimate the production elasticities, causing in turn an over-estimate of technical progress if an equation such as (2.3) or (2.8) is implemented. If an input market is monopsonistic, then the reverse is likely to be true.

The direct econometric estimation of the aggregate production function is in principle free of all these above-mentioned biases because it does not require the traditional assumptions as part of the maintained hypothesis. However, it will be subject to the same biases if any of the traditional assumptions is kept as part of the maintained hypothesis of the econometric model.

2 Accounting for Economic Growth of the United States

What are the results of the growth accounting exercises conducted by the various authors for the United States using the conventional approach?

TABLE 2.1

Comparison of Growth Accounting by Authors and Time Periods

	Time Period	wK	K̇/K	wL	L̇/L	Scale	Ȧ/A	Y/Y	Ȧ*/A*
Abramovitz 1956	1869–1953	0.25	3.10%	0.75	1.53%		1.68%	3.51%	
Solow 1957	1909–1949	0.35	1.77%	0.65	1.09%		1.49%	2.90%	
Kendrick 1961	1889–1953	0.28	2.71%	0.72	1.72%		1.60%	3.60%	
Denison 1962	1909–1929	0.234	3.16%	0.689	employment 1.58% / hours quantity 4.34% / hours quality 0.38% / educ. 0.56% / women exp & util. 0.10% / age-sex comp. 0.01%	0.28%	0.28%	2.82%	0.92%
	1929–1957	0.225	1.88%	0.73	employment 1.31% / hours quantity −0.73% / hours quality 0.50% / educ. 0.93% / exp & util. 0.15% / age-sex comp. −0.01%	0.34%	0.59%	2.93%	1.70%
Denison 1967	1950–1962	0.144	struc & equip 3.74% / inventories 3.00%	0.786	1.42%: employ-ment 1.14% / hours of work 4.21% / educ. 0.62% / age-sex comp. −0.13%	0.30%	1.07%	3.32%	1.56%
Kuznets 1971	1889–1929	0.33	3.76%	0.67	1.74%		1.24%	3.70%	
	1929–1957	0.23	1.01%	0.77	0.53%		2.30%	2.95%	
	1950–1962	0.214	3.88%	0.787	0.80%		1.87%	3.36%	
Jorgenson & Griliches 1972 reply	1950–1962	0.414	4.09%: stock 3.14% / quality 0.70% / util. 0.25%	0.585	1.22%: stock 0.63% / quality 0.75% / util. 4.16%		1.03%	3.47%	1.76%
Kendrick 1973	1948–1966	0.247	3.48%	0.754	1.31%		2.30%	4.10%	
Denison 1979	1929–1976	0.151	struc & equip 2.17% / inventories 2.48%	0.81	1.27%: employ-ment 1.07% / hours quantity −0.63% / hours quality 0.31% / educ. 0.65% / age-sex comp. −0.13%	0.27%	0.89%	2.98%	1.49%
Denison 1985	1929–1982	0.154	struc & equip 2.44% / inventories 2.52%	0.804	1.33%: employ-ment 1.17% / hours quantity −0.64% / hours quality 0.29% / educ. 0.65% / age-sex comp. −0.13%	0.26%	0.76%	2.92%	1.33%
Jorgenson et al. 1987	1948–1979	0.399	4.04%: stock 1.04% / quality 2.98%	0.601	1.73%: hours 1.12% / quality 0.61%		0.81%	3.42%	2.37%

Notes to Table 2.1

wK = share of capital; $\dot{K}/K$ = rate of growth of capital;
wL = share of labor; $\dot{L}/L$ = rate of growth of labor;
$\dot{A}/A$ = estimated rate of technical progress;
$\dot{A}^*/A^*$ = estimated rate of technical progress in the absence of quality adjustments.

Abramovitz 1956

K series: capital stock—includes land, structure, producers' durable equipment, and net foreign assets.
L series: man-hours (labor force less estimated unemployment multiplied by standard hours).
Y series: real net national product.
Assumptions: constant-returns-to-scale, profit-maximization, and neutrality of technical progress.

Solow 1957

K series: Goldsmith's (1956) capital stock (excludes government, agriculture, and consumer durables), adjusted for percentage of labor force unemployed.
L series: man-hours.
Y series: real private non-farm GNP.
Assumptions: constant-returns-to-scale, profit-maximization, and neutrality of technical progress.

Kendrick 1961

K series: real net capital stock (by asset type) multiplied by the rate of compensation = real capital services.
L series: man-hours worked weighted by average hourly earnings in industry group.
Y series: real net product (Kuznets' concept—national security version, private and public consumption outlay and net investment plus national security expenditures).
Assumptions: constant-returns-to-scale, profit-maximization, and neutrality of technical progress.

Denison 1962

K series: privately owned capital.
L series: Office of Business Economics (OBE) estimates of persons engaged in production, adjusted for work hours (quantity and quality effects), education, experience and age and sex composition changes.
Y series: real national income.
Assumptions: degree of returns to scale=1.1, cost-minimization, and neutrality of technical progress.

Denison 1967

K series: nonresidential structure and equipment, inventories, dwellings, and international assets.
(Denison shows that the total contribution of capital amounts to 25% of the growth in real national income. This figure includes contributions from dwellings of 7.5%, international assets of 1.5%, nonresidential structures and equipment of 12.9%, and inventories of 3.0%. The contributions reported in Table 2.1 include only those from nonresidential structures and equipment and inventories, whereas the figures in Table 2.2 below represent Denison's estimate of the entire contribution of capital. Denison attributes 0.25% of the 3.32% growth in national income to growth in dwellings and 0.05% to growth in international assets.)
L series: employment, adjusted for hours worked, education, and age and sex composition changes.
Y series: real national income.
Assumptions: degree of returns to scale=1.1, cost-minimization, and neutrality of technical progress.

Jorgenson & Griliches 1967, 1972

K series: stock—consumers' durables, nonresidential structure, producers' durables, residential structure, nonfarm inventories, farm inventories, and land; utilization (of business equipment and structures)—approximated by the utilization of power-driven equipment in manufacturing.

L series: persons engaged, adjusted for changes in labor force composition by educational attainment and for effective hours per person.

Y series: real gross private domestic product.

Assumptions: constant-returns-to-scale, profit-maximization, and neutrality of technical progress.

Kuznets 1971

1889–1929 and 1929–1957 estimates derived from Lithwick 1967.
K series: structure and equipment in 1929 prices.
L series: man-hours.
Y series: real GNP (Department of Commerce version).
Assumptions: constant-returns-to-scale, profit-maximization, and neutrality of technical progress.

1950–1962 estimates derived from Denison 1967. The factor shares reported here are inferred from Kuznets' estimate of technical progress. Denison's rates of growth of real output and input as well as his estimate of the relative factor shares.

K series: Denison's (1967) capital stock.
L series: Denison's (1967) man-hours, excluding education adjustment.
Y series: real national income.
Assumptions: constant-returns-to-scale, profit maximization, and neutrality of technical progress.

Kendrick 1973

K series: real capital stock weighted by sectoral capital compensation.
L series: employment multiplied by hours worked, weighted by industry average rate of compensation.
Y series: real net national product, adjusted for government productivity improvement.
Assumptions: constant-returns-to-scale, profit-maximization, and neutrality of technical progress.

Denison 1979

K series: inventories, weighted average of an index of gross and net stock of nonresidential structure and equipment, dwellings, and international assets (Denison attributes 0.17% of the 2.98% growth in national income to growth in dwellings and 0.02% to growth in international assets).

L series: employment adjusted for hours (quantity and quality effects), education, and age and sex composition changes.

Y series: real national income.

Assumptions: degree of returns to scale=1.1, cost-minimization, and neutrality of technical progress.

Denison 1985

K series: inventories, weighted average of an index of gross and net stock of nonresidential structures and equipment, dwellings, and international assets (Denison attributes 0.20% of the 2.92% growth in national income to growth in dwellings and 0.06% to growth in international assets).

L series: employment adjusted for hours (quantity and quality effects), education, and age and sex composition changes.

Y series: real national income.

Assumptions: degree of returns to scale=1.1, cost-minimization, and neutrality of technical progress.

Jorgenson, Gollop, and Fraumeni 1987

K series: capital stock adjusted for changes in the composition of aggregate capital stock by asset class and legal form of organization.

L series: hours worked adjusted for changes in composition of hours worked by age, sex, education, employment class, and occupation.

Y series: aggregate real value-added.

Assumptions: constant-returns-to-scale, profit-maximization, and neutrality of technical progress.

In Table 2.1 we survey and summarize the results of growth accounting exercises by selected authors for the United States over different periods. All of the authors included in the survey except Denison (1962, 1967, 1979, 1985) assume constant returns to scale and profit maximization. Denison (1962, 1967, 1979, 1985) assumes that if all inputs increase by one percent, output increases by 1.1 percent, so that there are modest increasing returns to scale and cost minimization. All of the authors assume, at least implicitly, neutrality of technical progress.

From Table 2.1, it is apparent that the results of the various growth accounting exercises, even over identical or similar periods, differ significantly. For example, the estimates of technical progress for the prewar period range between 0.3 and 1.7 percent per annum and those for the postwar period range between 0.8 and 2.3 percent per annum. The lowest estimated rate of technical progress is the 0.3 percent by Denison (1962), for the period 1909–29. The highest estimated rate of technical progress is the 2.3 percent by Kuznets (1971) for the period 1929–57 and by Kendrick (1973) for the period 1948–66.

One potential source of the differences is the concept of aggregate real output used by the various authors: Abramovitz (1956) and Kendrick (1961, 1973) use real net national product; Solow (1957) uses real private non-farm gross national product; Denison (1962, 1967, 1979, 1985) and Kuznets (1971) for the period 1950–62 use real (net) national income; Jorgenson and Griliches (1967, 1972) use real gross private domestic product; Kuznets (1971) for the periods 1889–1929 and 1929–57 uses real gross national product; and Jorgenson, Gollop, and Fraumeni (1987) use real aggregate value added (approximately equivalent to real gross domestic product). Thus, the rates of growth of aggregate real output as used by the various authors may not be the same even over identical or similar periods.

Moreover, due to variations in the concept of aggregate real output, the measured share of labor also varies across studies—from approximately 0.6 if real output is measured gross of depreciation to approximately 0.8 if real output is measured net of depreciation. Under the maintained hypothesis of competitive profit maximization with respect to labor, the measured share of labor may be used as an estimate of the production elasticity of labor. Thus, aggregate real output may be estimated to grow by between 0.6 and 0.8 percent for every one-percent growth in labor input, other things being equal. Under the maintained hypothesis of constant returns to scale, the production elasticities sum to unity. Hence, the production elasticity of capital may be estimated as one minus the share of labor—between 0.2 and 0.4. For every one-percent growth in the capital input, aggregate real output may be expected to grow by between 0.2 and 0.4 percent, other things being equal.[12]

These growth accounting exercises may be further distinguished by whether they incorporate quality adjustments in the measurement of the quantities of inputs. Abramovitz (1956), Solow (1957), Kendrick (1961, 1973) and Kuznets (1971) do not make any quality adjustments to the quantities of measured inputs. Their estimates of technical progress range from 1.24 to 2.30 percent per annum with an average of 1.78 percent. Denison (1962, 1967, 1979, 1985), Jorgenson and Griliches (1967, 1972) and Jorgenson, Gollop, and Fraumeni (1987) all make quality adjustments to the quantities of labor and Jorgenson and Griliches (1967, 1972) and Jorgenson, Gollop, and Fraumeni (1987) make quality adjustments to the quantities of capital as well. Their estimates of technical progress range from 0.28 to 1.07 percent per annum with an average of 0.78 percent. Studies without adjustments for the changes in the quality of inputs, which include most of the early ones, have yielded much higher estimates of technical progress. On average, the estimated rate of technical progress in these studies is more than 100 percent higher than that found in studies with quality adjustments.

For the purpose of comparison, we have also computed, for those studies with quality adjustments, approximate implied estimates of the rate of technical progress if quality adjustments were *not* made. These implied estimates are reported in the last column of Table 2.1. In every case, we obtain significantly higher estimates of the rate of technical progress, as expected. The average estimate is 1.59 percent per annum, comparable to that obtained for the studies with no quality adjustments. This is consistent with the view that changes in the qualities of the inputs may be considered as one of the forms in which technical progress is materialized.

It is also apparent, from Table 2.1, that Denison's (1962, 1967) estimates of technical progress for the periods 1929–57 and 1950–62 are lower than those of Kuznets (1971) and Jorgenson and Griliches (1972) because of his assumption of increasing rather than constant returns to

---

[12]Denison (1962, 1967, 1979, 1985) assumes that the degree of returns to scale is 1.1. Thus, the sum of his estimates of production elasticities of capital, labor and land should sum to 1.1. Under the assumption of cost minimization, the production elasticities may be estimated as the respective cost shares multiplied by 1.1. Thus, the production elasticities of capital, labor, and land in Denison 1985 may be estimated as 0.17, 0.88, and 0.05, respectively.

scale. Under constant returns to scale, Denison's estimate would have been almost identical with those of the other authors.

In Table 2.2, we present the estimated (implied) contributions of the three sources of growth—capital, labor, and technical progress—for the various studies, with and without quality adjustments. On average, the proportion of economic growth explained by technical progress, with quality adjustment of the inputs, is 25 percent. This proportion increases to 52 percent in the absence of quality adjustments of the inputs. In other words, when changes in the quality of inputs are attributed to technical progress, the estimated contribution of technical progress is on average doubled.

The proportion of economic growth explained by growth in the capital input is, on average, 28 percent with quality adjustment and 22 percent without quality adjustment. Quality adjustments thus increase the proportion of economic growth attributable to capital.

Estimates of the *combined* contribution of capital and technical progress from the various studies are also presented in Table 2.2. The estimated combined contributions of capital and technical progress, in the absence of quality adjustments of inputs, range from 59 to 91 percent. With quality adjustment of inputs, the estimates range from 35 to 79 percent. Once again, quality adjustments lower the estimated contribution of technical progress and raise the estimated contribution of capital.

It can be debated whether quality adjustments to the inputs should be considered part of technical progress. To the extent that the origin of the quality improvements is unknown or unpredictable, a good case can be made that they belong in the "residual" just as much as other forms of technical progress. Ultimately, of course, it depends on the objective of the particular growth-accounting exercise whether quality adjustments should be considered as part of technical progress.

We conclude, from this necessarily brief and cursory survey, that major differences exist in the estimates of technical progress and in the decomposition of economic growth into its sources. We find that adjustments of the quality of inputs reduce the estimates of the rate of technical progress and the latter's contribution to economic growth by approximately one-half and that increasing returns to scale also reduce the estimates of the rate of technical progress and the latter's contribution to economic growth. We shall investigate the impact of relaxing the traditionally maintained assumptions on the estimates of technical progress and of the contributions of the different sources of economic growth.

TABLE 2.2
Contributions of the Different Sources of Growth (Percent)

	Time Period	With Quality Adjustment					Without Quality Adjustment				
		Capital	Tech. Progress	Sub-total	Labor	Scale	Capital	Tech. Progress	Sub-total	Labor	Scale
Abramovitz 1956	1869–1953						22%	48%	70%	33%	
Solow 1957	1909–1949						21%	51%	72%	24%	
Kendrick 1961	1889–1953						21%	44%	65%	34%	
Denison 1962	1909–1929	26%	10%	36%	54%	10%	26%	33%	59%	32%	10%
	1929–1957	15%	20%	35%	54%	12%	15%	58%	73%	16%	12%
Denison 1967	1950–1962	25%	32%	57%	34%	9%	25%	47%	72%	19%	9%
Kuznets 1971	1889–1929						34%	34%	68%	32%	
	1929–1957						8%	78%	86%	14%	
	1950–1962						25%	56%	81%	19%	
Jorgenson and Griliches 1972 reply	1950–1962	49%	30%	79%	21%		40%	51%	91%	8%	
Kendrick 1973	1948–1966						21%	56%	77%	24%	
Denison 1979	1929–1976	15%	30%	45%	46%	9%	15%	50%	65%	26%	9%
Denison 1985	1929–1982	19%	26%	45%	46%	9%	19%	46%	65%	26%	9%
Jorgenson, Gollop & Fraumeni 1987	1948–1979	47%	24%	71%	30%		12%	69%	81%	20%	

NOTE: The capital and labor shares are assumed to be fixed by some authors and variable by others. In studies in which the factor shares are variable the average share over the sample period is used in our calculations. Thus, the contribution from the different sources may not sum to exactly unity because of approximation errors.

3 The Meta-Production Function Approach

In Boskin and Lau 1990, a new framework for the analysis of productivity and technical progress, based on the direct econometric estimation of an aggregate meta-production function,[13] that does not require the traditionally maintained assumptions, is introduced. This new approach enables the separate identification of not only the degree of returns to scale and the rate of technical progress in each economy but also their biases, if any. The estimated aggregate meta-production function is then used as the basis for a new measurement of technical progress as well as a new assessment of the relative contributions of capital, labor and technical progress to economic growth.

The new approach to the estimation of aggregate production functions from pooled time-series and cross-section data is based on the Lau and Yotopoulos (1989) modification of the concept of the meta-production function, introduced by Hayami and Ruttan (1970, 1985), through the use of time-varying, country-, and commodity-specific augmentation factors. The basic assumptions for this approach are:

(1) All countries have access to the same technology, that is, they have the same underlying aggregate production function $F(.)$, sometimes referred to as a meta-production function, but may operate on different parts of it. The production function, however, applies to standardized, or "efficiency-equivalent," quantities of outputs and inputs, that is:

$$(2.10)\quad Y_{it}^* = F(K_{it}^*, L_{it}^*),\quad i = 1,\ldots,n;$$

where Y_{it}^*, K_{it}^* and L_{it}^* are the "efficiency-equivalent" quantities of output, capital and labor respectively of the ith country at time t, and n is the number of countries. The assumption of a meta-production function implies that $F(.)$ does not depend on i (but may depend on t).

(2) There are differences in the technical efficiencies of production and in the qualities and possibly definitions of measured inputs across countries. However, the measured outputs and inputs of the different countries may be converted into standardized, or "efficiency-equivalent," units of inputs by multiplicative country- and output- and input-specific time-varying augmentation factors. The "efficiency-equivalent" quantities of output and inputs of each country are not directly observable. They are, however, assumed to be linked to the measured quantities of outputs, Y_{it}'s, and inputs, K_{it}'s and L_{it}'s, through possibly time-

---

[13]The term "meta-production function" is due to Hayami and Ruttan (1970, 1985). See also Lau and Yotopoulos 1989 and Boskin and Lau 1990.

varying, country-and commodity-specific augmentation factors $A_{ij}(t)$'s, $i = 1, \ldots, n; j = O, K, L:$

(2.11) $Y_{it}^* = A_{i0}(t)Y_{it};$

(2.12) $K_{it}^* = A_{iK}(t)K_{it};$

(2.13) $L_{it}^* = A_{iL}(t)L_{it}; \quad i = 1, \ldots, n.$

These assumptions require some explanation. Together they imply that the aggregate production function is the same in all countries in terms of standardized, or "efficiency-equivalent", units of outputs and inputs. Moreover, measured inputs of any country may be converted into equivalent units of measured inputs of another country. For example, one unit of capital in country A may be equivalent to two units of capital in country B; and one unit of labor in country A may be equivalent to one-third of a unit of labor in country B. These conversion ratios may also change over time. In terms of the measured quantities of outputs and inputs, the aggregate production functions of any two countries are *not* necessarily the same. However, in terms of "efficiency-equivalent" units, the production functions are identical across countries.

We note that in terms of the *measured* quantities of outputs, the production function may be rewritten as:

(2.14) $Y_{it} = A_{i0}(t)^{-1}F(K_{it}^*, L_{it}^*), \quad i = 1, \ldots, n;$

so that the reciprocal of the output-augmentation factor $A_{i0}(t)$ has the interpretation of the possibly time-varying level of the technical efficiency of production, also referred to as output efficiency, in the ith country at time t.

There are many reasons why these commodity augmentation factors are not likely to be identical across countries. Differences in climate, topography, and infrastructure; differences in definitions and measurements; differences in quality; differences in the composition of outputs; and differences in the technical efficiencies of production are some examples. The commodity augmentation factors are introduced precisely to capture these differences across countries. The commodity augmentation factors are assumed to have the constant exponential form with respect to time. Thus:

(2.15) $Y_{it}^* = A_{i0}\exp(C_{i0}t)Y_{it};$

(2.16) $K_{it}^* = A_{iK}\exp(C_{iK}t)K_{it};$

and

(2.17) $L_{it}^* = A_{iL}\exp(C_{iL}t)L_{it}; \quad i = 1, \ldots, n;$

where the A_{i0}'s, A_{ij}'s, C_{i0}'s, and C_{ij}'s are constants. We shall re-fer to the A_{i0}'s and A_{ij}'s as *augmentation level* parameters and c_{i0}'s and c_{ij}'s as *augmentation rate* parameters. For at least one country, say the ith, the constants A_{i0} and A_{ij}'s can be set identically at unity (or some other arbitrary constants), reflecting the fact that "efficiency-equivalent" outputs and inputs can be measured only relative to some standard. Econometrically this means that the constants A_{i0}'s and A_{ij}'s cannot be uniquely identified without some normalization. With-out loss of generality we take the A_{i0} and A_{ij}'s for the United States to be identically unity. Subject to such a normalization, it turns out that these commodity augmentation level and rate parameters can in fact be estimated simultaneously with the parameters of the aggregate production function from pooled intercountry time-series data on the quantities of *measured* outputs and inputs. Thus, it is actually pos-sible to answer the question of how many units of labor in country B is equivalent to 1 unit of labor in country A at some given time t empirically.[14]

(3) The wide ranges of variation of the inputs resulting from the use of intercountry time-series data necessitate the use of a flexible func-tional form for $F(.)$ above. In addition, a flexible functional form is also needed in order to allow the possibility of non-neutral returns of scale and technical progress.[15] In this study, the aggregate meta-production function is specified to be the transcendental logarithmic (translog) func-tional form introduced by Christensen, Jorgenson, and Lau (1973). For a production function with two inputs, capital (K) and labor (L), the translog production function, in terms of "efficiency-equivalent" output and inputs, takes the form:

$$(2.18) \quad \ln Y_{it}^* = \ln Y_0 + a_K \ln K_{it}^* + a_L \ln L_{it}^* + B_{KK}(\ln K_{it}^*)^2/2$$
$$+ B_{LL}(\ln L_{it}^*)^2/2 + B_{KL}(\ln K_{it}^*)(\ln L_{it}^*).$$

By substituting equations (2.15) through (2.17) into equation (2.18), and simplifying, we obtain equation (2.19), which is written entirely in terms of observable variables:

---

[14]These country- and commodity-specific augmentation level and rate parameters can be used as the basis for an international as well as intertemporal comparison of productive efficiencies. See Boskin and Lau 1990.

[15]For example, if the meta-production function $F()$ is chosen to be the Cobb-Douglas form, then the returns to scale will be neutral with respect to the inputs. Moreover, the commodity augmentation factors cannot be separately identified and thus the technology will be indistinguishable from one with neutral technical progress. For this last point, see, for example, Lau 1980.

(2.19) $\ln Y_{it} = \ln Y_0 + \ln A_{i0}^*$
$$+ a_{iK}^* \ln K_{it} + a_{iL}^* \ln L_{it}$$
$$+ B_{KK}(\ln K_{it})^2/2 + B_{LL}(\ln L_{it})^2/2$$
$$+ B_{KL}(\ln K_{it})(\ln L_{it}) + c_{i0}^* t$$
$$+ (B_{KK}c_{iK} + B_{KL}c_{iL})(\ln K_{it})t$$
$$+ (B_{KL}c_{iK} + B_{LL}c_{iL})(\ln L_{it})t$$
$$+ (B_{KK}(c_{iK})^2 + B_{LL}(c_{iL})^2 + 2B_{KL}c_{iK}c_{iL})t^2/2$$

where A_{i0}^*, a_{iK}^*, a_{iL}^*, and c_{i0}^* are country-specific constants. We note that the parameters B_{KK}, B_{KL} and B_{LL} are independent of i, that is, of the particular individual country. They must therefore be identical across countries—thus supplying the common link among the aggregate production functions of the different countries. This also provides a basis for testing the first maintained hypothesis of this study, namely, that there is a single aggregate meta-production function for all the countries. We note further that the parameter corresponding to the $t^2/2$ term for each country is not independent but is completely determined given B_{KK}, B_{KL}, B_{LL}, c_{iK},m and c_{iL}. This provides a basis for testing the second maintained hypothesis of the study, namely, that technical progress may be represented in the constant exponential commodity-augmentation form.

Equation (2.19) is the most general specification possible under our maintained hypotheses of a single meta-production function and constant exponential commodity-augmentation representation of technical progress. Conditional on the validity of equation (2.19), the traditionally maintained hypotheses of growth accounting—constant returns to scale, neutrality of technical progress, and profit maximization with competitive output and input markets can be tested.

In addition to the aggregate meta-production function, we also consider the behavior of the share of labor costs in the value of output: $w_{it}L_{it}/p_{it}Y_{it}$, where w_{it} is the nominal wage rate and p_{it} is the nominal price of output in the ith country at time t. Under competitive output and input markets, the assumption of profit maximization with respect to labor, which is a necessary condition for overall profit maximization, implies that the elasticity of output with respect to labor is equal to the share of labor cost in the value of output:

(2.20) $$\frac{w_{it}L_{it}}{p_{it}Y_{it}} = \frac{\partial \ln Y_{it}}{\partial \ln L_{it}}$$
$$= a_{iL}^* + B_{KL}\ln K_{it} + B_{LL}\ln L_{it}$$
$$+ (B_{KL}c_{iK} + B_{LL}c_{iL})t$$

In other words, the parameters in equation (2.20) are identical to the corresponding ones in equation (2.19). If we do not maintain the hypothesis of profit maximization with respect to labor, the parameters in equation (2.20) do not necessarily have to be the same as those in the aggregate meta-production function. This provides a basis for testing the hypothesis of profit maximization with respect to labor.

4 The Data and the Statistical Model

Boskin and Lau (1990) use data from the Group-of-Five (G-5) countries: France, Federal Republic of Germany, Japan, the United Kingdom, and the United States. The period covered is from 1957 to 1985 except for West Germany and the United States, data for which begin in 1960 and 1948 respectively. The aggregate real output of each country is measured as the real Gross Domestic Product (GDP) in 1980 prices. Labor is measured as the number of person-hours worked. The share of labor in the value of output is estimated by dividing the current labor income (compensation of employees paid by resident producers) by the current GDP of each country. Capital is measured as utilized capital. These data are converted into U.S. dollars using 1980 exchange rates. Time is measured in years chronologically with the year 1970 being set equal to zero. A detailed explanation of the variables and the data sources is given in Boskin and Lau 1990.

We introduce stochastic disturbance terms ϵ_{1it}'s and ϵ_{2it}'s into the first-differenced forms of the natural logarithm of the aggregate production function and the labor share equation, respectively.[16] We assume:

$$(2.21) \quad E \begin{bmatrix} \epsilon_{1it} \\ \epsilon_{2it} \end{bmatrix} = 0, \ \forall i, t;$$

and

$$(2.22) \quad V \begin{bmatrix} \epsilon_{1it} \\ \epsilon_{2it} \end{bmatrix} = \Sigma, \ \text{a constant, nonsingular matrix,} \ \forall i, t :$$

and the stochastic disturbance terms are uncorrelated across countries and over time. In the first-differenced form, our stochastic assumptions amount to saying that the influence of the stochastic disturbance terms is permanent—they raise or lower the production function and the labor share permanently until further changes caused by future stochastic disturbance terms.

---

[16]It turns out that the model without first-differencing yields a Durbin-Watson statistic that is close to unity for the labor share equation, indicating serious misspecification. The model with first-differencing yields reasonable values for the Durbin-Watson statistics. We therefore adopt the first-differenced model.

Under the further assumption of joint normality of the stochastic disturbance terms, we can estimate the system of two equations consisting of the production function and the labor share equation, first differenced, and its various specializations under the different null hypotheses by the method of nonlinear instrumental variables (see, e.g., Gallant and Jorgenson 1979). The list of instrumental variables used in the estimation is also given in Boskin and Lau 1990.

5 Summary of the Major New Findings

The major new findings of Boskin and Lau (1990) are that (1) the traditionally maintained hypotheses of constant returns to scale in capital and labor, neutrality of technical progress, and profit maximization with competitive output and input markets can all be rejected and (2) in the postwar period technical progress can be represented in the purely capital-augmenting form rather than the often assumed neutral or labor-augmenting form; (3) the elasticity of output with respect to measured capital input is much lower than the usual factor-share estimate based on the assumptions of constant returns to scale and profit maximization; and (4) returns to scale are not fixed but variable and as a consequence of (3) above have been decreasing rather than constant. All of these findings have implications for the measurement of technical progress and growth accounting. We discuss these findings in turn.

5.1 Tests of Hypotheses

First, the basic maintained hypotheses of the meta-production function approach adopted by Boskin and Lau (1990) are tested.[17] They consist of (1) the aggregate production functions of all five countries are identical in terms of "efficiency-equivalent" inputs, that is, there is a single aggregate meta-production function for all countries; and (2) technical progress can be represented in the commodity-augmentation form with each augmentation factor being an exponential function of time, conditional on the single meta-production function hypothesis. Neither hypothesis can be rejected at any level of significance. The nonrejection of these two maintained hypotheses lends empirical support to the validity of the aggregate meta-production function with commodity augmentation factors approach.

Next, the three major hypotheses traditionally maintained for aggregate production function or growth accounting studies—constant re-

---

[17]Readers interested in the details of these statistical tests may consult Boskin and Lau 1990.

turns to scale, neutrality of technical progress, and profit maximization with competitive output and input markets—conditional on the validity of our maintained hypotheses of a single meta-production function and exponential commodity augmentation factors are separately tested. It is found that all of these hypotheses can be separately rejected at their assigned levels of significance. The results of this series of tests suggest that the traditional assumptions may not be valid, at least for the countries and time periods under study.

After establishing the validity of the assumptions of the meta-production function approach and the lack of validity of the traditional assumptions, Boskin and Lau (1990) proceed to examine the nature of technical progress. Specifically, Boskin and Lau (1990) test whether: (1) augmentation level parameters are identical across countries separately for capital and labor; (2) technical progress can be represented by two sets of augmentation rates; and finally (3) technical progress can be represented by a single set of augmentation rates for output or an input. It is found that the hypotheses of identical capital and labor augmentation level parameters across countries cannot be rejected. This implies that in the base year (1970), the "efficiencies" of measured capital and labor were not significantly different across countries.[18] However, the hypothesis of equal augmentation level parameters across countries must be interpreted carefully because differences in definitions and measurements, in addition to differences in the underlying qualities, will also show up as differences in the estimated augmentation level parameters.

It is also found that the null hypothesis that technical progress can be represented by a single (instead of three) set of augmentation rate parameters cannot be rejected. Technical progress in these "one-rate" models may be identified as Harrod-neutral, Solow-neutral, and Hicks-neutral respectively. Again, at the assigned level of significance, the hypotheses of zero output and capital rates and zero capital and labor rates[19] are rejected. However, the hypothesis of zero output and labor rates cannot be rejected. Boskin and Lau (1990) conclude that technical progress can be represented by a single set of augmentation rates for capital, that is, *technical progress is capital-augmenting*.

A final hypothesis on the nature of technical progress is that of identi-

---

[18]However, because the definitions of the capital stocks are the least inclusive for Japan and the United States and the most inclusive for West Germany and the United Kingdom, it also implies that the efficiencies of capital are highest in Japan and the United States, followed by France and then West Germany and the United Kingdom in the base year.

[19]This hypothesis is in fact identical to that of neutrality.

cal capital augmentation rate parameters across countries. This hypothesis, conditional on the maintained hypotheses of the study, identical capital and labor augmentation level parameters, and zero output and labor augmentation rate parameters, can be rejected at the assigned level of significance. In fact, the five countries fall into two groups: France, West Germany, and Japan all have capital augmentation rates in the range of eleven to thirteen percent per annum whereas the United Kingdom and the United States have capital augmentation rates in the range of six to eight percent per annum. The hypothesis of equal augmentation rate parameters across countries must also be interpreted carefully because they may reflect *changes* in the definitions, measurements (e.g., depreciation rates, deflators, and their errors, if any), and improvements in the quality of complementary inputs over time, in addition to changes in the underlying quality of the inputs. Moreover, one cannot in general associate an improvement in the quality of an input with an increase in its augmentation factor. For example, an increase in the number of individuals who are computer-literate may show up as an augmentation of capital (an increase in the effective number of computers) rather than labor. Better roads may also show up as an augmentation of capital (an increase in the effective number of vehicles).

Finally, the Cobb-Douglas production function hypothesis, which implies constant production elasticities of capital and labor, is also tested. As expected, it can be rejected at any level of significance.

5.2   The Estimated Aggregate Meta-Production Function

Boskin and Lau (1990) synthesize the results of the hypothesis testing and impose the restrictions implied by the hypotheses that are not rejected at the assigned levels of significance, namely: identical augmentation level parameters for capital and labor and zero augmentation rate parameters for output and labor.[20] The estimated capital augmentation rate parameters are statistically significant and positive for all countries. Japan has the highest rate—12.8 percent per annum—followed by France (12.1 percent), West Germany (11.0 percent), and the United Kingdom (7.2 percent). The United States has the lowest—6.9 percent per annum.[21] As mentioned previously, the estimates of augmentation level and rate parameters should be interpreted carefully. For example, an increase in computer literacy may be reflected as an augmentation

---

[20]See Table 6.1 in Boskin and Lau 1990.

[21]The t-ratios for the estimated country-specific capital augmentation rate parameters (for the hypotheses of zero capital augmentation rates) are: France, 5.7; West Germany, 5.3; Japan, 4.4; the United Kingdom, 4.9; and the United States, 5.7.

of capital (an increase in the effective number of computers). It may be further noted that the estimated capital augmentation rates for France, West Germany, and Japan are very similar in magnitude. This similarity is consistent with the assumption of equal access to technology for these industrialized countries.

5.3 Production Elasticities

The estimates of labor production elasticities range between 0.50 for Japan and 0.55 for West Germany (see Boskin and Lau 1990, Table 6.2). For the year 1970, our estimates of the labor production elasticities are somewhat higher than the actual shares of labor costs in total output for France and Japan and somewhat lower for West Germany, the United Kingdom, and the United States. (This finding suggests that labor may possibly be paid more than its marginal product in the latter countries.) On the whole, our estimated production elasticities of labor do not differ substantially from the actual labor shares. The estimates of the production elasticities of capital range between 0.21 for the United States and 0.29 for Japan in 1970 and are much lower than those estimated from the more conventional factor-share method under the assumptions of constant returns to scale and profit maximization with competitive markets. Given the values of the actual labor shares, the factor-share estimates of the production elasticities of capital would have been between 0.4 and 0.5, approximately twice our estimated production elasticities of capital.

Moreover, in theory, for a concave production function, the production elasticity of capital can be increasing as well as decreasing with respect to capital, depending on the nature of the production technology (but must lie between zero and unity). For the five countries under study, however, the elasticities of capital turn out to be decreasing with respect to both capital intensity and time (see Boskin and Lau 1990). A low production elasticity of capital does not necessarily imply a low capital share. Given decreasing returns to scale (see below) and that the production elasticity of labor is close to the actual labor share, there is actually a surplus that accrues to capital or other fixed factors such as land as the residual claimants to output.

5.4 Returns to Scale

As the hypotheses of homogeneity and constant returns to scale in production have both been rejected, not only is the degree of returns to scale not unity but it is also not a fixed constant—it depends on the quantities of capital and labor and time. Since the estimated production elasticity of labor is approximately the same as the share of labor

TABLE 2.3

Average Annual Rates of Growth of Real GDP, Capital, and Labor

Country	Period	GDP	Capital Stock	Utilized Capital	Labor Force	Employment	Labor Hours
France	57–85	0.039	0.045	0.044	0.007	0.004	−0.003
West Germany	60–85	0.029	0.041	0.039	0.001	−0.002	−0.005
Japan	57–85	0.068	0.092	0.093	0.011	0.011	0.007
United Kingdom	57–85	0.024	0.031	0.031	0.005	0.001	−0.002
United States	48–85	0.031	0.032	0.034	0.018	0.017	0.016

and the estimated production elasticity of capital is much lower than the share of capital implied by the factor share method under the assumptions of constant returns to scale and profit maximization, returns to scale, which are equal to the sum of the capital and labor elasticities, must be less than unity, or equivalently, decreasing. This is indeed found to be the case. At the 1970 values of the measured inputs of each country, statistically significant decreasing returns to scale are found for all five countries. The estimated degrees of returns to scale range between approximately 0.75 and 0.79. The finding of decreasing returns to scale may possibly be attributed to omitted factors of production such as land, public capital stock (in the case of Japan and the United States), human capital, R&D capital stock, natural resources, and the environment.

5.5 Is Capital-Augmenting Technical Progress Capital-Saving or Labor-Saving?

One interesting question is whether capital-augmenting technical progress is also capital-saving, in the sense that the (cost-minimizing) demand for capital relative to labor,[22] at a given quantity of output and prices of capital and labor, is reduced as a result of the technical progress. It can be shown that capital-augmenting technical progress is capital-saving if and only if the elasticity of substitution between capital and labor is less than unity in absolute value (Boskin and Lau 1990). It turns out that the estimated elasticities of substitution for all of the countries lie between 0.25 and 0.48 in absolute value (Boskin and Lau 1990, Table 6.2). Thus, technical progress has been capital-saving rather than labor-saving in all of the countries. One implication of the capital-saving nature of technical progress is that structural unemployment in the aggregate economy is unlikely to be technologically induced. In-

---

[22]Recall that the hypothesis of profit maximization has been rejected.

TABLE 2.4

Alternative Estimates of Average Annual Rates of Technical Progress

Country	Conventional Estimates	Our Estimates
France	.019	.029
West Germany	.013	.024
Japan	.016	.038
United Kingdom	.012	.017
United States	.007	.015

stead, new technology makes a given quantity of capital go further as a complementary input to labor.

6 Alternative Estimates of Technical Progress and Growth Accounting

6.1 Alternative Measurements of Technical Progress

Our first application of the estimated aggregate meta-production function is to use it to compute alternative estimates of the average annual rates of technical progress, or equivalently, the rates of growth of total factor productivity, without relying on the assumptions of constant returns to scale, neutrality of technical progress and profit maximization. In Table 2.3, we present a summary of the data on the outputs and inputs of the five countries over the sample periods. It shows that Japan had the highest average annual rate of growth of real GDP and the United Kingdom the lowest. Japan also had the highest average annual rate of growth of capital stock and United Kingdom the lowest. The United States had the highest rate of growth of the labor force and West Germany the lowest.

In Table 2.4, we present our estimates of the average annual rates of technical progress based on our estimated aggregate meta-production function[23] and then compare them with estimates obtained with the same data but using the conventional method. (Thus, the data on the capital and labor inputs are *not* adjusted for improvements in quality.) We observe significant differences between the two alternative sets of estimates of technical progress. Our estimates are much higher, partially reflecting our finding of a lower production elasticity of capital and hence decreasing returns to scale for the five countries (see Table 2.5), and

---

[23]For a detailed discussion of how the estimate of the rate of technical progress is computed from the estimated aggregate meta-production function, see Boskin and Lau 1990.

TABLE 2.5

Estimated Parameters of the Aggregate Production Functions
(at 1970 Values of the Independent Variables)

	Capital Elasticity	Labor Elasticity	Degree of Local Returns to Scale	Rate of Local Technical Progress	Actual Labor Share	$\dfrac{\partial \ln(K/L)}{\partial \ln A_K(t)}$	Elasticity of Substitution
France	0.251 (8.241)	0.506 (5.945)	0.758 (9.565)	0.030 (12.937)	0.489	−0.690 (−4.699)	0.310 (2.108)
West Germany	0.234 (9.395)	0.519 (7.792)	0.752 (11.944)	0.026 (9.299)	0.532	−0.521 (−3.952)	0.479 (3.633)
Japan	0.292 (8.056)	0.498 (6.399)	0.791 (9.752)	0.037 (9.012)	0.435	−0.639 (−8.018)	0.361 (4.537)
United Kingdom	0.242 (8.884)	0.513 (7.111)	0.755 (11.138)	0.017 (7.753)	0.597	−0.749 (−9.797)	0.251 (3.283)
United States	0.205 (10.426)	0.548 (5.766)	0.753 (8.189)	0.014 (7.070)	0.614	−0.561 (−7.749)	0.439 (6.063)

NOTE: Numbers in parentheses are t-ratios.

show somewhat greater dispersion. However, the rankings of the countries by the average annual rate of (realized) technical progress change only marginally, with France and Japan trading first and second places.

Our estimate, based on the new method, of the average annual rate of technical progress in the United States for the postwar period (1948–85) of 1.5 percent per annum is twice our estimate, based on the conventional method, of 0.7 percent, for the same period. This difference may be attributed to the relatively low estimate of the production elasticity of capital obtained under our new approach, and the resulting decreasing returns to scale. In order to compare the estimate based on our new method with those of other authors, we need, first of all, to standardize the sample period, which is taken to be, broadly speaking, the postwar period. Denison's (1985) estimate for the period 1948–82 may be calculated to be 0.8 percent per annum;[24] Jorgenson, Gollop, and Fraumeni (1987) also estimate the rate of technical progress to be 0.8 percent per annum for the period 1948–79. However, both of these estimates of the average annual rate of technical progress are *net* of the adjustments for improvement in the qualities of the inputs. If the estimated contributions from the quality improvements of the inputs are included as part of technical progress, Denison's (1985) estimate of technical progress increases to 1.2 percent per annum and Jorgenson, Gollop, and Fraumeni's (1987) estimate to 2.4 percent per annum. Moreover, if Denison's (1985) estimate is further adjusted to incorporate the constant returns to scale assumption, it would have further increased to approximately 1.5 percent per annum. However, since we find decreasing returns to scale of approximately 0.75, if both the Denison and Jorgenson-Gollop-Fraumeni estimates are adjusted to reflect the same degree of returns to scale, their estimates would have been approximately 2.0 and 2.6 percent, respectively. By comparison, our estimate of 1.5 percent is actually lower, even though it might have been expected, a priori, to be higher! Thus, estimates based on our new method, especially those of $\dot{A}(t)/A(t)$, are *genuinely different* from those of the other authors.

The rate of local technical progress *realized* in each period is defined as the rate of growth of real output, holding inputs constant. Boskin and Lau (1990) find that the rate of local technical progress, given the rate of capital augmentation, declines with the level of capital and time but rises with the level of labor for each of the five countries. Thus, even though the rates of capital augmentation are exogenous, the rates

---

[24]Denison (1985) does not give a separate estimate for the period 1948–82. The 0.8 percent estimate is derived from his data.

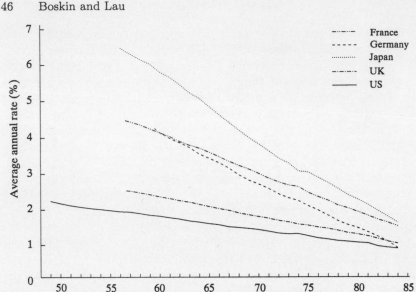

FIGURE 2.2 Local Technical Progress

of technical progress *realized* depend on the quantities of capital and labor and to that extent may be regarded as endogenous. The rates of local technical progress are plotted against time for each country in Figure 2.2. It is interesting to note that, for every country, the local rate of technical progress declines with time. In fact, the rates of local technical progress show strong signs of equalization, *over time*, despite significant differences in the rates of growth of the inputs and in the rates of capital augmentation across countries. This decline in the rates of local technical progress, given that the rates of capital augmentation are constant (by assumption) may be largely attributed to a declining production elasticity of capital with respect to additional capital. However, the equalization in the local rates of technical progress realized, which depend on the quantities of inputs as well as time, should not be confused with the equality of the rates of capital augmentation, which are assumed fixed and exogenous in this study, or with the concept of *convergence in technology*, discussed in Boskin and Lau (1990).

6.2 Alternative Accounts of Economic Growth

A second application of the estimated aggregate meta-production function is to use it to assess the relative contributions of the three sources of growth—capital, labor and technical progress—again without relying on the assumptions of constant returns to scale, neutrality of technical progress and profit maximization. In Table 2.6, we present two alterna-

TABLE 2.6
Relative Contributions of the Sources of Growth

Country	Capital	Labor	Technical Progress
This Study			
France	28	−4	76
West Germany	32	−10	78
Japan	40	5	55
United Kingdom	32	−5	73
United States	24	27	49
Conventional Estimates			
France	56	−4	48
West Germany	69	−11	42
Japan	72	5	23
United Kingdom	55	−6	51
United States	47	30	23

tive sets of estimates of the relative contributions of the different sources of growth for each of the five countries, first using the estimated aggregate meta-production function of Boskin and Lau (1990) and secondly using the conventional approach of subtracting the factor-share weighted sum of rates of growth of the inputs from the rate of growth of output, again based on the same data. Table 2.6 shows, according to the meta-production function, that over the period under study, technical progress is by far the most important source of economic growth, accounting for half or more (three quarters for the European countries), and capital is the second most important source of economic growth (except for the United States). Labor accounts for less than 5 percent except for the United States. These results may be contrasted to those of the conventional approach which identify capital as the most important source of economic growth (more than 45 percent), followed by technical progress (between 23 and 51 percent).

Our estimated contribution of capital of 24 percent for the United States may also be compared to the 19 percent estimated by Denison (1985) and the 47 percent estimated by Jorgenson, Gollop, and Fraumeni (1987). Jorgenson, Gollop, and Fraumeni (1987) adjust the quantity of the capital input for improvements in quality. If, however, the contribution of the improvements in the quality of the capital input is attributed to technical progress, the Jorgenson, Gollop, and Fraumeni (1987) estimate of the contribution of capital is reduced to only 12 percent. Our estimated contribution of technical progress of 49 percent may also be compared to the 26 percent estimated by Denison (1985) and the 24 percent estimated by Jorgenson, Gollop, and Fraumeni (1987). If, however, the contributions of the improvements in the quality of inputs are attributed to technical progress, Denison's estimate is increased to 46 percent and the Jorgenson, Gollop, and Fraumeni (1987) estimate to 69 percent. Thus, once the adjustments to the quality of inputs are taken into account, our results of the growth accounting exercises, using the new method, are not qualitatively different from those of others, even though the conventional method under the traditional assumptions may have been expected to attribute a higher proportion of economic growth to capital and a correspondingly lower proportion to technical progress.

It is interesting to note that the estimated *combined* contributions of capital and technical progress are virtually the same under both approaches as are the contributions due to labor. By either the new or the conventional method, capital and technical progress combined account for more than 95 percent of the economic growth of France, West Germany, Japan, and the United Kingdom. In the United States, where the labor force grew more rapidly than in other countries during this period, they still account for 70 percent of the economic growth. The comparable estimates (estimates without the adjustment of the quality of inputs) of the combined contributions of capital and technical progress for the United States, from the studies surveyed in Section 3, range between 59 and 91 percent. Thus, despite the differences in the underlying assumptions, the estimates of the combined contributions are not qualitatively different.

The reason why the estimates of the combined contributions of capital and technical progress are so similar is because the estimated contributions of labor are very similar by either approach—our estimated output elasticities with respect to labor are not that different from those obtained by the factor share method. Thus, the estimated combined contributions of the other factors—capital and technical progress—must also be very similar. However, our approach yields much lower output elasticities with respect to capital than those obtained by the factor share method under the constant returns to scale assumption. Thus, our estimated contributions of the remaining factor, technical progress, or the "residual," must be correspondingly higher. Another way of understanding our results is to observe that our low estimated production elasticities of capital imply decreasing rather than constant returns to scale and thus the estimated rates of technical progress must be higher to be consistent with the same rates of growth of real output and inputs.

The effect of the assumption of neutrality of technical progress on our measurement of technical progress and on growth accounting may be illustrated with reference to the case of the United States. For the United

TABLE 2.7

The Effect of Non-Neutrality on the Estimates
of the Rate of Technical Progress

(percent)

Country	Initial Inputs as Reference	Final Inputs as Reference
France	3.3	2.4
West Germany	2.8	2.1
Japan	4.7	2.8
United Kingdom	1.9	1.5
United States	1.7	1.3

States, the estimated average annual rate of technical progress is 1.5 percent by our new method. However, if we measure technical progress holding inputs constant at their levels in 1948, the rate of technical progress would have been estimated at 1.7 percent per annum. If we hold inputs constant at the levels of 1985, the rate of technical progress would have been estimated at 1.3 percent per annum. The proportion of U.S. economic growth which can be attributed to technical progress may be estimated at 56 percent and 43 percent respectively. Similar results are obtained for the other four countries and are presented in Table 2.7. Thus, technical progress does not improve productivity uniformly and its measurement over long periods of time depends on the base or reference levels of the measured inputs.

7 Concluding Remarks

We have introduced and implemented a new method of analyzing technical progress and economic growth based on the concept of an aggregate meta-production function, using pooled time series data from the Group-of-Five countries for the postwar period. We have found that the empirical data are inconsistent with the traditionally maintained hypotheses of constant returns to scale, neutrality of technical progress and profit maximization under competitive conditions, at the aggregate, national level. Instead, we find sharply decreasing local returns to scale and capital-augmenting technical progress. All of the traditional hypotheses are, however, customarily maintained in the application of the conventional method of measuring the rate of technical progress and of growth accounting.

With our new approach, we have obtained alternative estimates of the rates of technical progress as well as alternative decompositions of economic growth into its sources—capital, labor and technical progress—that are independent of the conventional assumptions. We have found much higher and somewhat more dispersed rates of *realized* technical progress. We have also found that technical progress is by far the most important source of economic growth of the industrialized countries in our sample, accounting for half or more, in contrast to the growth accounting studies of Denison and Griliches and Jorgenson and his associates.

We have also found that technical progress is capital-augmenting rather than the more frequently assumed neutral or labor-augmenting. Capital-augmenting technical progress has important implications. It means the aggregate production function can be written in the form:

$$(2.23) \quad Y_t = F(A(t)K_t, L_t)$$

Thus, the benefits of technical progress are higher the higher the level of the capital stock. A country with a low level of capital stock relative to labor will not benefit as much from technical progress as a country with a high level of capital relative to labor. Capital and technical progress are thus complementary. Moreover, capital-augmenting technical progress implies that the rate of realized technical progress, defined as the growth in real output holding inputs constant,

$$(2.24) \quad \frac{\partial \ln Y}{\partial t} = \frac{\partial \ln F(A(t)K, L)}{\partial \ln K} \times \frac{\dot{A}(t)}{A(t)}$$

depends on the elasticity of output with respect to measured capital as well as the rate of capital augmentation. The former in turn depends on the actual quantities of capital and labor. In this study, the rate of capital augmentation, $\dot{A}(t)/A(t)$, is taken to be exogenous and equal to a constant for each country, but the rate of realized technical progress, $\partial \ln Y/\partial t$, which varies with the quantities of capital, labor and time, through the elasticity of output with respect to capital, is endogenous. Moreover, the elasticity of capital can be increasing as well as decreasing with respect to capital (so long as it lies between zero and unity), depending on the production technology. For the five countries under study, however, the elasticity of capital is decreasing with respect to capital as well as time. Thus, with a constant rate of capital augmentation, the degree of local technical progress can be expected to continue to decline over time.

The consequence of capital-technology complementarity can be readily appreciated from our empirical results. Consider France, West Germany, and Japan. They have almost the *same* estimated rate of capital augmentation of between eleven and thirteen percent *per annum*. However, according to our estimates in Table 2.4, Japan has the highest average annual rate of (realized) technical progress, followed by France and then West Germany, in the same order as their respective rates of growth of capital stock (see Table 2.3). This is precisely the *complementarity* of capital and technical progress at work. Thus, an increase in the saving rate which results in a higher level of capital formation may also bring about an acceleration in the rate of economic growth in the short and intermediate runs.

However, we should emphasize that our finding of capital-augmenting technical progress does not necessarily mean that the quality of labor has not improved over time, or that all the investments in human capital have gone to waste. As mentioned earlier, improvements in the quality of labor may manifest themselves in the form of capital-augmenting technical progress.

A further implication of capital-augmenting technical progress, given that the elasticity of substitution between capital and labor has been found to be less than unity, is that technical progress is capital-saving rather than labor-saving, in the sense that the desired capital-labor ratio for given prices of capital and labor and quantity of output declines with technical progress. Capital-augmenting technical progress is thus less likely to cause structural unemployment through the technological displacement of workers. In fact, given that F_{KL}, the cross-partial derivative of output with respect to capital and labor, is positive, capital-augmenting technical progress is likely to enhance employment, in the intermediate and long runs.

The results of our growth accounting exercise identify technical progress as the most important source of economic growth. While this finding may be reminiscent of the findings of a large unexplained "residual" in early studies of economic growth (see Table 2.1), such superficial similarities may be misleading. Our finding is quite distinct from those of the earlier studies on at least two counts. First, the earlier studies typically assume constant returns to scale, neutrality of technical progress, and profit maximization with competitive markets. None of these assumptions are maintained in our study. Second, while technical progress is assumed to be exogenous in our model, as in the earlier studies, we have found it to be capital-augmenting, so that it is complementary to the capital input. Thus, the effect of technical progress on real output depends on the size of the capital stock—technical progress does a country with a high level of capital stock much more good than a country with a low level. This capital-technology complementarity, which implies a positive interactive effect of capital and technical progress, distinguishes our results from the earlier studies. It would be wrong to interpret our finding to mean that capital is not an important source of economic growth. In addition to its direct contribution, capital also enhances the effect of technical progress. Capital and technology are inextricably intertwined and are both indispensable ingredients for economic growth.

However, we have not made explicit adjustments for the quality of capital or labor, as were done by Denison (1962, 1967, 1979, 1985) and Jorgenson, Gollop, and Fraumeni (1987). Instead, we allow any trend of improving input quality to be captured by the rates of capital and labor augmentation themselves. Thus, what we attribute to technical progress includes what others may attribute to the improvement in the qualities of the inputs.

Appendix

Estimated Parameters of the Aggregate Production Function and the Labor Share Equation
(First-Differenced Form)

Parameter	Estimate	t-ratio
Aggregate Production Function		
a_K	0.178	4.430
a_L	0.521	1.974
B_{KK}	−0.056	−3.975
B_{LL}	−0.001	−0.004
B_{KL}	0.030	1.690
c_{FK}	0.121	5.733
c_{GK}	0.110	5.314
c_{JK}	0.128	4.432
c_{UKK}	0.072	4.922
c_{USK}	0.069	5.663
R^2	0.855	
D.W.	1.818	
Labor Share Equation		
B_{KLF}	−0.145	−2.030
B_{KLG}	−0.164	−1.653
B_{KLJ}	−0.061	−1.647
B_{KLUK}	0.070	0.503
B_{KLUS}	−0.028	−0.449
B_{LLF}	0.148	0.687
B_{LLG}	0.261	1.689
B_{LLJ}	−0.399	−4.317

Parameter	Estimate	t-ratio
B_{LLUK}	0.210	1.741
B_{LLUS}	0.090	0.572
B_{*FLt}	0.010	2.561
B_{*GLt}	0.011	2.043
B_{*JLt}	0.013	3.267
B_{*UKLt}	−0.002	−0.467
B_{*USLt}	0.001	0.432
R^2	0.188	
D.W.	1.797	

References

Abramovitz, M. 1956. Resource and Output Trends in the United States Since 1870, *American Economic Review*, 46, pp. 5–23.

Arrow, K. J. 1964. Optimal Capital Policy, the Cost of Capital, and Myopic Decision Rules, *Annals of the Institute of Statistical Mathematics*, 16, pp. 21–30.

Arrow, K. J., H. B. Chenery, B. S. Minhas and R. M. Solow. 1961. Capital-Labor Substitution and Economic Efficiency, *Review of Economics and Statistics*, 43, pp. 225–50.

Boskin, M. J. 1988. Tax Policy and Economic Growth: Lessons from the 1980s, *Journal of Economic Perspectives*, 2, pp. 71–97.

Boskin, M. J. and L. J. Lau. 1990. Post-War Economic Growth of the Group-of-Five Countries: A New Analysis, Technical Paper No. 217, Stanford, CA: Center for Economic Policy Research, Stanford University (mimeographed).

Boskin, M. J., M. S. Robinson and A. M. Huber. 1989. Government Saving, Capital Formation and Wealth in the United States, 1947–1985, in R. E. Lipsey and H. S. Tice, *The Measurement of Saving, Investment, and Wealth*, Chicago: University of Chicago Press, pp. 287–356.

Christensen, L. R., D. W. Jorgenson and L. J. Lau. 1973. Transcendental Logarithmic Production Frontiers, *Review of Economics and Statistics*, 55, pp. 28–45.

David, P. A. 1972. Just How Misleading Are Official Exchange Rate Conversions? *Economic Journal*, 82, pp. 979–90.

David, P. A. and T. van de Klundert. 1965. Biased Efficiency Growth and Capital-Labor Substitution in the U.S., 1899–1960, *American Economic Review*, 55, pp. 357–94.

Denison, E. F. 1962. United States Economic Growth, *Journal of Business*, 35, pp. 109–21.

Denison, E. F. 1967. *Why Growth Rates Differ: Post-War Experience in Nine Western Countries*, Washington, D.C.: Brookings Institution.

Denison, E. F. 1979. *Accounting for Slower Economic Growth: The United States in the 1970s*, Washington, D.C.: Brookings Institution.

Denison, E. F. 1985. *Trends in American Economic Growth, 1929–1982*, Washington, D.C.: Brookings Institution.

Denison, E. F. and W. Chung. 1976. *How Japan's Economy Grew So Fast*, Washington, D.C.: Brookings Institution.

Denison, E. F. and R. P. Parker. 1980. The National Income and Product Accounts of the United States: An Introduction to the Revised Estimates for 1929–80, *Survey of Current Business*, vol. 60, no. 12, pp. 1–26.

Gallant, A. R. and D. W. Jorgenson. 1979. Statistical Inference for a System of Simultaneous, Nonlinear, Implicit Equations in the Context of Instrumental Variables Estimation, *Journal of Econometrics*, 113, pp. 272–302.

Goldsmith, R. 1956. *A Study of Saving in the United States*, Princeton: Princeton University Press.

Griliches, Z. and D. W. Jorgenson. 1966. Sources of Measured Productivity Change: Capital Input, *American Economic Review*, 56, pp. 50–61.

Hayami, Y. and V. W. Ruttan. 1970. Agricultural Productivity Differences Among Countries, *American Economic Review*, 60, pp. 895–911.

Hayami, Y. and V. W. Ruttan. 1985. *Agricultural Development: An International Perspective*, revised and expanded ed., Baltimore: Johns Hopkins University Press.

Hulten, C. R. and F. C. Wykoff. 1981. The Measurement of Economic Depreciation, in C. R. Hulten, ed., *Depreciation, Inflation, and the Taxation of Income from Capital*, Washington, D.C.: The Urban Institute Press.

Jorgenson, D. W. and Z. Griliches. 1967. The Explanation of Productivity Change, *Review of Economic Studies*, 34, pp. 249–83.

Jorgenson, D. W. and Z. Griliches. 1972. Issues in Growth Accounting: A Reply to Edward F. Denison, *Survey of Current Business*, vol. 52, no. 5, Part II, pp. 65–94.

Jorgenson, D. W., F. M. Gollop, and B. M. Fraumeni. 1987. *Productivity and U.S. Economic Growth*, Cambridge, MA: Harvard University Press.

Kawagoe, T., Y. Hayami, and V. W. Ruttan. 1985. The Intercountry Agricultural Production Function and Productivity Differences Among Countries, *Journal of Development Economics*, 19, pp. 113–32.

Kendrick, J. W. 1961. *Productivity Trends in the United States*, Princeton: Princeton University Press.

Kendrick, J. W. 1973. *Postwar Productivity Trends in the United States, 1948–1969*, New York: Columbia University Press.

Kuznets, S. S. 1965. *Economic Growth and Structure*, New York: Norton.

Kuznets, S. S. 1966. *Modern Economic Growth: Rate, Structure and Spread*, New Haven: Yale University Press.

Kuznets, S. S. 1971. *Economic Growth of Nations*, Cambridge, MA: Harvard University Press.

Kuznets, S. S. 1973. *Population, Capital and Growth*, New York: Norton.

Landau, R. 1990. Capital Investment: Key to Competitiveness and Growth, *The Brookings Review*, Summer, pp. 52–56.

Lau, L. J. 1980. On the Uniqueness of the Representation of Commodity-Augmenting Technical Change, in L. R. Klein, M. Nerlove and S. C. Tsiang, eds., *Quantitative Economics and Development: Essays in Memory of Ta-Chung Liu*, New York: Academic Press, pp. 281–90.

Lau, L. J. 1987. Simple Measures of Local Returns to Scale and Local Rates of Technical Progress, Working Paper, Stanford, CA: Department of Economics, Stanford University (mimeographed).

Lau, L. J. and P. A. Yotopoulos. 1989. The Meta-Production Function Approach to Technological Change in World Agriculture, *Journal of Development Economics*, 31, pp. 241–69.

Lau, L. J. and P. A. Yotopoulos. 1990. Intercountry Differences in Agricultural Productivity: An Application of the Meta-Production Function, Working Paper, Stanford, CA: Department of Economics, Stanford University (mimeographed).

Lindbeck, A. 1983. The Recent Slowdown of Productivity Growth, *Economic Journal*, 93, pp. 13–34.

Lithwick, N. H. 1967. *Economic Growth in Canada: A Quantitative Analysis*, Toronto: University of Toronto Press.

Musgrave, J. C. 1986. Fixed Reproducible Tangible Wealth in the United States: Revised Estimates, *Survey of Current Business*, vol. 66, no. 1 (January).

Shephard, R. W. 1953. *Cost and Production Functions*, Princeton: Princeton University Press.

Solow, R. M. 1956. A Contribution to the Theory of Economic Growth, *Quarterly Journal of Economics*, 70, pp. 65–94.

Solow, R. M. 1957. Technical Change and the Aggregate Production Function, *Review of Economics and Statistics*, 39, pp. 312–20.

Summers, R. and A. Heston. 1988. A New Set of International Comparisons of Real Product and Price Levels: Estimates for 130 Countries, 1950–85, *Review of Income and Wealth*, 34, pp. 1–25.

3

What Is "Commercial" and What Is "Public" About Technology, and What Should Be?

RICHARD R. NELSON

The papers in this volume are concerned with: effective commercialization as a lure and reward for innovative effort, what leads to effective commercialization, and the implications of all this for technical advance and economic growth. This paper attempts to set commercialization of technology into a broader context that recognizes that technology has public aspects as well as private.[1]

It is organized as follows. Section 1 lays out the private and public aspects of technology in modern capitalist economies, and their connections. In Section 2, I consider the means companies use to commercialize the technology they create, and the ways that initially proprietary technology goes public. Section 3 is concerned with the expressly public aspects of the capitalistic innovation system, particularly with the roles of universities. Section 4 addresses the question: what should be private and what should be public about technology?

1 The Private and the Public Aspects of Technologies in Capitalist Economies

Virtually all contemporary accounts of how technical change proceeds in capitalist economies are based on Joseph Schumpeter's in his *Capitalism, Socialism, and Democracy*. For-profit firms, in rivalrous competition,

---

[1]Some of the material presented in this paper was published earlier in Dosi et al. 1988.

are the featured actors. The context within which they operate is set, on the one hand, by the laws and ethos of capitalism that enable firms to keep proprietary, at least for a while, the new technology they create, and on the other hand by public scientific and technological knowledge. The latter lends problem solving power to industrial R&D. The former enable firms to profit when their R&D creates something the market values. Indeed, given that its rivals are induced by this context to invest in R&D, a firm may have little choice but to do so also. The result is significant competition in innovation, generating a bountiful flow of new products and processes.

This volume is focused on the proprietary or commercial aspects of technology. If prospects for commercialization are inadequate, investments in industrial R&D will not be forthcoming. But technology has a public aspect as well as a private one.

Schumpeter recognized this clearly. While in his model of technical advance the lure and reward for private innovative efforts reside in a temporary monopoly over the new product or process, he stressed that in the general run of things that monopoly is temporary. Sooner or later the company's proprietary technology will be imitated or invented around. He did not see this as a problem, but rather as a principal mechanism through which the benefits of successful innovation are spread widely, and as a factor preventing firms from being able to rest on their laurels.

Schumpeter saw less well the consciously public parts of the capitalist innovation system, parts which were important even in his day, and which have become even more so since then. There were and are technical societies, industry associations, and other mechanisms through which new technical information is shared, and which can be used to pursue it cooperatively. Universities play a major role as repositories of public scientific and technological information, and as a prime locus of work aimed to enhance public knowledge. Public monies, as well as private, serve to fund the system.

Both the private and public, the commercial and collective, aspects of technology play an essential role in its advance. Together they set up technical advance as a cultural evolutionary process.

Again Schumpeter saw and stated the matter clearly: "The essential point to grasp is that in dealing with capitalism we are dealing with an evolutionary process." Empirical research on innovation amply supports Schumpeter's point. Technical advance inevitably proceeds through the generation of a variety of new departures in competition with each other and with prevailing practice. The winners and losers are determined in actual contest. Many contemporary modelers ignore this, treating technical advance as if it proceeded with much more accurate ex-ante calculation and before the contest agreement on winners than is the case. Sideÿ Winter and I (1982) have argued that such models not merely oversimplify, but fundamentally mis-specify, how technical advance proceeds under capitalism, which is through an evolutionary process in the sense above.

Of course the process through which technical advance proceeds in capitalist economies differs in various respects from evolutionary processes in biology, although on reflection, some of the apparent differences may not be real. Thus technology occasionally makes "big jumps." This is inconsistent with traditional concepts of evolution in biology, but not with more modern notions of punctuated equilibrium. Also, it is clear that innovation is far from a strictly random process; rather, efforts to advance technology are carefully pointed in directions that innovators believe to be feasible and potentially profitable. However, here again the difference with biological evolution may not be sharp if one recognizes the possibility that selection has operated on genes to make viable mutations more likely than would be the case were mutation strictly random.

I propose that the feature that most sharply distinguishes the evolutionary processes through which technology advances from biological evolution is that new findings, understandings, and generally useful ways of doing things do not adhere to their finder or creator and their well-defined descendants, but are shared among contemporaries, at least to some extent. In many cases the sharing is intentional, in others despite efforts to keep findings privy. But in any case that the new technology ultimately goes public means that technology advances through a "cultural" evolutionary process. The capabilities of all are advanced by the creation or discovery of one. This is fundamentally different than in biological evolution.

Some economists, recognizing this phenomenon, have attempted to analyze it in terms of a trade-off between incentives for innovation, which are seen as hinging on the ability of firms to make new technology proprietary, and the extent to which an innovation is used where it is efficient to use it, which may depend on the degree to which new technology is public.[2] While there is something to this point of view, it misses several important matters. First, the going public of new technology not only increases society's ability to use it in its present form, but also widens the range of parties who are in a position to further improve it, variegate it, more generally contribute to its advance. While analysts have

---

[2]Perhaps the best exposition of this point of view is contained in Nordhaus 1969.

argued the case both ways, I maintain that by and large the experience is that technical advance proceeds much more rapidly when a considerable number of parties are engaged in competitive efforts, than in a context where one or a few parties are in a position to control developments.

Second, for-profit firms on their own financial bottom are not the only actors in the innovation game. There are public institutions and public monies as well, and these can step in where profit incentives are weak.

Third, simple analyses of this genre do not recognize that technology and technical advance can be unpacked into different dimensions or aspects. It matters what aspects are kept proprietary or privy, and what aspects are left or made open to the public. The analytical problem here is reflected in the failure of many economists, and other scholars, to see into the complex relationships and overlays between technology and science which are characteristics of the modern system.

Technology at any time needs to be understood as consisting of both a set of specific practices, and a body of generic understanding that surrounds these and provides interpretations of why they work as they do.[3] The distinction I am drawing here might seem to be between a science and its applications. However, it is a mistake to think of manufacturing technology as simply applied science. In many cases where it is known that a method works, it is not well understood how. The technique may have been discovered initially through trial and error learning, or even accident. And it may take a long time before a solidly grounded understanding of why and how it works is developed.

Further, in almost all technologies a considerable portion of generic understanding about how things work, key variables affecting performance, the currently binding constraints and promising approaches to pushing these back, stems from operating and design experience with products and machines and their components, and analytic generalizations reflecting on these. These understandings may have only limited grounding in any fundamental science, standing, as it were, largely on their own bottom. This is not quite what philosophers of science tend to mean when they talk of a "science."

However, the philosophers of science may be myopic about what a science is. A number of observers have noted that many modern fields of inquiry that call themselves sciences do not fit the classic mold. Thus fields like computer science, metallurgy, and pathology, are basically about this kind of understanding, and reflect attempts to make it "more

---

[3]Dosi (1982) refers to this body of understanding as a technological paradigm.

scientific." The various disciplines that call themselves "engineering" have a similar cast. These "applied sciences" tend to lie between the body of extant technique and simple experiential understanding, and the more fundamental sciences. They represent attempts to go deeper into the understanding of a technology than simple practitioner's knowledge, and often they reach down into fundamental physics, chemistry, and biology to provide a basis for such understanding. However, they by no means have been able to encompass and codify all of generic knowledge. And to a considerable extent they have an analytic structure of their own which is not completely reducible to the underlying fundamental sciences.

In much of economic theory it is presumed that technology is a latent public good, in the sense of having the capacity to benefit many parties, and inexpensive (if not literally costless) to teach and learn compared with the cost of invention or discovery in the first place. It is this view of technology and technical advance that lies behind the notion that intellectual property rights or secrecy are needed if there is to be private incentive for invention and innovation, but that when technology is proprietary there are major "dead weight" costs.

Now the notion that technology is a latent public good is a reasonable first approximation if the focus is on generic knowledge. Generic knowledge tends to be germane to a wide variety of uses and users. Such knowledge is the stock-in-trade of professionals in the fields, and there tends to grow up a systematic way of describing and communicating such knowledge, so that when new generic knowledge is created anywhere it is relatively costless to communicate to other professionals. Indeed, generic technology knowledge not only has strong latent public good properties; it may be difficult to prevent such knowledge from being manifestly a public good even if one tries to keep it privy. As I shall show in some detail shortly, rival firms have a variety of ways to ferret out the generic aspects of a competitor's technology, even if the specific details of particular products and process may remain beyond their ken.

On the other hand, the body of particular extant technique is more of a mixed bag. Some of practiced technique is widely applicable and easily learned by someone skilled in the art, if access were open. But scholars like Keith Pavitt (1987) and Nathan Rosenberg (1976, 1982) have argued persuasively that many prevailing industrial techniques that operate effectively in a given establishment can be transferred to another only with considerable cost, even if the original operator is open and helpful. Partly that is because of the complexities involved and the high cost of teaching and learning.[4] Partly it is because significant modifications may need to be made if the technology is to be effective in a somewhat different context. In these cases, technology, or at least a particular practice, is more like a private good than a latent public one.

So, in a sense the evolution of technology involves the co-evolution of a public good—generic knowledge—and a collection of private goods— specific practices. One might presume, therefore, that corporate R&D efforts would focus tightly on creating new techniques, with public institutions like universities doing most of the work advancing generic knowledge.

Note that this division of labor largely obviates the "trade-off" problem. However, while this is a tolerable first approximation, it misses three important points. First, a large part of generic knowledge is created as a by-product of creating new products and processes. Thus a portion of the trade-off problem remains. Second, in order to tap into an advancing field of generic knowledge one often needs to do work in it oneself. Third, in complex technologies which have strong underlying applied sciences, corporate and university scientists share a common training and belong to the same professional associations. They also share many interests. Because of these facts corporate R&D is an important source of public generic knowledge, in some fields professors invent, and in some technological areas the lines between what corporations do and what goes on in universities are blurry, not sharp.

2 How Firms Appropriate Returns to Their Innovative Efforts, and How They Ferret Out What Their Competitors Have Done

Schumpeter never was explicit about just how a firm that invested in R&D established and protected a proprietary edge. Economists and historians of technology writing since his time have recognized a variety of means, and their work suggests that different ones are operative in different technologies and industries. However, until my colleagues and I designed a survey, there was no systematic map of the terrain. The details of our questionnaire have been reported elsewhere (in particular, see Levin et al. 1987). Here I simply will briefly recount our findings.

To oversimplify somewhat, we distinguished three broad classes of means through which firms are able to appropriate returns to their innovations—through the patent system, through secrecy, and through

---

[4]For good studies of the cost of technology transfer, see Teece 1977 and Mansfield et al. 1981.

various advantages associated with exploiting a head start—and asked our respondents in different lines of business to score on a scale from one to seven the effectiveness of these for protecting product and process innovations. I will focus on product innovations.

There were significant cross-industry differences regarding the means rated most effective for appropriating returns to product innovation. However, in most industries, a head start and such associated advantages as ability to move down the learning curve ahead of one's competitors, or getting ahead in sales and service, were rated as the most effective means. Included here were such industries as semiconductors, computers, telecommunications, and aircraft. In some of these industries patents also were rated as reasonably effective, in others not. But clearly in many industries where innovation is important, the gains to an innovator come largely from getting in early and exploiting that advantage.

An interesting feature of most of the above industries is that imitation of a new product is time consuming and costly even if it isn't protected by patents. In some, the products take the form of complex systems. Our respondents from industries producing aircraft and guided missiles, canonical complex systems, reported that it would cost a competent imitator three-fourths or more of what the innovator invested to come up with something comparable, that considerable time would be involved as well, and that it did not matter much whether or not there were patents. Producing complex systems effectively requires that many components and details be got right, and this is difficult to learn to do even if one has a model to take apart, or a blueprint to follow. These industries, and others like semiconductors, also involve complex production processes with tooling and equipment often finely tuned to product design. Simply getting the production line in place and running right can yield the inventor a substantial lead over potential followers.

While only a few industries reported patents to be their most effective instrument for protecting product innovation, there are two groups that did.[5] One consists of industries where chemical composition is a central aspect of design: pharmaceuticals, industrial organic chemicals, plastic materials, and synthetic fibers. The other consists of industries producing products that one might call devices: air and gas compressors, scientific instruments, power-driven hand tools, etc. In both of these cases the composition of the products is relatively easy to define

[5] Our finding that patents are important in only a few fields is consistent with the earlier findings of Scherer et al. (1959) and Mansfield et al. (1981).

and delimit. This means that they are relatively easy to reverse engineer and imitate; thus patent protection is essential if the innovator is to be able to reap profits. Happily, the same conditions seem to be conducive to ability to draw patents that can be enforced.

Secrecy was rated an effective means of protecting product innovation in virtually no industries. On the other hand, secrecy was scored as quite effective in many industries in protecting process innovations.

Note that the results of our questionnaire provide a quite different picture of how firms gain returns from their product innovations than is often presumed by lay persons. Gaining returns from innovation in most cases is not so much a matter of establishing property rights or guarding secrets. Rather, profit comes from getting ahead of one's competitors, and exploiting that advantage. But with time competitors can catch on, if they make the required investments.

This picture is confirmed by the answers to the probes in our questionnaire about how firms acquire knowledge about new products developed by their competitors. The means included doing independent R&D or reverse engineering, trying to get information from employees of the innovating firms and perhaps hiring them away, patent disclosures and publications of various sorts, and open technical meetings.

The responses suggest that monitoring outside technological developments generally is an active and costly business. In most industries the means of monitoring judged most effective was either doing independent R&D (presumably while attending to clues about what one's competitors are doing) or reverse engineering. The industries that gave these means low scores almost invariably were those that do little R&D themselves, and hence don't have the capabilities to employ them. Conversely, virtually all R&D intensive industries rated one or both very effective as a means of learning about (and presumably mastering something comparable to) one's competitors' innovations. It is apparent that in these industries the fact that viable firms have active R&D efforts serves to bind them together technologically, as well as to advance the frontiers.

Those industries that reported reverse engineering to be effective also tended to report that they often learned a lot from conversations with scientists and engineers of the innovating firms. Some reported that they make a practice of trying to hire some away. It is apparent that in the United States in many industries communication between scientists and engineers and interfirm flow of R&D personnel both act to keep generic knowledge relatively public. Publications and open technical meetings also were rated effective sources of information in a wide range of industries.

It thus appears that, contrary to common beliefs, firms do not keep tight controls on all information about their new technology, and in some cases they seem actively to divulge information. How come? In the first place, the very staking of claims involves the release of information. That is one of the intents of the patent system, and where patents are effective, they also make knowledge public. Where patents are not effective, but aggressive use of a head start advantage is, firms have strong incentives to stake their claims through advertising, open meetings, and a wide variety of other ways, in addition to patenting. They need to attract customers, and this means they need to tell them about their new wares, but this means telling something to their competitors too. Also, as von Hippel (1988) has documented, engineers in different firms tend to help each other out by sharing and trading information.

Further, the divulging of certain kinds of information does not significantly undermine a company's real proprietary edge. Where new products are patentable and patents are effective, as in pharmaceuticals, it does not hurt a company to divulge generic information if it has the patent. Letting generic knowledge won in R&D go free does not handicap a firm from reaping handsomely from its product innovation, if it has a significant head start on production and marketing of the product in question, and the capacity to take advantage of that lead.

The way I have described the matter suggests that what firms have great difficulty in keeping privy is the generic aspects of the new technology that they have created, rather than the specific practical implementations. I think that is correct. Not just patents, but the location and organization-specific nature of the details of a firm's particular product and process technologies often make them difficult to directly imitate, even if another firm wanted to do just that. On the other hand, the general understanding behind those particular products and processes are very difficult to keep privy for very long, in an industry where most firms have sophisticated and alert engineers and scientists, and these tend to communicate with each other. Advances in generic understanding won through corporate R&D efforts thus go public quite quickly. As the data presented above indicate, this does not mean that rival firms get a free ride. They must do their own R&D in order to create their own specific ways of implementing this new knowledge.

3 The Expressly Cooperative and Public Components of Innovation Systems

Until recently most accounts by economists of how technical advance occurs in capitalist economies focused almost exclusively on the role of for-profit firms (and sometimes private inventors), on rivalry, and on proprietary technology. The limits of proprietary grip were recognized and some of the significance of that understood, but voluntary sharing of technology was seldom mentioned. Above I have begun to develop the point that the public aspects of technology are there not simply because of unavoidable holes in proprietary shields, but because in many cases firms deliberately leak and share knowledge.

In contrast with the orientation of economists, virtually all sociologists and many historians writing about technical advance, while recognizing rivalry of business firms, have tended to stress the social and professional aspects of the process. The key actors in the stories they tell are engineers or applications oriented scientists with a wide shared base of technical and scientific knowledge. The competition is among ideas and approaches. Some approaches prove better than others, or their proponents more skilled in pushing them forward and getting them to work. When that happens, all have learned something, although disagreements are not necessarily put away. The process moves on.[6]

This point of view of course is not necessarily antithetical to the Schumpeterian one. Both capture part of what is going on. Advancing technology is a matter of both business and professional competition. And it is a matter of business and professional cooperation as well.

Recently writings by economists have begun to pick up on interfirm technology sharing and R&D cooperation. However, there still is inadequate attention to the roles of public finance, of professional associations, and of universities and other expressly public and cooperative components of the innovation systems. And where there is attention it often is rather simpleminded.

Here, I want to focus on the role of universities. Within the United States, university science and engineering, and our science-based industries, grew up together.[7] Chemistry took hold as an academic field at about the same time that chemists began to play an important role in industry. The rise of university research and teaching in the field of electricity occurred as the electrical equipment industry began to grow up in the United States. In both cases the universities provided the industry with its technical people, and many of its ideas about product and process innovation.

The situation is dynamic not static. Academic research was very important to technological developments in the early days of the semi-

---

[6]For good examples of such studies see Constant 1980 and Hounshell 1984.
[7]See, e.g., Noble 1977, Thackray 1982, or Rosenberg 1985.

conductor industry, but as time went by, research and development in industry increasingly separated itself from what the academics were doing. As I will document in a moment, at the current time academic biology and computer science are very important sources of new ideas and techniques for industry. The latter is a new field, and the former is experiencing a renaissance. On the other hand, technologies associated with complex product systems or production processes, like aircraft and aircraft engines, telecommunications, and semiconductor production, involve much that the academics do not do, and mostly do not know about in any detail.

In our questionnaire, my colleagues and I asked our respondents to score on a scale from one to seven the relevance of various fields of basic and applied science to technical advance in their line of business. We also asked them to score, on the same scale, the relevance of university research in those fields. I propose that a high score for a science on the first question signals the importance of university training in that field, and a high score on the second indicates the relevance of what academic researchers are doing.

A large number of scientific fields received a relatively high score on the first of these questions. A much smaller number scored highly on the second. Recall my earlier remark that a number of fields of science are expressly concerned with the scientific underpinnings of various technologies. It is interesting that the fields of university research that were scored as highly relevant by a considerable number of industries tended to be of this sort. Computer science, material science, and metallurgy head the list. Academic research in chemistry, a more conventional science, also was scored as relevant by a number of industries, which is not surprising to anyone familiar with the long-standing close links between academic and industrial chemistry. University research in the engineering disciplines also received a high relevance score from a number of industries. Industries for which these fields are important look to universities for new knowledge and techniques as well as for training.

On the other hand, while physics as a field of science was scored as relevant by a large number of industries, academic research in physics was scored as relevant by only a few. This confirms the importance of training in physics to corporate research, and also signals that what the professors are doing in the way of research is at some distance from applications. Now the fact that academic research in a field like physics or mathematics was not scored by our respondents as relevant to technical advance by no means implies that literally. Historians of science and technology know better. Rather, the responses suggest that the impact occurs over the long run, and is indirect, through influences on the applied sciences and engineering disciplines, which are the fields the industrialists are hooked into.

Industries where technological advance is being fed significantly by academic research naturally look for close links with university scientists and laboratories where that work is being done. In the application oriented sciences and engineering disciplines research is to a large extent motivated to facilitate technical advance. Many academics are strongly interested in industrial practices. Knowledge of prevailing technology, its strengths, and its weaknesses, helps them to define their own research agendas.

In fields like electrical engineering and biochemistry, the lines between what academics do and what at least some corporate scientists and engineers do may be blurred, not sharp. Thus, some research scientists and engineers in companies like IBM, AT&T, and DuPont, engage in research that is virtually indistinguishable from what their academic counterparts are doing. While in some cases this research is motivated by beliefs that a particular new product or process innovation will come of it, in many instances the reason is that this research is needed if the company is to stay abreast of developments in the relevant scientific fields, from wherever these developments may come (see, e.g., Rosenberg 1988). On the other hand, in fields like biochemistry and electrical engineering, academics often come up with patentable inventions.

4 What Should Be Public and What Should Be Private?

The lines between what is private and what is public about technology are blurred and shifting. A good portion of the debate about technology policy is about where these lines are drawn, or about redrawing them. I wish to conclude this essay by considering some of the current issues under debate.

At the present time in the United States there are many voices raised pressing the case for better enforcement of intellectual property rights. The complaint, of course, is that foreign firms are stealing American-created technology, but the argument is not simply about American interests. It also is that, if innovators cannot reap returns, innovative R&D will disappear. I think there is something to that position. But the problem is the broad one of enabling an innovator somehow to reap returns, not the narrower one of the adequacy of intellectual property rights. Patents are the principal means of appropriating returns in only a few industries, and many of the loudest complaints are not coming from these. What has been happening, I believe, is that generic public knowledge has grown in power in many fields, and this has enabled sophisticated companies quickly to find the generic technological basis of a competitor's innovation, and to come up with an equivalent. In many of these industries, patents never were important. A head start was, and now the advantage a head start gives is diminished. There is a problem here, but it is not necessarily a problem of intellectual property rights.

Another current issue is where to encourage firms to cooperate in generic research, and what greater cooperation would buy them. If I am right about the enhanced power of generic knowledge, it is of increased importance that firms tap into its source, which I have suggested requires that they themselves work at that endeavor. And cooperative generic research programs seem to some an attractive vehicle for doing this. However, it is important to recognize that the advantage that can be gained by participating firms will not be gained unless they individually build something proprietary from what they have learned cooperatively. Cooperative generic research may (or may not) be a superior substitute for individual uncoordinated basic research effects, but is a complement not a substitute for private efforts at product development.

While a moment ago I characterized much of the current argument about intellectual property rights as superficial, there are some quite deep questions about what should be patentable and what should not, about the scope of allowable patent claims, criteria for judging whether a patent has been infringed, etc. Yet there is very little careful analysis behind the key decisions affecting these variables. Thus as a result of a series of court decisions and congressional votes, in the United States substances found in nature in dilute or non-pure form now are patentable in pure form. Not simply the process which creates these pure forms but the natural substance itself has been ruled patentable. More generally, there has been a tendency to allow quite broad patent claims. At the present time the scope of several basic patents in biotechnology is under contest. At stake is whether the further development of this technology is to be the province (subject to licensed extension) of a single or small number of parties, or if it is to be open to a large number. While one can argue what ought to be both ways, it is clear that society can influence which way it is by decisions about the scope of proprietary patent rights.

A number of years back there was a change of policy regarding who owned patents emanating from university research financed by the National Institutes of Health and the National Science Foundation. The old rule had been that rights rested with the government, and the general policy was open licensing of these patents. Under the new policy, patents are owned by the researcher. While these rights tend to be given over to the university, the researcher now can profit handsomely. A number of observers have argued that a consequence has been less open communication among academics in fields of commercial relevance.

I am concerned that successful university entrepreneuring in biotechnology and computer science has caused a swing in beliefs about the appropriate roles of universities, so as to diminish emphasis on their mission as guardians of public knowledge. I believe that such a swing would be both ineffective and harmful. If I am right, in most fields of technology what gives advantage to a company is a particular manifestation of a generic technology that is tailored to its own particular circumstance, products and processes. In most fields university scientists are not very good sources of proprietary capabilities. Computer programs and products of biotechnology currently are exceptions. But it may be a mistake to glorify this and try to make the exception the rule. To try to make universities more like industrial labs will tend to take attention away from their most important functions, which are as a major source of new public technological knowledge, and as society's most effective vehicle for making technological knowledge public.

Technology is a public good as well as a private one. It is important to preserve both aspects.

References
Constant, E. W. 1980. *The Origins of the Turbojet Revolution*. Baltimore: Johns Hopkins University Press.

Dosi, G. 1982. Technical Paradigms and Technological Trajectories, *Research Policy*, vol. 111, pp. 147–162.

Dosi, G., C. Freeman, R. Nelson, G. Silverberg, and L. Soete. 1988. *Technical Change and Economic Theory*. London: Printer Publishers.

Hippel, Eric von. 1988. *The Sources of Innovation*. New York: Oxford Press.

Hounshell, D. A. 1980. *From the American System to Mass Production, 1800–1932*. Baltimore: Johns Hopkins University Press.

Levin, R., A. Klevoick, R. Nelson, and S. Winter. 1987. Appropriating the Returns to Industrial R&D, *Brookings Papers on Economic Activity*, 3.

Mansfield, E., M. Schwartz, and S. Wagner. 1981. Imitation Costs and Patents: An Empirical Study, *Economic Journal*. December.

Nelson, R., and S. Winter. 1982. *An Evolutionary Theory of Economic Change*. Cambridge: Harvard University Press.

Noble, D. 1977. *America by Design*. New York: Knopf.

Nordhaus, W. D. 1969. *Invention, Growth, and Welfare: A Theoretical Treatment of Technological Change*. Cambridge: M.I.T. Press.

Pavitt, K. 1987. On the Nature of Technology. Inaugural lecture given at the University of Sussex, 23 June 1987.

Rosenberg, N. 1976. *Perspectives on Technology.* Cambridge: Cambridge University Press.

Rosenberg, N. 1982. *Inside the Black Box: Technology and Economics.* Cambridge: Cambridge University Press.

Rosenberg, N. 1985. The Commercial Exploitation of Science by American Industry. In Clark, K., R. Hayes, and C. Lorenz, *The Uneasy Alliance: Managing the Productivity-Technology Dilemma.* Boston: Harvard Business School Press.

Rosenberg, N. 1988. Why Do Companies Do Basic Research (With Their Own Money)? Manuscript, Stanford University, January 1988.

Scherer, F. M. et al. 1959. *Patents and the Corporation.* Privately printed, Boston.

Schumpeter, J. A. 1950. *Capitalism, Socialism, and Democracy.* New York: Harper.

Teece, D. 1977. Technology Transfer by Multinational Firms: The Resource Cost of International Technology Transfer, *Economic Journal*, June.

Thackray, A. 1982. University-Industry Connections and Chemical Research: An Historical Perspective, in *University-Industry Research Relationships.* Washington, D.C.: National Science Board.

4

Successful Commercialization in the Chemical Process Industries

RALPH LANDAU AND NATHAN ROSENBERG

The story of technological innovation in Chemicals and Allied Products (SIC 28) is an anomalous one. It is the high technology industry about which the general public probably has the least knowledge. Yet, judged by criteria that are generally regarded as socially and economically significant, this sector might well be ranked at the top of the industrial scale.

The most common criterion for "high tech" is an industry's expenditure upon research and development.[1] Chemicals and Allied Products is near the very top of the list when industries are ranked in terms of the total amount of industry funds spent on R&D and virtually at the top in terms of the share of total R&D that is financed by private funds (see Table 4.1). Only the petroleum and food industries exceed the chemical industry's 98 percent figure for the share of R&D that is privately (i.e., nongovernmentally) funded, but these two industries spend far less in total dollars.

Another widely accepted criterion is the willingness to spend money on scientific research, especially basic research. Here again this industry receives top grades. A far larger share of R&D expenditures in Chemicals

The authors wish to acknowledge the invaluable contributions of Ashish Arora and Alfonso Gambardella to the preparation of this paper.

[1] A broader definition, encompassing all of the costs associated with innovation, might also include design, market development, worker training, plant startup, and initial operating losses. These expenses might properly be called intangible capital, and are almost always financed by equity capital—the most expensive kind. (See Bernheim and Shoven, "Comparing the Cost of Capital in the United States and Japan," this volume.)

TABLE 4.1

The Major R&D Investment Industries: 1990 Estimates

R&D Expenditures (in billion $)

Industry	Total	Privately Financed	Percentage Privately Financed
1. Aerospace	24.26	6.88	28
2. Electrical Machinery Communications	16.19	10.10	62
3. Machinery	14.96	13.07	87
4. Autos, Trucks, Transportation	10.86	9.35	86
5. Chemicals	10.76	10.54	98
6. Professional and Scientific Instruments	4.24	3.58	84
7. Petroleum Products	1.88	1.86	99
8. Food and Beverages	1.52	1.52	100
9. Rubber Products	1.43	1.10	77
TOTAL	86.10	58.00	

SOURCE: Battelle Memorial Institute.

NOTE: Total R&D estimated at $138.7 million, of which all industrial R&D is $100.1 billion (66% comes from companies and the rest from government) so that the above are the bulk of the investors in R&D.

and Allied Products consists of basic research, and basic research and applied research together represent a much larger share of total R&D for this industry than for any other industry (see Table 4.2). It is tempting to conclude that this industry has received so little public attention because its performance has been, in certain respects at least, so exemplary! A more important reason may be the low visibility of an industry most of whose output consists of intermediate goods rather than final product.[2] It is, as a result, difficult for the public to understand how pivotal this industry is in the provision of food, clothing, health, and shelter. Its low visibility may also account for the lack of awareness that it is one of only two major high-tech industries that has consistently maintained a strong competitive position in international trade, when competitiveness is measured by a positive balance of trade. Some sense of the diversity of activity in this sector is conveyed by stating that its output includes paints, pharmaceuticals, soaps and detergents, perfumes and cosmetics, fertilizers, pesticides, herbicides and other agricultural chemicals, solvents, plastics, synthetic fibers and rubbers, dyestuffs, photographic supplies, explosives, antifreeze, and many other kinds of chemicals.

---

[2]Only about one-quarter of this industry's products go directly to the consumer.

TABLE 4.2

Percentage Composition of the R&D Expenditures in the Six
Industries of the U.S. Economy Where R&D is Mostly Concentrated

	Chemicals and Allied Products			Non-electrical Machinery			Electrical Machinery			Automobile and Other Transp. Equipment			Aeronautics and Missiles			Scientific and Professional Instruments		
	BR	AR	D	BR	AR	D	BR	AR	D	BR	AR	D	BR	AR	D	BR	AR	D
1965	12.9	38.8	48.3	2.1	12.9	85.0	4.6	13.5	81.9	3.0		97.0	1.4	14.3	84.3	na	na	na
1966	13.0	39.0	48.0	2.1	13.0	85.0	4.0	12.0	84.0	3.0		9.0	1.0	14.0	85.0	na	na	na
1967	12.4	37.8	49.7	2.0	13.2	84.8	3.4	12.9	83.7	na	na	na	1.3	12.8	85.9	na	na	na
1968	12.5	37.1	50.4	1.9	12.8	85.2	3.3	13.7	83.0	na	na	na	1.2	11.9	86.9	na	na	na
1969	12.7	38.5	48.8	1.3	15.2	83.5	3.1	15.0	81.9	na	na	na	1.1	10.2	88.7	na	na	na
1970	12.0	38.7	49.3	1.3	15.2	83.5	3.3	15.0	81.7	na	na	na	1.2	9.6	89.2	na	na	na
1971	13.2	38.8	47.9	1.1	14.0	84.9	3.2	15.3	81.5	0.7	8.4	90.8	1.1	9.4	89.5	2.2	11.4	8.4
1972	12.9	39.5	47.6	1.2	13.6	85.2	2.9	16.3	80.8	0.6	7.9	91.6	1.1	8.5	90.4	1.9	11.7	86.4
1973	10.8	40.8	48.4	1.1	13.6	85.3	3.3	15.6	81.1	0.3	6.2	93.5*	1.0	10.1	88.9	2.4	10.5	87.0
1974	11.3	39.8	48.9	1.0	13.0	86.0	3.3	15.6	81.1	0.4	6.4	93.2*	1.0	11.6	87.4	2.2	11.0	86.8
1975	10.4	38.9	50.6	1.1	12.2	86.7	3.5	15.7	80.8	0.5		99.5*	0.8	11.2	88.0	1.2	9.6	89.2
1976	10.1	41.0	48.9	1.6	11.3	87.1	2.9	17.4	79.8	0.3		99.7*	0.9	10.5	88.6	1.7	11.6	86.7
1977	10.3	41.7	48.0	1.5	11.2	87.3	3.0	17.1	79.8	0.4		99.6*	0.8	10.7	88.5	1.6	13.0	85.4
1978	na	na	na	na	na	na	na	na	na	na	na	na	na	na	na	na	na	na
1979	9.1	41.6	49.3	1.3	13.1	85.5	3.0	15.4	81.6	na	na	na	1.1	10.9	88.0	na	na	na
1980	na	na	na	na	na	na	na	na	na	na	na	na	na	na	na	na	na	na
1981	10.1	42.5	47.4	1.9	18.4	79.7	2.7	17.0	80.3	na	na	na	1.1	12.4	86.5	1.1	99.8	
1982	na	na	na	na	na	na	na	na	na	na	na	na	na	na	na	na	na	na
1983		46.0	5.0	1.4	1.5	84.0	3.1	16.5	80.4	na	na	na	1.1	25.0	73.9	na	13.4	na
1984	8.4	37.4	54.2	1.6	14.0	84.4	3.0	16.7	80.4	na	na	na	1.5	19.1	79.3	na	13.7	na

SOURCE: Percentages calculated on data published by the National Science Foundation, *R&D in Industry*, various years.

NOTES: BR=Basic Research; AR=Applied Research; D=Development; na=not available.

*Does not include other transportation equipment.

It also follows that the boundary line between chemicals and other industries is often difficult to draw and must, to some degree, be arbitrary. This vagueness is even apparent in the SIC classification, which has to refer to "Allied Products" in addition to "Chemicals" in defining SIC 28. Petroleum products and rubber might easily have been included in this classification, since they are industries where chemical processing activities are central.

In each of these industries some kind of chemical transformation takes place, as contrasted with other industries cited in Table 4.1 where the transformations are primarily physical in nature. High pressure and/or high temperature reactions and/or continuous-flow processing are essential defining characteristics, and have been so since about the time of the First World War. Hence, if the criterion for inclusion was that chemical transformations were a central part of the productive activity, a strong case could also be made for the inclusion of steelmaking, aluminum, paper and pulp, glass, and foodstuffs. Indeed, in terms of the central role played by chemical processing in generating technological innovations for other industries, the industry bears some interesting similarities to the role played by the machine tool industry in the nineteenth century. In both cases the industries concerned served as the locus of innovative activities and were responsible for the diffusion of new technologies and other spillover effects to a wide range of other industries which were involved in similar kinds of productive activities (Rosenberg 1963).

Not only are the boundaries of the industry very imprecise; in addition, there is a great deal of variety within its boundaries. The nature of the innovation process as well as the trajectories along which it moves, are distinctly different between organic and inorganic chemicals; and the pharmaceutical sector has a research process, a class of final products, and a regulatory environment that sets it distinctly apart from all other sectors of the industry.

How this industry became such an important part of the modern industrial society is not only an interesting story in its own right, but as the first science-based industry it has numerous lessons to teach more recent science-based industries.

1   A Brief History of the Origins of the Modern Chemical Industry

The manufacture of chemicals received its impetus in the nineteenth century from the rapidly expanding requirements of the textile industry, but soon established itself as an innovative sector in its own right. Britain had achieved an initial dominance in chemicals because of its early leadership in textiles, and developed an enormous LeBlanc soda ash capacity. However, as the continental countries began to catch up, the newer Solvay process undermined British superiority. The British failure to match the new competitive processes and, even more important in the long run, the failure to develop new products, facilitated the eventual emergence of German leadership in the industry. Part of the reason for the British failure may have been that industrialization had come earlier in that country, and hence more capital had been sunk in plants embodying a now-obsolete technology. More importantly, technical education and scientific research in Britain were not accorded the status and government promotion that applied in Europe, particularly Germany.

As early as the 1850s, most of the really important research was done on the continent where the education of chemists was more systematic and pervasive. This was particularly evident in the evolution of the synthetic organics industry which revolutionized dyestuffs and eventually opened the road to plastics, synthetic fibers, and modern pharmaceuticals. Still another reason may have been that British investors found opportunities for lower risk and higher financial returns overseas, especially in their empire, as well as in the more traditional fields of iron, coal, steel, and textiles;[3] whereas the German investors, as latecomers on the industrial scene, had to look elsewhere for new opportunities and found them in domestic industries based more heavily on scientific research and newer technologies. The Germans appear, of necessity, to have adopted longer time horizons than the British in their industrial undertakings.

The Germans were particularly skilled in applied scientific research. Close relations were fostered between industry and the universities, and industrialists sent their children to study science in the universities far more frequently than did their British counterparts. German industry in the latter part of the nineteenth century established important research laboratories to complement academic research, whereas no such laboratories seem to have existed in Britain until the First World War.

Close study of the evolution of new chemical discoveries in the second half of the nineteenth century suggests that there were many contributions in basic discoveries and scientific understanding from outside Germany, including important contributions from Russia, France, and England. The Germans may actually have been less inventive at first in

---

[3]Late twentieth-century America over-invested in commercial real estate at the expense of productive equipment for perhaps similar reasons.

this regard than has previously been thought. Nevertheless, they became especially skilled in the production and marketing of organic chemicals derived from coal, perhaps partly because of their determination to become militarily more self-sufficient. Lacking an empire and possessing therefore a relatively smaller domestic market, they devoted great energy to improving their export capability. Because initially the several German states were very supportive of the early dyestuffs manufacturers, many new firms were formed, and the intense competition between them not only forced prices down but also led to frantic searches for new colors and differentiated products. This also demanded close coordination with customers' needs, at which the Germans excelled, and the development of better processes to make the basic dyestuffs—all essentially batch in character.

An example of this process is found in the history of the first dye which was discovered in England by William Henry Perkin in 1856 (mauve). It was synthesized from coal tar residues which were abundant by-products of the large illuminating gas industry. Perkin was in turn a student of a very famous chemist, Dr. August Wilhelm von Hofmann. Hofmann was himself a student and assistant to one of the most famous German chemists, Justus Liebig (Beer 1959). In the history of the dyestuff and organic chemical industry more generally, von Hofmann occupied as central a place as William Walker was later to achieve in the United States for chemical engineering and modern chemicals manufacture. He was a singularly attractive and creative figure, who had moved to England in 1845 to teach chemistry. Liebig had preceded him in a triumphal tour of England in 1842, at a time when Liebig's ability to popularize chemistry by showing its usefulness to industry and agriculture filled a vacuum not provided by distinguished British chemists such as Dalton, Davy, and Faraday. His book on chemistry as applied to agriculture was very popular with the English landed gentry. Englishmen who wanted training in chemistry went to Germany, especially Göttingen and Giessen. Their enthusiasm for the German methods of teaching chemistry led them upon their return to wish to establish educational facilities modeled on German universities, such as Liebig's Giessen.

However, the attempts to found the Royal College of Chemistry, begun in 1845, proved to be unsuccessful. It was eventually folded into the Museum of Practical Geology in 1865 as its chemical department. At the same time, Hofmannn received an extremely attractive offer to return to Germany. He had become disappointed with the unprogressiveness of the British dye industry, the backward state of organic chemical education, and the lack of sympathy on the part of business, the government, and the very conservative banks. In Germany, a very different atmosphere prevailed. There he could invigorate chemical education and systematic research and help turn out more chemists for industry, since there were insufficient academic positions for them. This episode reveals some of the major weaknesses that continued to plague Britain's scientific and technical education even well into the earlier part of the twentieth century. Britain under Queen Victoria was becoming preoccupied with empire. Its great universities, Oxford and Cambridge, neglected science education in favor of the classics, languages, religion, and literature. Students went into the armed forces, the foreign service, law, politics, the clergy, and medicine. In a fashion consistent with the aristocratic nature of British culture, sons of industrialists neglected science and shunned industry. Although Britain was the first country to experience an industrial revolution, it would be fair to say that it never experienced a bourgeois revolution (Cannadine 1990, Wiener 1981).

Germany, too, was under aristocratic influence, but of a more militaristic and autocratic character than the British. The Germany that was united by Otto von Bismarck in 1871 was dominated by the landed aristocracy of Prussian Junkers. Nevertheless, learning and scholarship were long respected in Germany. Even today eminent German industrialists (particularly in the chemicals industry) seek a part-time position of some sort in a university so that they may be addressed as "Herr Professor Doktor." There was also a persisting authoritarian tradition, involving command by the state and obedience by the citizenry. The German government has always played a large role in the chemical industry as it became increasingly important to the economy and to national security. The British government, by contrast, took no initiatives to strengthen industry or its infrastructure (Grant, Patterno, and Whitston 1988), and favored free trade. In the 1860s and 1870s Britain had almost every comparative advantage in the new chemical industry. It had the largest supplies of high quality coal in Europe, and the largest and most successful textile industry. It knew how to make dyes. It was rich. But it let its advantages slip away.

It did not take the Germans long to comprehend the significance of Perkin's discovery, and because they were willing and able to risk more capital investment in production, distribution, and management, and exhibited a firm determination to take the initiative in dyestuffs manufacturing, they forged ahead, initially by copying the British techniques. The similarities with the recent Japanese emergence to positions of technological leadership are suggestive. The Germans, realizing that copying alone could not sustain for long the competition from the originators, soon developed research capabilities of their own and went on to improve, invent, and manufacture a number of other dyes, as well as some pharmaceuticals (which involve related chemistry). This effort provided a great spur to industrial development throughout the German economy and several major companies, like Bayer, BASF,[4] and Hoechst, emerged to compete intensely with each other. This vigorous competition among large German firms was a key factor in the many advances made in organic chemical synthesis in the late 1800s and the first third of the twentieth century. In Alfred Chandler's terminology, they drove down the price of dyes and the chemically related pharmaceuticals by using economies of scope as well as of scale (Chandler 1990a, 1990b).

By the end of the 1880s, the large German plants were producing more than 500 different dyes and pharmaceuticals at unit costs far below those of smaller competitors. Although England continued to manufacture dyestuffs, by 1873 its relative decline had begun, and it never regained its earlier prominence. Thus, the Germans came to excel in synthetic organic chemistry, based on coal tar and its derivatives. By 1913, German companies produced 140,000 tons of dyes (72 percent of which came from the big three), 10,000 tons came from Switzerland, and only 4,400 tons from Britain. This rapidly growing German domination of dyestuffs helped to propel that country into the position of the strongest continental industrial power. The parallels to the Japanese strategy in electronics in recent decades are striking.

Across the Atlantic, the American chemical industry in the last half of the nineteenth century grew rapidly, primarily in inorganic chemicals and explosives. By the time of the First World War it had already achieved a large dollar volume, of the same order of magnitude as the German chemical industry, reflecting the much larger size of the American market. As an example, when Allied Chemical was formed after the First World War in 1920, it was a bigger company than the soon-to-be-formed ICI or I.G. Farben. It was during this phase of the construction and operation of large plants for chlorine, caustic soda, soda ash, sulfuric acid, phosphates, etc., that American companies developed skills in the manufacture of large volume chemicals on a continuous basis. However, the know-how to do so was empirical in nature and specific to the company involved.

The First World War, of course, by cutting off American imports of German dyestuffs and other fine chemicals, forced the development of an

---

[4]Heinrich Caro, the first BASF research director, had made a fortune earlier in an English dyestuff plant, and returned to Heidelberg in 1866.

indigenous organic chemical industry. At the same time, however, the Germans, who had been preparing for war for some time, had started research on a process to achieve self-sufficiency in a key chemical like synthetic ammonia for fertilizer. This was developed by Fritz Haber, who was located at the Technische Hochschule in Karlsruhe. His fundamental chemical breakthrough in synthesizing ammonia from atmospheric nitrogen came to the attention of Carl Bosch at BASF. Bosch had both chemical and mechanical engineering training at Leipzig and thus, in a sense, was a completely self-contained chemical engineer, the first important one in Germany. Bosch was able to develop a research laboratory specializing in designing and manufacturing high-pressure equipment needed to make the process work and then to design full scale plants. As a result of this, the BASF group became highly skilled in high-pressure technology and, to this day, still maintains an outstanding capability in this area as well as in process engineering more generally. Both Haber and Bosch received Nobel Prizes for their work and helped render Germany militarily strong during the First World War. Here, too, their strengths in dyestuff manufacture permitted rapid conversion to the manufacture of explosives.

2 The Growth of the American Chemical Industries in the Twentieth Century

An understanding of the present state of this industry, including in particular an appreciation of the reasons for its commercial success in the international arena, requires more historical perspective. America's considerable success in this sector has to be understood against the background of international differences in natural resource endowments, institutions, the size and speed of growth of the domestic market for its products and the resulting ability to exploit scale economies, and the gradual working out of path-dependent phenomena. In fact, the industry provides an excellent illustration of the limited explanatory value of analytical approaches, such as neoclassical growth theory, that ignore the importance of path dependence.

The rise of the American industry to international pre-eminence owes much to several different interrelated developments, which will be elaborated on in what follows:

1. The early and continuing pre-eminence of chemical research within the American industrial context.

2. The emergence of the distinctly American discipline of chemical engineering, supported by a university system that was highly

TABLE 4.3

Number of Car Registrations in the United States

1900	8,000	1950	40,339,077
1910	458,300	1960	61,671,390
1920	8,131,522	1970	89,243,557
1930	23,034,753	1980	121,600,843
1940	27,165,826	1985	131,864,029

SOURCE: Highway Statistics Summary to 1985 (U.S. Department of Transportation) and Historical Statistics of the United States.

accommodating to the needs of a newly emerging industry, and aided by the development of the unusual institution of the independent process-design firm. By contrast with the prominent role played by the university system in America, some aspects of which will shortly be described, it appears that, except for two that appeared in the 1930s in Britain, there were no departments of chemical engineering—or even extensive curricula in chemical engineering—outside of the United States before the Second World War.

3. The movement toward exploitation of the benefits of large-scale production, even in the early years. American chemical processing technology has shown a willingness to lead in the direction of construction of plants of large size, whenever such benefits appeared attractive, throughout the entire twentieth century.

4. The expansion to large-scale continuous processing and modernization of the petroleum refining industry after the First World War, necessitated by Henry Ford's introduction of the assembly line manufacture of cheap automobiles just before the war, and the rapid diffusion of the automobile in American society after that war (Table 4.3).

5. The cutoff of organic chemical imports from Germany in the First World War, which forced the United States to develop its own supplies, and thus vastly accelerated a growth process that was already at work.

6. The conversion of the organic chemicals sector from a coal base, which was—on a worldwide basis—the dominant feedstock up until the Second World War, to a petroleum based petrochemical industry that soon grew to enormous dimensions around the world and especially in the United States. The United States, as it happened, had vast oil and gas resources, and the American chemical engineering discipline had matured in a direction that was distinctly well suited to developing a large scale manufacturing technology that was appropriate to the new feedstocks.

The rebirth, in the two decades after the Second World War, of a powerful European chemical industry, once again provided strong competitive pressure on American firms and fostered continuing innovation in the postwar years. This competition was further enhanced by many new entrants from other industries into chemicals manufacture.

3 Research in the Chemical Process Industries

This sector has been the most research-intensive sector of the American economy throughout the twentieth century. Indeed, it is generally agreed that systematic industrial research had its historical origins in this industry, specifically in the German chemical industry in the second half of the nineteenth century (Beer 1959). As science became institutionalized in American industry through the establishment of research laboratories inside industrial firms, chemicals accounted for a far larger share of these research labs than any other industry. For the entire period 1899–1946, the chemicals industry accounted for more than 25 percent of all new laboratory foundings. If other industries that are heavily involved in chemical processing are included (e.g., petroleum products, glass, and rubber) these sectors together accounted for almost 40 percent of all industrial research labs founded between 1899 and 1946. If research intensity is measured by the employment of scientific personnel (scientists and engineers) expressed as a percentage of total employment, occasional surveys conducted by the National Research Council indicated that the chemical sector's research intensity was more than twice as great as any other sector between 1921 and 1946 (Mowery 1981).

Innovation in chemicals is more directly dependent upon prior findings in scientific research than is the case in most other high tech industries. In chemicals, especially organic chemicals, (which, of course, comprise many other classes of products than dyestuffs) the development of new products depends heavily upon the findings of scientific experiments performed at the laboratory level. The initial stages of the development of new and differentiated polymers, for instance, depend upon the laboratory combination of individual molecules (monomers) to form a single or multiple composite molecule (polymers). Depending upon the length, the shape and the chemical properties of the individual monomers, one may create materials with different chemical and physical properties such as plastics, resins, synthetic rubber and fibers, films, foams, etc. Of course, the role of laboratory research declines at later stages of the development process when chemical engineering becomes the fundamental discipline for translating the bench-scale reactions to the full industrial manufacturing scale. But, because of the highly scientific nature of chemical engineering itself, the best results in commercialization occur when chemical engineers work closely with chemists at every stage of the development process, beginning with the laboratory (Landau 1980).[5] Thus, the particular nature of the products and the production process in chemicals accounts for the significance of scientific research at the early stages of the innovation development cycle, which also sets closer ties later on between science and production than in most other industrial realms. Ralph Gomory (this volume) has identified what he calls the "ladder process" of breakthrough innovation, which is dominated by scientists building a product around a new idea. However, as he also points out, the later stages are dominated by manufacturing engineers making incremental but cumulatively massive improvements to existing products ("Challenges," Council on Competitiveness, June 1989, p. 1).

We have written a fuller account of this "cyclic" process, in Gomory's phrase, in several recent publications (Landau and Rosenberg 1990, 1991) wherein we lay particular emphasis on learning-by-doing, as reflected in learning or experience curve behavior. For this article, we wish only to illustrate its importance by showing, in Figure 4.1, a real case history for a high density polyethylene process practiced by a major international company. Initially, the gas-phase pilot plant (1975) produced at lower cost than the older solution processes, as practiced on a large scale. Nevertheless, production costs of the older technology declined more rapidly over the subsequent decade than the new pilot plant. However, when the company, despite this, took the risk and built a full-scale gas plant (1985), its production costs turned out to be decisively lower than the older, solution process technologies. Suppose that they had decided not to take this risk, and rest on the continuing improvements of their solution process plant—the conservative path. Another competitor, also experimenting with the gas-phase process in a pilot plant, but not possessing a full scale investment in a solution plant, might well have decided to go forward with the gas-phase process, and leapfrogged the earlier entrant. The learning curve is thus not at all a simple or obvious phenomenon, but an integral part of a corporate technological strategy. The existence of such risks underlines the findings of Bernheim

---

[5]Valuable sources on these issues include the Pimentel report on chemistry and the Amundsen report on frontiers in chemical engineering research, both issued by the National Research Council's Board on Chemical Sciences and Technology. See *Opportunities in Chemistry*, 1985 (Pimentel Report) and *Frontiers in Chemical Engineering: Research Needs and Opportunities*, 1988 (Amundsen Report). Washington, D.C.: National Academy Press.

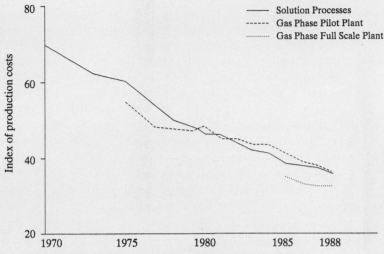

SOURCE: Industry data. Index derived from data in which input figures
for each year were expressed in 1988 prices.

FIGURE 4.1 High Density Polyethylene: Cost of Production

and Shoven (this volume) that risk premiums are an essential component
of the critically important cost of capital, especially when technological
change is rapid and internationally driven, as this article illustrates.

In chemicals, therefore, product innovation continues to remain intimately dependent upon the specific prior finding of basic and applied
research, and on the learning-by-doing cycles. In other high tech industries the links between scientific research and new product design and
development are less intimate, as is suggested by the much larger share
of D in their total R&D expenditures.

4 The Emergence of Chemical Engineering

The development of the discipline of chemical engineering was associated, to a quite striking degree, with a single institution: M.I.T.[6]

---

[6]We are particularly indebted in this section to the excellent article by John W. Ser-
vos, "The Industrial Relations of Science: Chemical Engineering at M.I.T., 1900–
1939", *Isis*, Vol. 71, No. 259, December, 1980, pp. 531–49, and to an unpublished
manuscript by James W. Carr, retired Exxon Engineer. Additional material comes
from *75 Years of Progress—A History of the American Institute of Chemical Engineers, 1908–1983*, by Terry S. Reynolds, American Institute of Chemical Engineers:
New York, 1983; W. J. Furter, ed., *History of Chemical Engineering*, published by
the American Chemical Society as *Advances in Chemistry*, Series 190, Washington,
D.C., 1980; and Roger L. Geiger, *To Advance Knowledge—The Growth of American
Research Universities 1900–1940*, Oxford University Press: New York, 1986.

Although Germany had, in the late nineteenth century, established a strong tradition of chemical research with a highly effective university base, the findings of the chemist were subsequently more or less handed over to mechanical engineers, and the plants designed and operated by companies that developed their own know-how. In Germany, separate roles were maintained for the chemist and the engineer which, in America, were partly fused into the new discipline of chemical engineering. This led, in Germany, to a highly artificial division of labor, inasmuch as mechanical engineers had little understanding of the chemical transformations involved in the manufacture of diversified chemicals, and were ill-trained to evaluate the numerous trade-offs that were inherent in the design of efficient chemical processing equipment. As M.I.T.'s Warren K. Lewis astutely observed of the German scene:

Details of equipment construction were left to mechanical engineers, but these designers were implementing the ideas of the chemists, with little or no understanding of their own of the underlying reasons for how things were done. The result was a divorce of chemical and engineering personnel, not only in German technical industry but also in the universities and engineering schools that supplied that industry with professionally trained men (Lewis 1953, pp. 697–98).

The essence of chemical engineering is a cluster of integrative skills that are applied to the design of chemical processing equipment. But there is much more to it than that. The chemical engineer has, at the center of his or her activities, the examination and synthesis of different technologies from the point of view of their comparative cost. The work and decision making of the chemical engineer is inherently economic as much as it is engineering, since it involves the explicit consideration of innumerable trade-offs in determining optimal design.

Moreover, it is clear from what has already been said that success in the commercialization of chemical processing innovations has depended critically upon the productivity gains realized through an improvement process that takes place after an innovation is introduced in the market. An integral part of this process of cumulative improvement has involved the exploitation of economies of large scale production and therefore a movement toward larger scale plants.

Historically, success in the commercialization of new technologies in this sector has turned upon the ability to make the transition from small scale, batch production to large scale, continuous processing plants. The benefits of larger scale have been so pervasive in this sector that chemical engineers have developed and employed a "six-tenths rule," which is regularly invoked, that is, capital costs increase by only 60 percent of the increase in rated capacity.

A distinctive characteristic of the American chemical processing scene even in its earliest years was the continuous pressure toward the exploitation of larger size, and the alacrity with which American firms moved in that direction. One authoritative study, discussing the American situation shortly before its entry into the First World War, has referred to "... the American attitude to the size of chemical works, which was, in short, to build a large plant and then find a market for the products" (Haber 1971, p. 176). It would seem plausible to infer that such an attitude developed at the time because the relevant markets were, as a matter of fact, both large and growing rapidly.

The acquisition of these skills involved an educational curriculum vastly different in content from that offered to a chemist or to a mechanical engineer. It took many years before these things were understood and properly sorted out.

What first appeared in a lecture series by Prof. Lewis M. Norton at M.I.T. in 1888 was essentially descriptive industrial chemistry, in an era of primarily heavy inorganic chemical manufacture (sulfuric acid, soda ash, chlorine, etc.) plus a few coal-tar-based organics. Industrial chemists focused on the production of a single chemical from beginning to end, and saw few unifying principles between the manufacture of different chemicals. Mechanical engineers focused primarily on the machinery, as they have since then. Applied chemists broke industrial processes into their chemical steps, with little attention to production methods. Industrial companies alone possessed the know-how to design and operate large scale plants, largely by empirical methods.

Not long afterward (1903), Prof. Arthur A. Noyes, an M.I.T. graduate and a Ph.D. from Leipzig, sought to convert M.I.T. into a science-based university including a graduate school oriented toward basic research. In order to train young people properly for careers in industry, he believed that graduates should be trained also in the principles of the physical sciences, using the problem-solving method, in which he pioneered. In 1903, he succeeded in establishing a German-style Research Laboratory of Physical Chemistry, although he also found himself personally financing about half of it! It produced the first Ph.D.'s from M.I.T. not long after. The Graduate School of Engineering Research was also established in 1903.

However, a different vision possessed Prof. William H. Walker, working closely with Arthur D. Little, who had already established his chemical consulting firm. They rather emphasized that M.I.T. should remain a school of engineering and technology, but should train the builders and leaders of industry by focusing on applied sciences. A sharp controversy between Noyes and Walker had begun that was to last for many years. Walker believed that learning physical science principles was not enough. Only through understanding of problems drawn from industry could a student move from theory to practice. In particular he felt that, in this way, the problems of scale, involving measurements in terms of tons rather than grams, could be understood by the young engineer, given the constraints imposed by materials, costs, markets, product specifications, safety and the many other considerations which up to then were solely the province of industrial firms. Furthermore, the large and growing domestic market fitted in well with America's tradition of mass production rather than the older craft traditions of European origin. Although he was primarily involved in chemistry and chemical engineering, his vision embraced other engineering disciplines as well.

Acting on his conception, Walker reorganized the languishing program in industrial chemistry, converting it from its heterogeneous collection of courses in chemistry and mostly mechanical engineering into a unified program in chemical engineering. As time passed, it was increasingly based on the study of "unit operations," a concept first used by Arthur D. Little in 1915, but practiced several years before. The central idea was to reduce the vast number of industrial chemical processes into a few basic steps such as distillation, absorption, heat transfer, filtration, evaporation, reaction, and the like. Hence, Walker and Little were developing their own scientific analysis of the principles of chemical engineering, based on unit operations but, unlike Noyes, they involved new conceptual frameworks not used or needed by chemists studying their basic science in the laboratory, because chemists did not concern themselves with how one designs industrial size plants and equipment.

In fact, what was emerging here was chemical engineering in the sense of a unifying discipline with direct applications, and guidance, for a very wide range of individual products and processes. An engineer trained in terms of unit operations could mix and match them as necessary in order to produce a wide variety of distinct unit processes. Such an engineer was much more flexible and resourceful in his approach to problem solving. He was also better equipped to take techniques and methods from one branch of the industry and to transfer them to other branches. This capability was especially valuable in the innovative process—particularly as new materials and new products emerged.

To emphasize his disagreement with Noyes' approach, Walker succeeded in establishing the Research Laboratory of Applied Chemistry in 1908. (At around the same time he was one of the founders of the American Institute of Chemical Engineers.) It was intended to obtain research contracts with industrial firms and thus to provide not only income but experience in real problems for both faculty and students. It would also serve as the basis for graduate studies in chemical engineering. It would involve close mutually beneficial links between M.I.T. and industry. In this era, it must be remembered, few American firms had their own research facilities (GE in 1900 and DuPont in 1902 were among the pioneers in establishing their own research laboratories). Walker had observed that the rapid growth of the German chemical industry was based upon close cooperation between industry and academics, and the existence of industrial research laboratories. Thus, his conclusions became one of the principal steps in the full internationalization of the chemical industry.

To further establish industrial linkage, Walker and his younger colleague, Warren K. Lewis, again with Arthur D. Little's support, founded the School of Chemical Engineering Practice in 1916. This gave students access to the expensive industrial facilities required to relate classroom instruction in unit operations to industrial practices, but still under the supervision of a faculty member.

As between Noyes and Walker, it was Walker's view that eventually prevailed. After the war, the rapid expansion of the undergraduate chemical engineering program paralleled and derived additional momentum from the rapid expansion of the American chemical industry—a momentum that had been accelerated by the war itself and the loss of German chemical imports. Thus, the first true link between the discipline of chemical engineering and the growth of the chemical industry appeared. It is a link that has remained close ever since. By the end of 1920 the formation of the first free-standing chemical engineering department in the world was assured, and Warren K. Lewis became its first chairman.

Under the leadership of Lewis and his colleagues, the chemical engineering department became an intellectually powerful center during this period, cemented by the publication of the pioneer text in 1923: "Principles of Chemical Engineering," by Walker, Lewis, and William H. McAdams (Walker, Lewis, and McAdams 1923). At the same time, Lewis developed a new kind of relation with industry which proved to have an enormous impact on the chemical process industries. The industry which proved the key to many future developments was petroleum refining.

5 Chemical Engineering and the Petroleum Industry

This industry had been broken up in 1911 under an antitrust judgment that split the Standard Oil Company into a group of competitive companies which did little in the way of research, development, or engineering. Before the First World War, the introduction of the assembly line at the Ford plant brought into existence the first automobiles aimed at a mass market. General Motors had been founded by W. C. Durant in 1908 and provided a wider range of vehicles. Although the war interrupted the diffusion of the automobile, the oil companies foresaw that gasoline demand would boom after the war. They did not feel that they were adequately equipped to develop and install improved continuous processes for large-scale production. The two largest refiners were Standard Oil Company (Indiana) and Standard Oil Company (New Jersey)—now Amoco and Exxon. In building up an R&D capability, Exxon hired Frank A. Howard, a patent attorney for Amoco (who also coordinated their innovative efforts). Their new Development Department with Howard as its head came into existence in 1919.

One of Howard's first acts was to engage the best consultant for Exxon that he could find, and that was M.I.T.'s Warren K. Lewis. A partnership was formed that would have a profound influence on Exxon as well as on chemical engineering. Lewis' first efforts were to provide precision distillation and to make the process continuous and automatic. Batch processes were clearly inadequate for the rising market demand for gasoline. However, in bringing about the transition from batch to continuous process technologies, the petroleum industry was also inventing the technologies upon which the petrochemical industries of the future would be based.

Furthermore, by 1924 Lewis helped increase oil recovery by the use of vacuum stills. From 1914–27, the average yield of gasoline rose from 18 percent to 36 percent of crude throughput. This work and his earlier bubble tower designs became refinery standards. At the same time, course work at M.I.T. was altered to embody these new concepts and their design principles.

By 1927, Howard began a series of agreements with the German chemical giant, I.G. Farben Industrie. Howard wanted access to their work on hydrogenation and synthetic substitutes for oil and rubber from coal. He felt this might increase gasoline yields and could promote Exxon's activity in chemicals. To exploit the application of these technologies to Exxon's purposes, Howard needed a whole new research group, particularly a group that was knowledgeable about chemical process technology. Again, Lewis was consulted and he in turn introduced Robert Haslam, head of the Chemical Engineering Practice School, to Howard. Haslam, on leave from M.I.T., formed a team of fifteen M.I.T. staffers and graduates who set up a research organization at Baton Rouge. Six of these later rose to positions of eminence in petroleum and chemicals. Haslam himself left M.I.T. and became vice president of the Exxon Development Department in 1927.[7]

In this one step, by exploiting the German technology for American refining use, Exxon had overcome much of the large gap that had so long existed between applied and innovative research in the petroleum industry, including introducing the use of chemical catalysis in petroleum refining. Much of what took place in modern petroleum processing until the Second World War originated in Baton Rouge, and it was basically the M.I.T. group that was primarily responsible for it. With the continuing advice of Lewis, and later (1935) of Edwin R. Gilliland, Baton Rouge produced such outstanding process developments as Hydroforming, Fluid Flex Coking and Fluid Catalytic Cracking (which ultimately became the most important raw material source for propylene and butane feedstocks to the chemical industry).

It will be apparent that Lewis had created a very different approach from the earlier Walker programs. Instead of bringing industry to the campus, as in the RLAC, he in effect brought the campus to industry. Unlike the Practice School, which also did this, he helped solve a number of the major problems of the industry. In so doing, he focused the discipline of chemical engineering on an overall, systems approach to the design of continuous automated processing of a large variety of products, first in petroleum refining but then, later, in chemicals. These developments provided the essential technological ingredients for the growth of the world petrochemical industry.

Thus, the earlier path along which America had progressed in designing equipment for a rapidly expanding petroleum industry had stimulated major learning experiences in continuous processing technology that were later transferred to the much more diversified canvas of the petrochemical industry. America's later commercial success in petrochemicals would have been a vastly different story had the country not enjoyed the learning benefits that flowed from her earlier experience with petroleum.

---

[7]By the mid-1920s, M.I.T. 's dominant influence throughout American industry was already apparent. At DuPont in 1926, for example, "...there were ten chemical engineers...in the Chemical Department. More than half of them had graduated from M.I.T." (Hounshell and Smith 1988, p. 279).

The rise of the petrochemical industry began with the entry of a few companies, particularly Union Carbide, Shell, Dow, and Exxon, not long before the Second World War. These firms soon encountered the need for chemical engineering skills as the scale of operations forced resort to continuous processing, just as it had previously in the petroleum refining industry. The techniques and skills formerly developed for the refining industry could now be applied to the problems of the petrochemical industry, as chemical raw materials began the epochal shift from coal-based to petroleum-based feedstocks. To be sure, the problems in manufacturing chemicals were different, and in many respects even more challenging: corrosion; complex product separations and purifications; diversified markets; toxic wastes and hazards; and many others.[8]

It is not commonly realized how far the transition to a petroleum base had already been carried even before the Second World War.

Between 1921 and 1939, the production of organic chemicals *not derived from coal tar* rose from 21 million pounds valued at $9.3 million to three billion pounds, with a value of $394 million. Coal tar chemicals production in 1939...still amounted to only 303 million pounds, valued at $260 million. The average 1939 price for petrochemicals was, in fact, thirteen cents per pound versus 87 cents for coal tar derived chemicals. The average price of noncoal tar chemicals had, over the period, been reduced by a factor of three, from a 1921 level of 43 cents per pound (Spitz 1988, pp. 67–68).

As the industry shifted to petroleum feedstocks in the interwar years and mastered the problems of large-scale, continuous process operations, the optimal size of plant often grew to exceed the market requirements of even the largest of western European countries. Since the European industry had relied much more heavily in its earlier years upon coal as the basic raw material, the transition to larger scale was impeded by skills, attitudes and educational preparation that had been developed under that coal-based industrial regime. European developments were also influenced by the determination of each country to maintain a capability for satisfying the requirements of its own domestic market.

Even in countries with a relatively large population, such as France and Great Britain, chemical firms planning new projects in the postwar period found it difficult to build a large enough plant that would have reasonably attractive

---

[8]For an examination of the ways in which modern chemical plant design evolved and differed from petroleum refining, see Ralph Landau, *The Chemical Plant* (Landau 1966).

economics. Substantial exports were needed to build such plants, but the products in question would not necessarily be salable in adjacent European countries, since potential purchasers were still averse to being dependent on supply from across the border (Spitz 1988, p. 348).

Building larger chemical processing plants is, however, much more than merely having assurance of access to sufficiently large markets. Such larger plants are necessarily also a product of technological innovations that make them feasible. In this respect it is much more common than it ought to be to assume that the exploitation of the benefits of large scale production is a separate phenomenon independent of technological change. In fact, larger plants typically incorporate a number of technological improvements, based upon the wealth of experience and insight into better plant design, that could be accumulated only through prolonged exposure to the problems involved in the operation of somewhat smaller plants. The building of larger plants must, as a result, often await advances in the technological capabilities in plant design, equipment manufacture, and process operation. Thus, the benefits of scale cannot be attained until certain facilitating technological conditions have been fulfilled.

Both as a conceptual matter and as a practical matter, it is not easy to disentangle the benefits of larger scale production from those achieved through introduction of improved equipment, improved design, or better "know-how," that is, better understanding of the technological relationships that are eventually embodied in the larger plant. Such later plants typically incorporate a large number of cumulative improvements and conceptual insights.

6 The Post-World War I German and British Position

The loss of the First World War, of course, weakened the German industry tremendously (although they had kept many of their technical secrets concealed from the allied investigating teams), and the numerous firms found themselves with large overcapacity particularly brought on by the recession of 1920–21. This brought about government encouragement for the major German companies to merge together to form the famous I.G. Farben Industrie.

Bosch was a central figure in this period. He noted that other countries which had been able to inspect German technology after losing the war were rapidly putting in ammonia plants of their own and would be creating severe competition for I.G. He therefore looked for other applications of their surplus high-pressure equipment, much of which was idle. What engaged him for this purpose was the hydrogenation of coal to create synthetic fuels. However, Germany in its weakened condition could not finance the extensive experimental work required nor did it have adequate raw materials, but it had the trained people. It was this situation that led to the arrangements with Standard Oil of New Jersey.

Other applications for the high-pressure equipment were found in the oxo-alcohol synthesis for detergent alcohols and the like. ICI, also encouraged to become a world class company by the British government, following this lead after the First World War inspection teams, developed methanol by similar techniques. This gave them high pressure experience and opened the door to their development of low density polyethylene, which initially required reaction pressures on the order of 30,000 pounds per square inch. It is noteworthy that BASF almost simultaneously was also able to develop such a high pressure polyethylene process. Experience does count!

What is notable in all of this is that, lacking indigenous raw materials, Germany was limited to coal as its base. This is an industry that has undergone a total transformation when its raw material base changes. Coal, alcohol, crude oil, or natural gas can all be transformed into many of the same organic chemicals. What proved decisive was that the cheapest feedstocks would come from petroleum and natural gas. When these materials became abundantly available in the United States, coal and alcohol chemistry were not economic for producing commodity organic chemicals. Thus, despite Europe's chemical industry being more advanced than that in the United States for a long time, the future of organic chemicals was going to be related to petroleum not coal, as soon as the four companies mentioned above turned their attention to the production of petrochemicals.

The European chemists of the earlier period concentrated on synthesizing molecules from available coal-derived feedstocks which largely involved aromatic chemicals. They therefore largely excluded industrial aliphatic chemistry. When the American oil and chemical companies recognized the value of petroleum hydrocarbons they not only came up with a number of industrially significant processes for aliphatic chemicals, but also discovered ways to make the original coal tar derived chemicals more inexpensively and in much larger quantities. European know-how greatly helped in this effort, particularly after the Second World War, when allied inspection teams combed German plants and laboratories to learn all they could find about German chemical knowledge and brought it back to the United States.

After the First World War the Dawes reparation plan forced the Germans once more to emphasize exports and this, of course, encouraged the German government to assist not only in the creation of I.G. but to turn a blind eye toward the creation of a European-wide system of cartels, cross licensing agreements, consortia and the like. Markets were shared, technology was carefully guarded, and the number of key producers in each country strictly limited. In addition, I.G. started a major search for autonomy in production of motor fuels, recognizing how important it would be in the next war. Thus, led by the now highly experienced high-pressure chemists and engineers, work began on the production of synthetic fuels by gasification of coal, hydrogenation of coal, and also on other related processes like carbonylation and hydroformylation. Ultimately the major triumph was the Fischer-Tropsch process. This plant was first built at Holten in the Ruhr shortly before the Second World War, and was developed at the Kaiser Wilhelm Institute for Coal Research in Muehlheim-Ruhr. This process was in fact a transition from coal tar organics to the present and future organic chemical industry, as it led to the ultimate emphasis on aliphatic chemicals that were made cheaply and better using petroleum and natural gas feedstocks. However, the cartel arrangements limited innovation among the members. The U.S. companies, because of American laws, were less inhibited and thus contributed to the competitive skills which created the great Second World War effort, and the dominance after. There was a tremendous burst of innovation between 1935 and 1950, which came from the loosening of the cartel bonds and the demands of the war.

The inspection teams found that the Germans had developed a competitive process and product for nylon, which DuPont had discovered before the war, and had also been engaged in the early phases of the development of a brand new category of polymers called polyurethanes by Otto Bayer in 1937, which ultimately was picked up by DuPont to make Spandex and Lycra fiber and by many American companies to develop the great polyurethane industry.

The reasons for the German chemical industry's being most advanced before 1939 had been its large coal and steel industry, its lack of protected colonial outlets, and its desire for self-sufficiency, which all combined together to favor rapid commercialization of new ideas based on chemistry from all over the world. The government's pro-cartel policy was also very important, especially after 1933. At first, the domination of chemists over engineers in Germany was a great strength in exploiting this situation but it became a drawback when chemical engineering became the key discipline for the design of large scale petrochemical and derivative plants. Another previous advantage was the better educational system in Germany and the high esteem in which technically trained people continued to be held. Another feature of the German system had been that much of their commercialization of innovation was due to superb development, because highly qualified scientists and engineers worked together and stayed together from the laboratory all the way through to the operation of the plant, rather than having a consecutive handing over of material from one step to the next. This is also the present Japanese technique for innovation (see Ken-ichi Imai, this volume).

7 The Path-Dependent Nature of the Industry's Growth

The foregoing account of the emergence of the American industry to a position of world leadership after the Second World War stresses its base in differences in natural resource endowments and the consequences that flowed, in a path-dependent way, from that initial difference. Thus, the abundance of a specific resource—petroleum—encouraged movement in a very particular technological direction. In addition, the parallel development of a huge automobile industry, uniquely suited to the large land mass of the country, also interacted in a path-dependent way with these resource differences to build and to reinforce an American comparative advantage in petrochemicals.

On the other hand, an important aspect of the discipline of chemical engineering is that it may offer ways of exploiting alternative lower-cost materials in the production of new or old products. The German Haber/Bosch process, the first great industrial milestone of chemical engineering, although developed by a chemist and a team of mechanical engineers, involved a new way of producing a very old product—ammonia. But it did so by *shifting* the underlying German resource base from a limited source—the byproduct ovens of the iron and steel industry—to an immensely abundant base—atmospheric nitrogen.

There is an interesting counterpoint to these examples. On the one hand, the huge United States endowment of petroleum gave rise to a whole set of path-dependent phenomena by shifting its industry to dependence upon a resource that was available in abundance. On the other hand, the Haber/Bosch process, emerging in the second decade of the twentieth century, was a supreme instance of a country developing a new technology that enabled it to *overcome* the shortage of a critical input—chemically bound nitrogen. Thus, it is safe to say that, in these matters, the sequence in which events unfold matters very much, and that many events in which we have a current interest can be accounted for only by recourse to path-dependency types of explanation.

In both countries, differences in resource endowments—abundance of petroleum and scarcity of nitrogen fertilizer—gave rise to research activities that had successful outcomes, but successful outcomes in solving very different classes of problems. In each case also the expanded technological capabilities gave rise to a wholly new set of consequences that cannot be understood outside of a sequence of events whose logic can only be grasped in path-dependent terms. Indeed, Germany's munitions- and food-producing capabilities during the First World War would have been vastly diminished—and twentieth century history vastly different as a consequence—had the sequence of events that led to the Haber/Bosch process not occurred.

But path-dependency is neither the whole story nor a simple story. What can be said is that Europe's early lead in the chemicals industry, in the late nineteenth and early twentieth centuries, did not provide the most effective path for leadership in the chemical processing technology that later came to dominate the twentieth century. Unfortunately for Germany, the dyestuffs industry, which they totally dominated in the late nineteenth and early twentieth centuries, locked them into certain methods and approaches that were far less suited for other branches of the chemical industry that became very important later on. In this sense the early leadership of the German chemical industry may have created serious impediments, through path dependence, to later success as the German industry and educational system developed in ways that did not conform to the requirements for commercial success later in the twentieth century. One major problem is that many dyestuffs are specialty products that are typically manufactured by batch processes. Warren K. Lewis has pointed out that

the very extent of the German success (in synthetic organic materials) deflected professional development in the wrong direction. The manufacture of dyestuffs was predominantly a set of small-scale operations, the equipment for which was blown-up laboratory apparatus, enlarged in dimensions. Similarly, the process was the batch operation of the laboratory. (Lewis 1953, p. 697)

Within the educational system, the German universities continued to produce students who were very capable research chemists but who could not design the system for manufacturing the products that they invented.

Although America was a distinct late-comer to the chemicals scene, its abundance of petroleum (and natural gas) deposits, as well as the experience that it gained in continuous processing methods in exploiting these deposits, opened up a development path that provided an excellent entry into the continuous processing technologies that became so central when the industry later moved to a petrochemical feedstock base. Even so, that can be only a part of the story: Opening up a path in no way guarantees accelerated movement along that path. To put it in modified Toynbeesque terms, challenges *sometimes* generate vigorous responses; but sometimes they also prove to be insurmountable barriers, or to lead to quite unexpected dead ends (Spitz 1988, pp. xiii, 26–29, 57–60). In the American case, the early exploitation of large petroleum deposits not only led to immediate commercial successes but also led to the acquisition of skills, methods and approaches with respect to process design that provided the basis for future success as well.

The differences with respect to the German experience were encapsulated in a private statement by a German chemical executive: "Chemists can't calculate." He was referring to the chemist's lack of comprehension of process design and its integration with product manufacture. The German refusal to learn this lesson until well after the Second World War, based upon a reluctance to dislodge the industry from the predominance of the chemists and to create space for the chemical engineer, undoubtedly played a large role in the decline of earlier unquestioned German leadership in the chemical industry. One pertinent example occurred in the mid-1950s, when a German chemical company chose to expand an initial plant for a modest production of a petrochemical product fourfold by building three more duplicate trains (that is, three more process lines), so as not to risk the problems of scaleup! However, the American-style chemical engineering was eventually adopted by the major German companies, and the universities moved in the same direction, albeit in a manner peculiarly suited to their own traditions (Landau 1958, pp. 64–68, 115).

By grafting the strengths of the American approach on to its traditionally strong chemical research, and greatly aided by a very favorable macroeconomic climate for industry, the German chemical industry has once again become a major factor in the world industry. Three of the world's top ten chemical companies are in West Germany, and three are from the United States. Nevertheless, the top ten chemical companies account for less than twenty percent of the industry's worldwide sales, compared with the approximately 60 percent of car sales which the top ten auto companies command. The Japanese chemical companies are only one-third to one-tenth the size of the large western companies. This seriously hampers their effort to become a major force outside Japan. On the other hand the English, French, Italians, Dutch, and Swiss have carved out very competitive industries of their own in recent years, but those developments are beyond the scope of this paper.

8 Expansion and Change in the Postwar Chemical Industry

A major factor in the postwar history of the chemical industry was the spectacular growth of demand. During the period 1950–70 the chemical market grew at annual rates of about 2.5 times the growth rate of GNP. Moreover, the growth of the market was essentially due to a high rate of growth of demand for relatively homogeneous products, often simple petrochemicals. Figures 4.2–4.5 illustrate the spectacular growth of four important chemicals—vinyl chloride, ammonia, polypropylene resins, and noncellulosic fibers—for this period. Table 4.4 reports the average percentage growth rate of annual output of 24 major chemical products in the United States, and the corresponding average annual percentage change in market price, for the 1960s (Lieberman 1987, p. 262). As can be seen, many of these products had annual growth rates higher than ten percent and almost all grew at a rate higher than five percent.

This had important consequences for the production of chemical products, and also established somewhat different patterns of innovation development in this field. In particular, immense new markets and large scale demand created the possibility for large scale expansion of the relevant production processes (Landau 1989, 1990b). The early phases of the innovation cycle in many chemical and petroleum-based products and processes had already occurred in the prewar era. Styrene and polystyrene, vinyl chloride and polyvinyl chloride, low-density polyethylene (high pressure), ethylene oxide and glycol by direct oxidation, alkylamines, methacrylates, and others, were first pioneered, on a small scale, before the Second World War. In the postwar period, the dramatic growth of demand, particularly the high growth of demand for relatively homogeneous bulk intermediate chemicals, created opportunities for scaling up chemical production processes to levels far higher than those of the prewar period.

At first, capacity increases were obtained without undertaking substantial transformations of the underlying characteristics of the production process and equipment design. Larger scale plants were initially built up by simply duplicating the equipment of a smaller scale plant (say, by building two reaction "trains" instead of one) (Spitz 1988:418).

This, however, did not lead to substantial reductions in production costs. The mere duplication of the internal equipment of a given plant

TABLE 4.4
List of Products and Selected Statistics

Product Name	Coverage Period	Numbers of Producers		Market Size in 1973 ($ million)	Total Numbers of Process Patents Over Coverage Period			Average percent Growth Rate of		Average Annual Percent Change in Market Place
		At Beginning of Period	At End of Period		By U.S. Producers	By U.S. non-Producers	By Foreign Firms	Annual Output	Cumulative Output	
Acrylonitrile	1960–73	4	4	143.1	56	40	220	14.0	18.1	−6.6
Aniline	1961–73	5	4	40.1	6	17	39	10.2	6.2	−5.2
Bisphenol	1959–73	3	4	51.5	11	6	51	13.9	16.3	−4.1
Caprolactam	1963–73	3	3	129.4	2	10	145	9.6	23.3	−8.0
Carbon disulfide	1963–73	5	4	29.8	3	7	37	2.8	4.6	−0.3
Cyclohexane	1959–73	2	9	105.7	40	17	70	12.3	13.5	−3.6
Ethanol	1959–73	5	6	108.1	4	24	119	2.5	7.1	−0.1
Ethylene glycol	1960–73	9	13	224.0	9	26	73	7.8	10.0	−3.9
Formaldehyde	1962–73	15	19	119.6	7	1	161	9.4	9.8	−4.7
Hydrofluoric acid	1962–73	10	9	135.6	6	24	65	6.9	8.2	+0.3
Isopropyl alcohol	1964–73	3	4	102.5	30	25	55	3.4	5.9	−0.7
Maleic anhydride	1960–73	4	7	42.6	25	32	90	11.9	13.5	−5.2
Methanol	1958–73	11	11	117.7	14	56	227	9.4	9.2	−6.6
Neoprene rubber	1960–73	1	2	147.6	2	0	3	3.1	8.3	−0.4
Pentaerythritol	1952–73	5	5	18.3	10	4	39	5.4	10.3	−3.3
Phenol	1959–73	8	11	171.5	17	46	124	9.7	10.2	−6.1
Sodium	1958–73	3	3	61.7	14	24	42	1.6	3.3	+0.1
Sodium chlorate	1963–73	7	9	21.6	16	8	42	6.2	9.3	−2.2
Sodium hydrosulfite	1964–73	6	4	29.6	1	4	18	4.6	3.5	+1.0
Sorbitol	1966–73	4	4	31.4	1	2	18	7.1	6.3	−0.4
Styrene	1958–73	7	12	419.0	45	53	88	10.9	12.5	−5.3
Trichloroethane	1967–73	3	4	48.0	3	4	14	6.1	10.0	−3.3
Vinyl acetate	1960–73	4	5	104.8	24	22	160	13.2	13.5	−6.2
Vinyl chloride	1962–73	12	10	222.1	7	14	110	14.4	14.7	−6.3

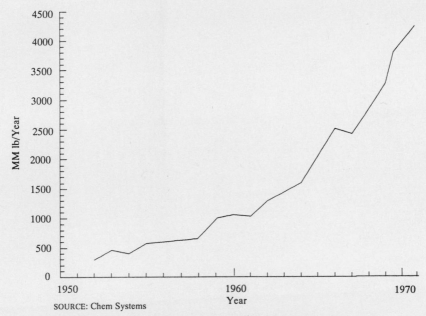

SOURCE: Chem Systems

FIGURE 4.2 U.S. Production of Vinyl Chloride Monomer

implied no reduction in average costs because of the higher total output levels. The cost of producing a given volume of final output using two reactors instead of one merely involved constant returns to scale, with no reduction in the average costs of production.

Systematic reduction of average costs did occur when producers realized that one could design larger scale plants by developing single "train" reaction system plants with the same capacity as two smaller reaction vessels. Single "train" reaction systems enabled the chemical companies to enjoy significant economies of scale which could not be exploited by simple duplication of the structure of the existing production process. Of course, single "train" plants entailed much higher fixed charges which could be absorbed only because of the high rate of growth of demand and the size of the chemical markets during the 1950s and 1960s.

Single "train" equipment design was first applied to the production of ethylene and ammonia, two of the most important petrochemical products (Spitz 1988, p. 425). As a result, the size of ethylene and ammonia plants grew at spectacular rates in this period. More generally, single "train" production processes induced dramatic increases in the

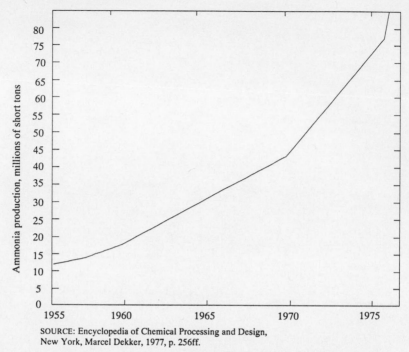

SOURCE: Encyclopedia of Chemical Processing and Design,
New York, Marcel Dekker, 1977, p. 256ff.

FIGURE 4.3   Annual World Ammonia Production

production of intermediate bulk chemicals, such as those documented in
Figures 4.2–4.5.

Thus, the drastic increase in the size of the chemical plants was
strictly related to the high level of the absorption rate of these products
in the market. Yet, it is essential to note that in general the output
capacity of each individual giant plant was well within the absorptive
capacity of the market. In the case of ethylene, for instance, each new
plant, even one of a billion pound capacity, corresponded to a relatively
small share of the total market (Spitz 1988, p. 457).

The opportunities for scaling up plant size and, more generally, for
enjoying economies of production through process development, induced
the chemical producers to focus their attention more carefully on the
production process. In the postwar period there were substantial needs
for better design and engineering of the production process. On the one
hand, this implied that chemical manufacturers paid increasing attention to process development by setting up internal departments aimed at
designing and engineering new and improved manufacturing processes.
On the other hand, the increases in capacity and the importance of

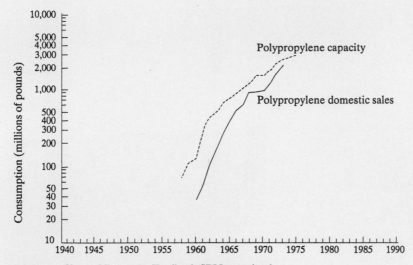

SOURCE: *Chemical Economics Handbook*, SRI International

FIGURE 4.4 U.S. Capacity and Domestic Sales
for Polypropylene Resins

manufacturing costs for the commercial success of a firm meant that a large market was emerging for improved processes for a set of relatively standardized, high volume products. Since process development for such products could be undertaken without simultaneously investing substantial funds in product R&D, there was a place for specialized contractors to design and engineer chemical production processes for the major chemical producers.

Indeed, specialized engineering firms (SEFs) entered this market and played a major role in this industry during the post World War II period. Chemical firms had subcontracted functions like procurement and installation to the SEFs even before the war, when design and process development was essentially carried out in-house. Chemical companies typically carried out their own process design, and used external contractors to handle construction, piping and mechanical work, electrical work, and other separate facets of the project. Petroleum companies typically farmed out most of the detailed design as well to SEFs. After the war, the chemical firms increasingly relied upon the SEFs to design, engineer and develop their manufacturing installations. In the 1960s nearly three-quarters of the major new plants were engineered, procured and constructed by specialist plant contractors (Freeman 1968).

There were various specific advantages accruing to the SEFs in

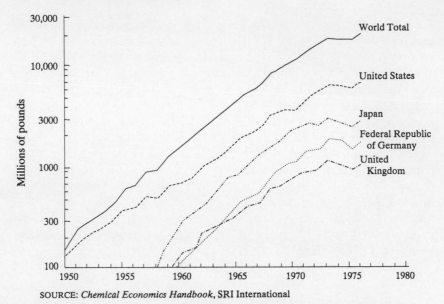

SOURCE: *Chemical Economics Handbook*, SRI International

FIGURE 4.5 Noncellulosic Figures: World Production, by Country

designing and developing chemical production processes. First, during
the 1920s and 1930s, while large chemical companies had concentrated
mainly upon product innovation and development, the SEFs had acquired an ability to handle sophisticated process design and development
work. In this, they had benefited greatly from their experience in the
petroleum sector, which had faced, earlier than the chemical industry,
problems of large scale processing and refining.[9]

Moreover, in the postwar period, the SEFs tended to be more successful than large chemical producers in attracting talented senior chemists
and chemical engineers. Their working environment was more congenial
to senior personnel, who enjoyed greater autonomy and control, and often also had financial stakes in the company (Landau and Brown 1965).
In addition, a most important source of advantage to the SEFs came from
their opportunity to exploit economies of specialization and learning by
doing. Once a major new process technology was developed, or the scaling up of a given production process was carried out, the SEFs could
reproduce that new technology, and/or larger scale production process,

---

[9]See Mansfield et al. 1977 (Chapter 3) for evidence on the increasing importance of
SEFs in process innovation during the post Second World War period. The enhanced
role of the SEFs was largely paralleled by the tendency of chemical firms to specialize,
to an even large extent than before, in product innovation.

for many clients. Such economies could not be accumulated by the chemical manufacturers, because of the SEFs' much more extensive experience with designing that particular plant many times for different clients. Furthermore, as they worked for many different clients, they accumulated useful information related to the operation of plants under a variety of conditions. This represented an opportunity for accumulating knowledge and specialization which was not available to the chemical producers, and helped the SEFs to design better plants for other customers.

The role of the SEFs had important consequences with respect to competition among chemical manufacturers. The most significant by far was their development of the complete technology and plant designs for the basic building blocks of the world chemical industry—olefins and aromatics (a distinctive achievement of American chemical engineering). American-designed ethylene cracking plants seemed to sprout everywhere and these in turn required technologies for manufacture of key intermediates for the chemical industries of many nations. Such technologies came from SEFs and from manufacturers in other countries who sought additional revenues outside their own domestic markets. Latecomers (abroad or domestic) to a particular chemical technology could benefit from their relations with the SEFs, which were able to provide them with the process know-how they had accumulated, at least in part, through their previous relations with earlier entrants. Moreover, the availability of such technology from the SEFs also permitted many new entrants into the industry from petroleum, paper, food, metals, and the like, both domestically and internationally. Vigorous competition ensued, and continues today. The effects on the American industry, including periods of overbuilding and resultant overcapacity, are detailed in (Landau 1989, 1990b).

9 The International Diffusion of New Technologies

The postwar maturing of chemical engineering and the growing skills at expanding the average size of the chemical plants and/or improving the chemical process technologies were primarily American achievements.

There were several reasons for this. Most specialized engineering firms were, to begin with, American firms, for reasons that have already been considered. After the war, most European countries encouraged the development of their domestic chemical industry, leading to a fragmentation into smaller national markets. Second, the European chemical industry had been disrupted by the war, whereas the American industry had remained intact. Not only did the U.S. chemical industry emerge unscathed from the physical devastation of the war; in addition, the events of the war involved a number of episodes that strengthened the industry at the beginning of the postwar period. This included crash programs involving new technologies and expanded facilities for such things as synthetic rubber, petroleum refining, chemicals, munitions, light metals, etc. And, of course, the most extensive crash program of all: The Manhattan Project that culminated in the successful development of the atomic bomb that brought the war in the Pacific to an abrupt halt. This project demonstrated the value of the mobility of skilled personnel for dealing with new situations. The new postwar entrants and the SEFs profited immensely from similar flows of people.

Third, there are important reasons why the American environment was most conducive to expanding the efficiencies of chemical production processes through enhancing the scale of the manufacturing sequence. The size of the U.S. market and high growth rates of demand for chemicals during the 1950s and 1960s played a major role. The market of each individual European country was far smaller than the American market, which reduced the incentive to bear the high fixed costs of giant chemical installations. Because of the high fixed costs of single "train" chemical processes with respect to the size of the European national markets, European chemical producers did not at first risk setting up giant chemical plants. Thus, their efforts to scale up chemical plants were for quite a period not comparable to that observed in the case of the American market.

In this way, the postwar chemical industry was one where American chemical firms dominated the world market for many years. Relatedly, the American market was the primary locus of attempts to develop new chemical production processes, and in particular to undertake substantial expansions of average plant size. Moreover, America possessed, at the same time, a large corps of chemical engineers who had for many years, through path-dependent sequences, acquired vast experience and expertise in designing large scale, continuous process plants.

The worldwide chemical market, at the same time, experienced successive waves of diffusion of new technologies, both process and product. Not only were many new technologies developed during the postwar period, but their diffusion throughout the world followed rapidly. Although the sources of chemical innovation were diverse, as indicated by Tables 4.5 and 4.6 (Landau 1989, 1990b),[10] which omit the com-

---

[10]In Table 4.5, the growing effect of environmental considerations on the chemical industry is apparent, by the omission of the important chlorinated pesticides which have been phased out. Now some of the newer ones are also likely to be replaced by more environmentally friendly products of biotechnological research.

TABLE 4.5

Major Postwar Commercial Chemical Developments:
Pesticides, Fungicides and Herbicides

Class	Product Example	Introduced	Company
Pesticides			
Phosphorodithioates	Thimet, Malathion	1953	American Cyanamid
Carbamates	Sevin	1956	Union Carbide
Synthetic Pyrethroids	Resmethrin	1967	NRDC,* Oxford
Fungicides			
Sulfanimides	Captan	1953	Chevron
Chlorthalonil	Bravo	1965	Diamond Shamrock
Systemic Fungicides	Benomyl	1967	DuPont
Herbicides			
Ureas	Monuron, Diuron	1950	DuPont
Triazines	Simazine, Atrazine	1955	Ciba-Geigy
Pyridinium Salts	Paraquat	1956	ICI
Nitroanilines	Treflan	1960	Eli Lilly
Thiocarbamates	Sutan	1962	Stauffer
Acetanilides	Lasso, Dual	1969	Monsanto, Ciba-Geigy
Glyphosate	Roundup	1974	Monsanto
Sulfonyl Ureas	Glean	1983	DuPont
Imidazolinones	Scepter	1985	American Cyanamid

*National Research & Development Council, UK.
SOURCE: S. Allen Heininger, Monsanto Company.

parable developments in pharmaceuticals, a major factor was the role played by the U.S. specialized engineering contractors as just described. The division of labor between the specialized engineering firms and the chemical manufacturers had important consequences for the diffusion of technology, both domestic and international. Although entry into manufacturing was regarded by some as financially more attractive, it was usually not feasible (Landau 1975, 1988). Hence, SEFs licensed extensively to chemical firms, including chemical firms in other countries. Of course, the wartime sharing of information among U.S. companies, the higher interfirm mobility of skilled personnel, and the increased frequency with which firms used licensing as a means of technology transfer, all contributed in significant measure to the rapid diffusion of technology. Nonetheless, the SEFs were major carriers of technological know-how, allowing a variety of producers to gain access to cost-reducing technology. In particular, by serving both U.S. and foreign companies, they represented an important means of diffusing the knowledge accumulated within the U.S. market to foreign producers as well, and in many cases also, in the reverse direction.

TABLE 4.6

Major Postwar Commercial Chemical Developments

Approx. Date	Product	Development	Company
Postwar	Low-pressure Polyethylene	Ziegler Chemistry	Karl Ziegler
Postwar	Phenol acetone, Cumene	Air oxidation	Distillers Co.; British Petroleum UK; Hercules
1953	Dimethylterephthalate	4-step air oxidation	Imhausen; Hercules
1953+	Ammonia	High Pressure synthetic gas—large single train	Pullman/Kellogg
1955	Maleic anhydride	High yield benzene oxidation	Halcon
1957	Irradiated polyethylene	Memory plastics	Raychem
thru-'57	Isocyanates-urethanes	Urethanes & foams (polyether polyols, one-shot foam, etc.)	Bayer; Houdry; Wyandotte
1958	High-density polyethylene	New catalysts	Montecatini-Natta; Phillips; Avisun; Amoco
	Polypropylene		
1958	α-Olefins and linear alcohols	New catalysts	Gulf; Ethyl; Conoco
1958+	Terephthalic acid	Air oxid. of p-xylene pure product	Halcon; Amoco
1959	Acetaldehyde	Vapor phase ethylene oxidation	Hoechst/Wacker
1960–70	Oxo alcohols	Improved catalysts	Exxon; ICI; Shell; Union Carbide
1960–70	Acetic acid, etc.	Oxidation of paraffins	Celanese
1960–70	Polycarbondates	Engineered plastics	GE; Bayer
1964	KA (cyclohexanol-cyclohexanone) oil (for nylon)	Cyclohexane oxidation Boric system	Halcon
1965	Acrylonitrile	Propylene ammoxidation	Sohio
1965	HMDA (for nylon)	Acrylonitrile electrohydrodimerization	Monsanto
1965+	Vinyl chloride	Oxychlorination of ethylene	Goodrich; Monsanto; PPG; Stauffer
1967+	Vinyl acetate	Ethylene+acetic+O_2, vapor phase	Bayer; Celanese; Hoechst; USI
1968	Acetic acid	High-pressure methanol+CO_2	BASF; DuPont

(TABLE 4.6, cont.)

Approx. Date	Product	Development	Company
1969	Acrylates	Propylene oxidation	BP; Celanese; Rohm & Haas; Sohio; Union Carbide
1969	Polyethylene terephthalate	Plastic bottles	DuPont
1969	Qiana (now abandoned)	From cyclododecane KA oil	DuPont; Halcon
1969+	Prophylene oxide, glycol, TBA	Epoxidation w/hydroperoxide	Arco/Halcon
1970	p-xylene	Recovery by adsorption	UOP
1970	Methanol	Low-pressure CO+H$_2$	ICI
1970	Aniline	Phenol+NH$_3$	Halcon; Mitsui
1970±	Ethylene oxide	Catalysts improvements	Halcon; Shell; Union Carbide
1970±	Polyphenylene oxide and Noryl polymers	Engineered plastics	GE
1972	Hexamethylene diamine (HMDA): for nylon	Butadiene+HCN	DuPont
1972	Styrene and propylene oxide	Epoxidation w/hydroperoxide	Arco/Halcon
1973	Acetic acid	Low-pressure methanol+CO	Monsanto
1974	Kevlar	Hi-tensile fiber	DuPont; Akzo
1974	Polypropylene	Vapor phase	BASF
1974+	Maleic anhydride	From butane	Amoco; Halcon; Monsanto; Denka
1977	Linear low-density polyethylene	Lower pressure	Union Carbide; later others
1978	Ethylene glycol (and vinyl acetate): now abandoned	Via acetoxylation	Halcon/Arco
1980	Acetic anhydride	Coal-based CO+methanol	Halcon; Tennessee Eastman
1981	Methacrylates	Isobutane or isobutylene-based	Mitsubishi, other Japanese companies; Halcon/Arco (development)
1985	Polypropylene	Improved catalysts and processing techniques (e.g., Catalloy)	Montedison/Himont; Shell; others

It is also important to note that the original developments of new chemical technologies, wherever developed, produced widespread benefits for world consumers. In this respect, the history of the world chemical industry during the postwar years is an excellent example of the effective functioning of the competitive market mechanism. New technologies were first developed, reducing the costs of producing chemical products. Diffusion mechanisms, generated by market incentives, made these technologies available to many producers, both in the United States and abroad. This generated worldwide price reductions for chemical products, old and new, which eventually resulted in passing on the benefits of new technology in the form of lower prices to consumer households.

Finally, it should be observed that, in the past fifteen years or so, there has been a substantial decline in the growth rate of chemicals. After the oil shock of 1973, the annual growth rate of the industry went from a double digit figure to about five percent (Krindl 1984, p. 12). There has been a corresponding decline in the relative importance of the production of large volumes of homogeneous chemicals as a major source of profitability for the industry. Many chemical companies have tended to place greater emphasis on the production of specialty chemicals, which have been characterized as products "... that are marketed and used on the basis of their performance rather than of their composition" (Krindl 1984, p. 12). As such they are more heterogeneous than the bulk compounds, and their development requires much more attention to the specific needs, increasingly diverse and sophisticated, of well-defined subsets of users. More attention is now being devoted to R&D, which has grown substantially since the early 1970s as a percentage of capital expenditures for the chemical industry as a whole (Figure 4.6).[11] Unlike traditional commodity chemicals, manufacturing costs here represent a much smaller proportion of the sales price (only about 20–25 percent, according to Krindl). In specialty chemicals, "... endless studies on manufacturing process improvements are generally not worthwhile... its success depends on flexibility, speed, and product reproducibility" (Krindl 1984, p. 13).

What this means is that, with increased worldwide competition in chemicals, emphasis has shifted, at least in part, to product innovation, and to the interactions between product innovation and marketing. Yet, in many of the best companies, technology push receives as

[11]A detailed account of the economics of the postwar industry as it developed into a worldwide competitive industry has just been published by Ralph Landau (Landau 1989, 1990b). Note Figure 4.4 therein, to illustrate the point above.

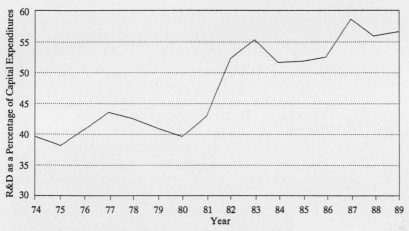

SOURCE: National Science Foundation & Chemical Mfrs. Assoc.

FIGURE 4.6   Chemical Industry R&D vs. Capital Expenditures
(R&D as a Percentage of Capital Expenditures)

much attention as product pull. The chemical industry today combines a large commodity-based, capital-intensive substructure with a proprietary research-based overlay. These businesses require different skills, but it is the *combination* that characterizes the best-managed companies and defines the conditions for competitiveness in the future.

These factors, together with the slower growth of markets (coupled with the general slowing down of world growth in the 1970s), would appear to account for the decline in the importance of the SEFs during the last decade. In the 1950s and 1960s the SEFs could focus on process development without undertaking substantial product R&D, which in chemicals is much more costly than process R&D. Because of the relative homogeneity of the bulk intermediate chemicals, the corresponding production processes could be developed in those years on the basis of well-defined features of the bulk product to be manufactured. For example polypropylene, which was virtually a commodity chemical for 30 years, has now entered the stage of high value added differentiated products, requiring much more specific "tailoring" of manufacturing technology to the diverse needs of end users. Nevertheless, many such differentiated products require flexible but continuous manufacturing systems, so that chemical engineering designers must become even more sophisticated in economics than before. This is one of the reasons that design engineering has been migrating from SEFs to the engineering departments of major chemical producers.

The heterogeneity of the specialty chemicals means that one cannot develop the production process unless one has an excellent grasp of the underlying features of the differentiated products to be manufactured. Thus, the decline of the role of the SEFs is smaller if one looks at the intermediate bulk products which are still produced in large quantities and where manufacturing costs are still a major determinant of competitiveness. But a slowdown in growth rates for commodity chemicals, and the increased integration of product and process development in specialty chemicals, has brought with it a decline in market opportunities for SEFs. And, along with their decline, one would expect to see a decline in the speed and extent of diffusion of technology as well. Whether this will in fact happen, however, will depend upon the ability of the industry to respond to the significant changes that have been taking place in the market for their products.

One major consideration is that the chemical industry of the future is likely to incorporate a wide range of new technologies originating in other sectors of the economy, and their eventual impact is impossible to discern at this point. The industry is likely to develop and make use of innovations in the form of new materials with special performance characteristics, as well as electronic components, equipment and instrumentation. New developments from biotechnology, as well as the extension and deepening of its own traditional product lines, are likely to play major roles in the future. In twenty years there may be few pure chemical companies, replaced by a mixed industry whose firms are part chemical, part electrical, electronic, biotechnological, mechanical, etc. It is a reasonable prospect that chemical engineers will continue to be employed in large numbers, not only in the more traditional product lines, but also in the newer specialty products requiring continuous processing techniques and astute scale-up design allowing for flexibility and quality control. The discipline has been moving intellectually in this direction, as the science orientation becomes even more important. Unit operations no longer represent the intellectual frontier, and chemical engineers are trained in the broader underlying concepts of transport phenomena, thermodynamics, kinetics, process design, and control theory.

In considering the future of the chemical process industries, and the role of the chemical engineer in helping to shape that future, certain key elements of past success may still serve as a guide. Those successes involved a creative approach to process design that was tightly linked to market demand and to the ongoing findings of R&D. This model may even serve as a paradigm for other engineering disciplines, which might well benefit from a more systemic approach to developing improved manufacturing technologies for a wide range of manufactured products,[12] where American companies are having substantial difficulties in competitive manufacturing of discrete mechanical goods. Paul Gray, past President of M.I.T., has recently emphasized that "In all too many cases, design engineers in America have given too little attention to quality, reliability and manufacturability in their designs. ... [W]alls... impede communication and cooperation between product designers and those who design and operate the manufacturing process." To alleviate this problem, M.I.T. has been designing a new form of educational program called Leadership in Manufacturing, which combines the engineering school and the Sloan business school. Perhaps it is no coincidence that a prominent feature of this program is a series of Practice Schools modeled on the Institute's unique and still flourishing Chemical Engineering Practice School.

It is tempting to ask what a new engineering discipline for manufacturing would look like, by analogy with the historic role played by chemical engineering with respect to the chemical industry. While its precise shape and content cannot yet be discerned, it should presumably have a strong computer science orientation, and include the integration of topics such as: object-oriented design; expert systems; linear programming; project scheduling; robotics; statistical design of experiments; statistical quality control; modeling and simulation; and concurrent design and manufacturing.

This paper has examined the circumstances shaping the past successes of the American chemical industry and, in particular, the central role played by the discipline of chemical engineering in determining those successes. It has also examined the growing internationalization of the industry and the changes in market conditions, and the diffusion of technologies, that have been associated with that growth. Other factors, such as macroeconomic environment and financial considerations have been neglected, and are crucial to future performance, but these are properly the subject of other papers in this volume. In closing, some

---

[12]Electrical power engineering and, more recently, nuclear engineering, have had a similar relationship with the electrical power industry. To some extent, modern electrical engineering and the semiconductor industry have evolved in a similar direction, as in Stanford's Center for Large Scale Integrated Circuits. During the Second World War, influenced by the pioneering leadership of Theodore von Karman at California Institute of Technology, aeronautical engineering became very tightly linked with the aerospace and defense industries. It is interesting to note that when Arthur A. Noyes left M.I.T. in 1919, he became one of the founders of Caltech, employing a model he originally conceived for M.I.T.!

relevant considerations that would appear to emerge from this industry's history are briefly identified.

10 Some Concluding Observations

Industrial development is, by its nature, a path-dependent phenomenon. The nature of such development cannot be adequately understood within the austere intellectual framework of neoclassical growth models. Steven Durlauf (this volume) finds from macroeconomic analysis that growth paths are distinct for each country, and that there are multiple equilibrium paths that depend on the ability of any country to manage its own affairs—for example, its skill in exploiting the findings of scientific research in a commercial context. Boskin and Lau (this volume), although employing a very different methodology, identify interactions between capital formation and technical change (in the sense that technology is capital-augmenting) that affect industrial efficiency in ways that have not been previously acknowledged in highly abstract models. Growth is a highly path dependent experience, and therefore history genuinely matters in understanding the rate and direction of development of specific industries (for example, aerospace is heavily dependent on government for research funding and purchases of military equipment; agriculture has always required government research). Comparative advantage is thus rightly seen as dynamic and ever-changing. The rise of the chemical engineering discipline in association with the petroleum and chemical industries and the importance of specialized engineering firms illustrates this dependence. What Durlauf, Boskin, and Lau point to is that a high productivity economy requires both a stable pro-investment policy and firms and industries that can manage technology successfully in the international marketplace (Landau 1990a, 1991).

The chemical industry has certain quite distinctive features, at least by comparison with other American industries. In this industry, science is vital from beginning to end and the breakthrough or ladder innovation (see Gomory, this volume) occurs simultaneously with cyclic improvements. In short, product and process innovation are fused. Ken-ichi Imai (this volume) states that this is essentially the Japanese system in certain other industries as well. In Imai's analysis of the Japanese innovation process, particular attention is drawn to the central importance of inter-organizational linkages, embedded in the larger Japanese system of multi-layered networks (including inter-firm, intra-industry, inter-industry, government-industry-university networks). At a theoretical level, Imai sees this as an *ex ante* coordinative mechanism, as distinct from the *ex post* coordinative mechanism represented by the price system in traditional economic theory. His position is also that the non-price information flows in this system provide an essential coordination mechanism and that, as the strategic importance of technology increases, the importance of this mechanism likewise increases.

There are many externalities and spillovers that characterize this and other industries. 75 percent of its output goes to other industries for further conversion and the products are used widely by plastics, apparel, rubber, and other industries. Reciprocally, other industries make equipment such as compressors, instruments, etc., which contribute to the growth of the chemical industry (Bernstein and Nadiri 1990).

The university-industry linkage is a profound factor in the growth of high-tech industries, as shown in the development of the chemical industry in both Germany and the United States. The lack of such a strong linkage in Great Britain was an important factor in the weak performance of the chemical industry in that country. The declining proportion of U.S. citizens receiving engineering degrees at U.S. universities may foretell a serious future insufficiency of trained personnel.

Wars have had decisive effects in shaping the direction of technological development. In Germany they have led to an emphasis on synthetic fuels, synthetic rubber, synthetic fertilizers and rocketry. In the United States, two world wars influenced the development of aeronautics, rocketry, electronics, and computers, as well as the rapid expansion of capability in chemicals that has already been discussed.

The history of the chemical industry raises complex questions concerning the appropriate role of government policy, although the exact nature of the "lessons" of history here is not very clear. The German government played a prominent role in the emergence of an industry-wide cartel, I.G. Farben, during the 1920s, an initiative that is hardly to be recommended. On the other hand, it seems clear that a successful chemical industry has required an educational establishment that was responsive to the peculiar requirements for skilled scientific and engineering personnel. The necessary educational infrastructure was provided in rather different ways in Germany and the United States, whereas the long lag in the development of such an infrastructure in Great Britain surely served to retard the development of the chemical sector in that country.

Examination of the history of the chemical industry emphasizes the economic importance of scale economies in the pursuit of competitive advantage, as well as the economic significance of learning experiences involving the use of new technologies. These phenomena were long neglected in models developed within the neoclassical tradition, and they would appear to justify much more attention on the part of both theoretical and empirical economists.

Changing rates of capacity utilization are often a neglected but highly important source of changes in productivity. An inappropriate interpretation of economic signals, such as occurred in the late 1960s and 1970s in response to initially low interest rates but visibly rising inflation rates, may result in considerable overcapacity with consequent low levels of productivity of capital. Furthermore, as Steven Durlauf has shown (this volume and *The Economist*'s editorial (*The Economist*, September 1989, pp. 13–14), low levels of investment in industries X and Y, which use the output of industry Z as an input into their productivity activities, may lead to reduced levels of productivity for existing capital in industry Z. Such phenomena may have played an important role in the productivity slowdown of the 1970s.

On the other hand, quality improvements in chemical products (and of many other high-tech products) are not adequately captured by ordinary productivity measures. As a result, the value of production, expressed in dollar terms, may understate the full extent of productivity improvement. Greater attention needs to be devoted to the attempt to measure the full significance of such understatement.

This industry everywhere has been one of the very first to encounter environmental restraints and challenges to its activities. Before Perkin's efforts in the 1850s, the sheer volume of waste products from the textile industry, and the resulting lawsuits, forced scientists to look for ways to utilize the by-products of industry more constructively. Its problems have continued right up to today, but it has now been joined by many other industries. Technology will also be the solution to the growing concern for sustainable growth.

The history of the chemical industry forcefully underlines the inadequacy of a neoclassical microeconomic approach that simply treats all industrial firms as homogeneous, maximizing agents, whose distinctive internal structure, characteristics, and history are not examined, or at least are not treated as central to the analysis. Such an approach virtually throws away the essential elements of the problem of success and failure in the commercialization of new technologies. These issues have now emerged explicitly in the business and management literature, most particularly in the work of Alfred Chandler and Michael Porter (Chandler 1990b, Porter 1990). The history of this industry shows that very skilled and educated leadership, as well as singularly creative individuals, matter a great deal. The German chemical companies have long had their managements and operations in the hands of scientifically trained or oriented people. This has also been largely true in the U.S. chemical industry, but not always in other technologically based industries. Greater attention to the manner in which firms acquire improved technical and commercial capabilities could make a major contribution to the microeconomic analysis of success and failure in the marketplace.

From a deeper philosophical point of view, the history described here illuminates a major change that has been occurring in the last two centuries. Until recently, aggressively led countries seeking rapid development have sooner or later resorted to plundering their neighbors or acquiring lands deemed to be rich in markets or resources. Past history has amply demonstrated the ultimate futility of such approaches, and the experience since the Second World War, as well as postwar economic research, has powerfully confirmed the existence of a far better alternative. Science and technology harnessed by an enlightened capitalistic democratic system can improve the standard of living of a country by higher growth rates without requiring either wars or colonies. It took Germany two world war losses to realize this, and the Japanese one, but they have shown that this path is the right one.

References

Beer, John J. 1959. *The Emergence of the German Dye Industry.* Urbana: University of Illinois Press.

Bernstein, J. I. and M. I. Nadiri. 1990. *A Dynamic Model of Product Demand, Cost of Production, and Interindustry R&D Spillovers.* Carleton and New York Universities working papers (November).

Cannadine, David. 1990. *The Decline and Fall of the British Aristocracy.* New Haven: Yale University Press.

Chandler, Alfred. 1990a. The Enduring Logic of Industrial Success, *Harvard Business Review* (March/April).

Chandler, Alfred. 1990b. *Scale and Scope.* Cambridge, MA: Harvard University Press.

Freeman, Christopher. 1968. Chemical Process Plant: Innovation and the World Market, *National Institute Economic Review* (August).

Frontiers in Chemical Engineering: Research Needs and Opportunities. 1988. (Amundsen Report). Washington, D.C.: National Academy Press.

Grant, Wyn, William Patterno, and Colin Whitston. 1988. *Government and the Chemical Industries: A Comparative Study of Britain and West Germany.* Oxford: Oxford University Press.

Haber, L. F. 1971. *The Chemical Industry 1900–1930.* Oxford: Oxford University Press.

Hounshell, David A. and John Kenly Smith, Jr. 1988. *Science and Corporate Strategy: DuPont R&D, 1902–1980.* Cambridge: Cambridge University Press.

Krindl, A. G. 1984. The Trend Towards Specialty Chemicals, *Chemical Engineering Progress* (October).

Landau, Ralph. 1958. Chemical Engineering in West Germany, *Chemical Engineering Progress*, Vol. 54, No. 7 (July).

Landau, Ralph. 1966. *The Chemical Plant, from Process Selection to Commercial Operation*. New York: Reinhold Publishing Co.

Landau, Ralph. 1975. The Chemical Process Industries in International Investment and Trade. In *U.S. Technology and International Trade*, April 1975, Washington, D.C.: National Academy Press, pp. 15–19.

Landau, Ralph. 1980. Chemical Industry Research and Innovation, in W. N. Smith and C. F. Larson, eds., *Innovation and U.S. Research*, ACS Symposium Series 129, American Chemical Society, pp. 44–45.

Landau, Ralph. 1988. Harnessing Innovation for Growth, *Chemical Engineering Progress*, July, pp. 31–42.

Landau, Ralph. 1989. The Chemical Engineer and the Chemical Process Industries: Reading the Future from the Past, *Chemical Engineering Progress*, September, pp. 25–39.

Landau, Ralph. 1990a. Capital Investment: Key to Competitiveness and Growth, *The Brookings Review*, Summer, pp. 52–56.

Landau, Ralph. 1990b. Chemical Engineering: Key to the Growth of the Chemical Process Industries, *AIChe Symposium Series 274*, Vol. 86, pp. 9–39.

Landau, Ralph. 1991. How Competitiveness Can Be Achieved. In *Technology & Economics*. Washington, D.C.: National Academy Press, pp. 3–46.

Landau, Ralph and David Brown. 1965. Making Research Pay. AIChE-I. Chem. E. Symposium Series No. 7, London Inst. of Chemical Engineers, pp. 35–43.

Landau, Ralph and Nathan Rosenberg. 1990. Chemical Engineering: America's High-Tech Triumph. In *American Heritage of Invention and Technology*, Fall, pp. 58–63.

Landau, Ralph and Nathan Rosenberg. 1991. Innovation in Chemical Processing Industries. In *Technology and Economics*, Washington, D.C.: National Academy Press, pp. 107–20.

Lewis, Warren K. 1953. Chemical Engineering—A New Science? In Lenox R. Lohr, ed., *Centennial of Engineering 1852–1952*. Chicago: Museum of Science and Industry.

Lieberman, Marvin. 1987. Patents, Learning by Doing, and Market Structure in the Chemical Processing Industries, *International Journal of Industrial Organization*, No. 5.

Mansfield, Edwin, et al. 1977. *The Production and Application of New Industrial Technologies*. New York: W. W. Norton and Co.

Mowery, David. 1981. *The Emergence and Growth of Industrial Research in American Manufacturing*. Doctoral dissertation, Stanford University.

Mowery, David and Nathan Rosenberg. 1989. *Technology and the Pursuit of Economic Growth*. Cambridge: Cambridge University Press.

Opportunities in Chemistry. 1985. (Pimentel Report). Washington, D.C.: National Academy Press.

Porter, Michael. 1990. *The Competitive Advantage of Nations.* New York: The Free Press.

Rosenberg, Nathan. 1963. Technological Change in the Machine Tool Industry, 1840–1910, *Journal of Economic History*, December.

Spitz, Peter H. 1988. *Petrochemicals: The Rise of an Industry.* New York: John Wiley & Sons, Inc.

Walker, William H., Warren K. Lewis, and William H. McAdams. 1923. *Principles of Chemical Engineering.* New York: McGraw-Hill.

Wiener, Martin J. 1981. *English Culture and the Decline of the Industrial Spirit, 1850–1980.* Cambridge: Cambridge University Press.

5

International Differences in Economic Fluctuations

STEVEN N. DURLAUF

1 Introduction

One of the most controversial areas of macroeconomic research over the last decade has centered on the question of the sources of movements in aggregate activity. Unlike the 1960s and 1970s, when aggregate volatility was largely ascribed to fluctuations in aggregate demand, most recent research has focused on the determination of aggregate supply. In particular, a school of thought has worked with real business cycle models, that is, models where the fundamental source of business cycle fluctuations is the evolution of productivity or technology over time. In this framework, aggregate activity is well represented as the realization of an economy characterized by flexible prices and complete markets. This paradigm further argues that there is a relatively small role for stabilization policy to improve macroeconomic performance in the sense that monetary and fiscal policies cannot affect the level of real activity through demand side effects except over the short run. In fact, in ascribing the fluctuations in real activity to exogenously given technology innovations, the real business cycle models suggest that demand side policies are largely irrelevant per se.

At the same time, empirical macroeconomics has changed the way

This research was generously supported by the program in Technology and Growth of the Center for Economic Policy Research. I thank Tim Bresnahan and Ralph Landau for many valuable discussions. Andrew Bernard and Suzanne Cooper have provided outstanding research assistance and Laura Dolly has provided outstanding secretarial assistance. All errors are mine.

in which economic growth and business cycles are conceptualized. A substantial body of research, most notably Nelson and Plosser 1982 and Campbell and Mankiw 1987, has concluded that aggregate real output in the United States contains a stochastic trend. Campbell and Mankiw (1988) have uncovered similar evidence for several major industrialized economies, including France, West Germany, and Japan. This means that some component of innovations to real activity will have long run effects on macroeconomic performance. Despite some question as to the magnitude of the unit root (see Cochrane 1988), there is widespread consensus that long run economic growth cannot be reduced down to a deterministic trend.

One critical implication of this empirical work is that trends and cycles cannot be treated as essentially distinct phenomena. Rather, they must be jointly conceptualized as the product of a common set of structural factors. For this reason, it is useful to study the behavior of overall economic fluctuations which contain both short run and long run factors.

The evidence of stochastic trends has been interpreted as supportive of some aspects of the real business cycle paradigm. King, Plosser, Stock, and Watson (1987) model technology as a stochastic trend driving the supply side. Authors such as Blanchard and Quah (1988), DeLong and Summers (1988), and Shapiro and Watson (1988) take stochastic as prima facie evidence of supply side factors in generating fluctuations. The idea behind this assumption is that a stochastic trend in output is possible only if the production possibilities of the economy are expanding across time. Therefore, stochastic trends are evidence of an erratically evolving technology for aggregate output. One feature of the real business cycle approach is that advanced industrialized economies are usually conceptualized as part of a large system of markets. The technology induced view of fluctuations suggests that there exist common sources to national fluctuations, unless one ascribes to the position that national technologies fail to converge even asymptotically. Such a view is clearly empirically untenable. As a result, the real business cycle paradigm suggests that there should be substantial interactions in fluctuations across countries.

An alternative perspective on aggregate fluctuations emphasizes the importance of domestic institutions in the determination of macroeconomic dynamics. This view, whose theoretical underpinnings are well exposited in Cooper and John 1988, argues that aggregate activity represents the realization of an extraordinarily complex set of interactions among many disparate firms and consumers. Further, these many economic activities exhibit powerful complementarities. Roughly speaking, a model exhibits complementarities when the decisions of one set of agents affect the decisions of another set of agents in ways that are not adjudicated through a complete set of markets. For example, if a firm makes a technological discovery that can be copied by other firms, a complementarity exists. Complementarities mean that there are feedbacks between agents which are not fully accounted for through the price system. In particular, high productivity by one agent positively affects the productivity of others. These feedbacks can dramatically affect the optimality of the market equilibrium.

A simple example of positive complementarities has been developed in Bryant 1985. Bryant considers a world of K agents who must each allocate an effort level e_k into a productive activity. This activity produces an aggregate output level Y equal to the minimum of the efforts of all agents:

$$Y = K \cdot \min(e_1, e_2, \ldots, e_k)$$

Now suppose that each agent receives $1/K$ of the output, agents enjoy leisure, and that the marginal utility of one unit of leisure is always exceeded by the marginal utility of one unit of consumption. In this case, any effort level which is identical across all agents constitutes an equilibrium. The decisions of $K - 1$ agents to simultaneously increase effort will create an externality for the Kth agent in the sense that his production opportunities will expand.

The basic insights of the Bryant model extend to general frameworks with positive complementarities. For sufficiently powerful complementarities, multiple equilibria exist. This idea has profound implications for macroeconomic theory. The microeconomics of a model no longer uniquely determine the macroeconomics. Domestic institutions become an essential determinant of the behavior of total output by helping to affect the selection of the equilibrium. The policy rules in this class of models assume a significant role in increasing output and stabilizing business cycles. Different choices of monetary and fiscal policies help cause individual agents to internalize the complementarities and collectively improve social welfare. This idea is a direct descendant of the Keynesian multiplier which illustrated how spillover effects across sectors could lead to large increases in output from relatively small changes in exogenous spending.

The purpose of this paper is to help understand some of the relationships between national economies and how different economies are linked. The paper is divided into two parts. First we consider some stylized facts about aggregate output behavior in different countries.

Our results strongly suggest aggregate activity is strongly affected by country-specific rather than common factors. We develop three pieces of evidence in support of this position. First, there exist country-specific components to long-term growth. Specifically, a permanent innovation to one country's output does not imply that other countries will experience the same innovation. Second, the degree of persistence in fluctuations is not uniform. The percentage of variance of output changes which can be attributed to long-run factors differs substantially across countries. Third, we find relatively little evidence of interactions in output fluctuations across countries. Together, these results call into question the strict productivity interpretation of economic fluctuations.

The second part considers some theoretical economic structures which are capable of generating these international disparities. We outline a model of dynamic coordination failure which captures a number of the complementarities which have been analysed in the literature. The novel feature of the model is its ability to generate dramatically different time series characteristics for economies with similar microeconomic institutions. In particular, stochastic trends are possible outcomes of the framework. The model demonstrates how private production incentives play an essential role in determining long-run growth. Government policy in promoting investment and production is a potentially essential factor in determining both business cycle behavior and long-run growth in the economy.

An appendix follows which discusses the interpretation of the time series techniques employed in the text as well as a formal model of multiple macroeconomic equilibria.

2 Some Stylized Facts About International Output Behavior

In neoclassical views of macroeconomic fluctuations, modern industrial economies are best thought of as collections of firms and consumers who efficiently act to maximize profits and utility respectively. In this framework, markets are complete and prices are flexible. The modern variant of this perspective, real business cycle models, argues that the major source of economic fluctuations, particularly in the long run, is productivity innovations which generate both business cycles and growth by altering the equilibrium levels of production and investment. These models imply that the demand side of an economy should have little effect on real variables, which results in a form of the classical dichotomy for 1980s macroeconomics where supply side factors determine the level of output and monetary policy determines the price level. A corollary of this perspective is that there is little role for government policy in affecting long-term growth. Long-term growth is a function of the rate of technical change. Demand side policies are therefore irrelevant.

The purpose of this section is to argue that there are fundamental differences across countries in their aggregate behavior over both the short and long terms. This means that historically, national experiences are unique. An implication of these results is that government policies are potentially essential components of both the short run and the long-run characteristics of an economy.

In this section, we provide evidence on three aspects of output fluctuations in OECD economies. First, we demonstrate that output fluctuations for these different economies are persistent. Persistence means that contemporaneous changes in real per capita GNP in each economy affect the long-run path of output. Evidence in favor of persistence suggests that a common set of forces determine both the business cycle and long-term growth. Second, we show that the persistent components of output in each economy are idiosyncratic in the sense that they cannot be explained by a common factor. This feature implies that countries can exhibit different long-run growth rates. Third, we show the importance of persistent output fluctuations in determining total output volatility differences across economies. This characteristic means that individual economies exhibit substantially different business cycle behavior. Collectively, our empirical results imply that country-specific factors play an important role in determining aggregate fluctuations.

3 Long-Run Characteristics of Postwar Economic Fluctuations

In order to describe long-run international output behavior, it is helpful to decompose per capita log output for country i at time t, $Y_{i,t}$, into trend and cycle components. We shall define the trend part of the series as the level of output forecastable in the long term based upon knowledge of the innovations in the history of the series plus the deterministic trend. One may decompose output into different parts corresponding to persistent and transitory components. Formally,

$$(5.1) \quad Y_{i,t} = C_{i,t} + \beta_i t + S_{i,t}$$

In this equation, $C_{i,t}$ denotes the mean reverting trend in the economy. This is the part of current GNP one expects to dissipate over time. On the other hand, deterministic trend in economy i, $\beta_i t$, is the part of the economy that is permanent and predictable from unchanging trends. It may be thought of as the part of long-run growth that is unaffected by the history of the economy. Finally, the stochastic trend, $S_{i,t}$, is the persistent part of the output that is affected by the realized history of the economy. It may be thought of as the part of the present that will continue into the indefinite future, after one has accounted for the deterministic trend.

This formulation distinguishes between two types of trend in growth. The deterministic trend represents the predictable part of economic growth. In traditional formulations of technical change, this term was designed to capture all long-run factors such as population growth that can be extrapolated into the future. On the other hand $S_{i,t}$ represents the stochastic component to long-term growth. This term captures those aspects of growth that are not predictable from the distant past. Events such as the invention of the automobile, which from a Schumpeterian perspective represents a fundamental source of business cycle fluctuations, would be captured by this term.

One can show that changes in the stochastic trend cannot be forecasted over time. This means in turn that $S_{i,t}$ is a random walk, that is,

$$(5.2) \quad S_{i,t} = S_{i,t-1} + u_{i,t}$$

where $u_{i,t}$ is an unpredictable or "white noise" process. Changes in output can therefore be decomposed into three components

$$(5.3) \quad \Delta Y_{i,t} = \Delta C_{i,t} + \beta_i + u_{i,t}$$

These three terms have a simple interpretation. The term $\Delta C_{i,t}$ represents the part of an output change which will eventually revert to zero. The term β_i captures the part of the change that was permanent and predictable from the past. The term $u_{i,t}$ represents that part of an output change that is permanent yet was unpredictable from the past.

The decomposition we have outlined highlights an important distinction between types of output changes. Permanent changes have a random as well as a predictable component. These aspects have different implications for the properties of $\Delta Y_{i,t}$. In particular, the deterministic component and stochastic components of trends may be analysed separately in order to understand the tendencies of economies to converge over time. The moving average representation of output changes represents output changes as the sum of a constant mean β_i of output changes with a weighted average of uncorrelated random variables ξ_{t-r} where the weights are designated as α_r.

$$(5.4) \quad \Delta Y_{i,t} = \beta_i + \sum_{r=0}^{\infty} \alpha_r \xi_{t-r}$$

The presence of a deterministic trend means that $\beta_i \neq 0$ whereas the presence of a stochastic trend means that the long-run effect of a par-

TABLE 5.1

Cross-Country Cointegration Tests 1950–1985

(Log Per Capita Output)

	France	West Germany	United Kingdom	Canada	USA
Japan	−1.99	−2.35	−1.70	−1.80	−1.00
France		−2.90*	−1.20	−1.09	−1.49
West Germany			−1.90	−2.30	−1.80
United Kingdom				−3.98**	−2.50
Canada					−2.720

$$Y_t = C + \beta X_t + \epsilon_t$$
$$\Delta \epsilon_t = \rho_1 \epsilon_{t-1} + \rho_2 \Delta \epsilon_{t-1} + U_t$$
$$H_0: \rho_1 = 0$$

SOURCE: Taken from Durlauf 1989c.   * Denotes significant at 10% level.

t = Statistics reported in table.   ** Denotes significant at 5% level.

Significance levels based on Engle-Granger 1987 described in text.

ticular ξ_t on the level of output is nonzero. Mathematically, this means that $\sum_{r=0}^{\infty} \alpha_r \neq 0$.

A natural test of the relevance of the real business cycle/efficient general equilibrium world view is the behavior of the trend components to national output. The idea is that if the long-run behavior of economies is determined by productivity, then there should be some relationship between the trends across different economies. In terms of equation (5.3) this would require that $u_{i,t}$ and $u_{j,t}$ be proportional, as long-run movements in one economy would be related to long-run movements in another. On the other hand, suppose that there is no tendency for the permanent components of output to converge across countries. This would mean that growth is determined by domestic economic conditions, as well as the magnitude of the aggregate production set.

In order to test for long-run convergence, it is necessary to introduce the idea of cointegration. Two time series, x_t and y_t, are said to be cointegrated if there exists a coefficient γ such that $x_t - \gamma y_t$ does not contain a persistent, that is, unit root, component. Roughly speaking, this means that if x_t and y_t both contain a unit root component, then this component is common to both series.

Cointegration is an ideal way of testing for the tendency of growth rates to be determined by a common long-run factor. We therefore consider the null hypothesis:

$$H_0 : Y_{i,t} \text{ and } Y_{j,t} \text{ are not cointegrated.}$$

Table 5.1 reports cointegration tests for six major industrial economies in the postwar period: Japan, France, West Germany, the United

Kingdom, Canada, and the United States. This table is taken from Durlauf 1989.[1] The data are taken from Summers and Heston 1988. All data are logs of per capita of real output and run from 1950–85.

The striking result of this table is the acceptance of the no cointegration hypothesis for fourteen of the fifteen country pairs. (If the null hypothesis were uniformly correct, one rejection out of fifteen would easily be consistent with random sampling error.) This means that the data are generally quite consistent with the view that there are country-specific determinants to the level of per capita output over the long term. The only rejection of the null hypothesis occurred for the United Kingdom and Canada; when the significance level is extended to ten percent, then France and West Germany appear cointegrated. The bottom line of this table is clear. It is extremely difficult to find evidence of convergence in national growth rates.

Notice that these results go further than rejecting long-run convergence. These results reject the presence of a common trend component in the growth rates of the various countries. This makes it particularly hard to attribute the sources of long-run growth movements to exogenously evolving technology shocks. Campbell and Mankiw (1988) found very similar evidence through a direct test of the number of unit root processes underlying the six economies. These authors were unable to reject the hypothesis that there are six distinct unit root components underlying six quarterly output series. The upshot of these results is that there is strong evidence of nationally unique components to long-term growth in the six economies we have examined.

In order to further understand the behavior of the trend components to the different national economies, an attempt was made to separate out the deterministic and stochastic components to national trends. In order to do this, observe that the cointegration tests in Table 5.1 check for either a deterministic or a stochastic trend. Cointegration can fail either because the means of output changes differ across countries or because permanent shocks to one country fail to be translated into permanent

---

[1] Our hypothesis tests are based upon an approach due to Engle and Granger (1987). If two times series are not cointegrated, then the residual $\epsilon_{i,t}$ in the regression

$$Y_{i,t} = C + \gamma Y_{j,t} + \epsilon_{i,t}$$

will contain a unit root. We therefore perform a second stage regression

$$\Delta \epsilon_{i,t} = \rho_1 \epsilon_{i,t-1} + \rho_2 \Delta \epsilon_{i,t-1} + \zeta_{i,t}.$$

Under the null hypothesis, ρ_1 will equal zero. Under the alternative, where $\epsilon_{i,t}$ is stationary, this will not occur. Therefore, H_0 may be formalized by computing a t-statistic for this regression.

TABLE 5.2

International Growth Rates–Annual, 1950–1985

(Standard Errors in Parentheses)

Canada	.024	Japan	.061
	(.005)		(.010)
France	.033	United Kingdom	.022
	(.006)		(.002)
West Germany	.039	United States	.019
	(.008)		(.003)

shocks for another. This distinction permits a decomposition of the cointegration tests so as to uncover more information about the failure of convergence in growth.

As equation (5.3) indicates, the deterministic growth rate will equal the mean of the first differences of the various log output series. Discrepancies in deterministic growth will therefore be revealed by different means for national output changes. Table 5.2 reports the mean growth rates for the six economies along with standard errors.

The results in Table 5.2 suggest that some of the failure of convergence in deterministic growth rates is attributable to the deterministic element. Over the entire sample, the growth rates of France, Japan, and West Germany are all substantially larger than Canada, the United Kingdom, and the United States. These differences are nearly statistically significant in the sense that the growth rates of the latter three countries are approximately two standard deviations away from the former group.

Similarly, it is possible to isolate the behavior of stochastic innovations in output growth.[2] This can be achieved by removing the deterministic trends in the two output series and then performing the cointegration tests. If distinct stochastic unit root components exist within the time series, then the detrended series will be distinct integrated processes. Table 5.3 reports the results of the detrended cointegration regressions. The results virtually mirror Table 5.1. With the exception of Canada and the United Kingdom, no pair of countries is cointegrated. The tests tend to be insignificant at the ten

---

[2]Formally, one modifies the first stage regression for the Engle-Granger test by adding a time trend:

$$Y_{i,t} = C + \beta t + \gamma Y_{j,t} + \epsilon_{i,t}$$

If the permanent components of $Y_{i,t}$ and $Y_{j,t}$ may be expressed as linear combinations of a time trend and a single unit root process, then the residuals in this regression will be stationary and the Engle-Granger test will apply.

TABLE 5.3
Cross Country Cointegration Tests
(Detrended Log Per Capita Output)

Country	France	West Germany	United Kingdom	Canada	USA
Japan	−1.790	−1.690	−0.920	−0.650	−2.910
France		−1.350	4.890	−0.040	−2.810
West Germany			−2.110	−2.020	−2.670
United Kingdom				−4.820**	−3.000
Canada					−2.720

$$Y_t = C + \beta X_t + \epsilon_t$$
$$\Delta \epsilon_t = \rho_1 \epsilon_{t-1} + \rho_2 \Delta \epsilon_{t-1} + U_t$$
$$H_0 : \rho_1 = 0$$

t = Statistics reported in table.   ** Denotes significant at 5% level.
Significance levels based on Engle-Granger (1987) described in text.

percent level. Somewhat striking are the extremely small statistics for almost all country pairs which do not include the United States. Some of these statistics matching the United States and various countries, most notably Japan and the United Kingdom, are significant at the ten–fifteen percent level. For the other country pairs, the values are extremely low.

The results of Tables 5.2 and 5.3 demonstrate that there is little evidence of convergence in growth rates among major industrialized economies in the postwar era. The holds true both for average output changes as well as random output fluctuations. The salient fact in comparing national growth rates is the extent to which growth is a domestic phenomenon.

4 Short-Run Characteristics of Postwar Economic Fluctuations

The discrepancies between national economic performance in the postwar period are also found in the behavior of output fluctuations over shorter horizons. These differences occur in two senses. First, the time series dynamics of output changes are different across countries. Second, there is little evidence that fluctuations in one country are associated with changes in another.

In order to compare output dynamics across countries, it is helpful to define a way of understanding the distribution of fluctuations according to periodicity. Let $\Delta Y_{i,t}$ denote the first difference of output in country i after the mean has been removed. As explained in the Appendix, $\Delta Y_{i,t}$ may be thought of as a weighted average of sines and cosines. Each sine and cosine of a given frequency ω makes some contribution to the total variance of the time series. The spectral density of $\Delta Y_{i,t}$ is a function which expresses the variance contribution of each frequency. The spectral density equals the infinite sum.

$$(5.5) \quad f_{\Delta Y_i}(\omega) = \sum_{j=-\infty}^{\infty} \sigma_{\Delta Y_i}(j) e^{-ij\omega}$$

In this sum, $\sigma_{\Delta Y_i}(j)$ denotes the autocovariance function of $\Delta Y_{i,t}$.

In order to interpret the spectral density, it is useful to recall that any time series can be expressed as the sum of an infinite number of orthogonal random variables multiplied by sines and cosines of various frequencies. The value of the spectral density at a particular frequency corresponds to the contribution of the random variable associated with the sine and cosine at that frequency to total variance. Sines and cosines of high frequency, that is, ω near π, represent shocks which tend to revert quickly whereas sines and cosines of low frequencies, that is, ω's near 0, represent shocks which tend to revert slowly. If $\omega = 0$, then the effect of the random term is the same for all elements of the time series. White noise has the feature that the variance contributions of all terms are equal. From our earlier discussion, equation (5.3) implies that if $\Delta Y_{i,t}$ is an uncorrelated process, the cyclical term $\Delta C_{i,t}$ equals zero. Hence if the first difference of output changes is white noise, the typical output fluctuation is essentially permanent.[3]

The extent to which a time series is dominated by high versus low frequency shocks in turn may be measured by the normalized spectral distribution function

$$(5.6) \quad F_{\Delta Y_i}(\lambda) = \frac{2 \displaystyle\int_0^{\lambda} f_{\Delta Y_i}(\omega) d\omega}{\sigma_{\Delta Y_i}^2}$$

which is the integral of the spectral density. The value of this function equals the percentage of the variance of $\Delta Y_{i,t}$ contributed by frequencies between 0 and λ. If the spectral distribution function is concentrated above the diagonal, then this is evidence that the shocks to output for a given country are persistent.

Figures 5.1 to 5.5 present the normalized spectral distribution functions of the six economies we have been examining. Each country is contrasted with the United States. The visual evidence is quite striking.

---

[3]See the Appendix for a more detailed discussion of the spectral distribution function.

Substantial differences exist in the behavior of output fluctuations across countries. The most notable difference is that the output dynamics of the United States, Canada and the United Kingdom are similar whereas the output dynamics of France, West Germany, and Japan are similar. In particular, for the first three countries output differences behave quite similarly to white noise. This means that output changes appear to persist indefinitely. For the latter three countries, the spectral distribution functions are all concentrated above the diagonal. This means that changes in output will expand across time before reaching their total impact on the long run level of output. This effect is the opposite of mean reversion.

In order to appreciate the differences in magnitude suggested by the diagrams, it is helpful to consider a specific frequency. The frequency value $\pi/4$ represents fluctuations whose period is eight years. The percentages of output fluctuations attributable to frequencies of equal or greater period for the different countries are as follows:

Canada	20%
France	57%
Germany	43%
Japan	55%
United Kingdom	10%
United States	18%

These point estimates are remarkably different. For example, over five times as great a percentage of output variance is attributable to cycles of eight years or more for Japan than the United Kingdom. The eight-year benchmark clearly illustrates that fundamentally different processes govern the output behavior of different industrialized economies. Output fluctuations in France, Germany, and Japan have been dominated by low frequency fluctuations over the postwar era. No such evidence exists for Canada, the United Kingdom, and the United States. If anything, relative to white noise, where the spectral distribution function equals .25, these countries are dominated by high frequency fluctuations.

In order to formalize the differences between national output movements across countries, two sets of tests were run. The first set of tests computed the Cramér-von Mises (CVM) statistic[4] for goodness of fit.

---

[4]Formally, the spectral deviations may be expressed as

$$U_T(t) = \sqrt{2} T^{\frac{1}{2}} \int_0^{\pi t} \left(\frac{I_T(\omega)}{\hat{\sigma}_x^2} - \frac{1}{2\pi} \right) d\omega; t \in [0, 1]$$

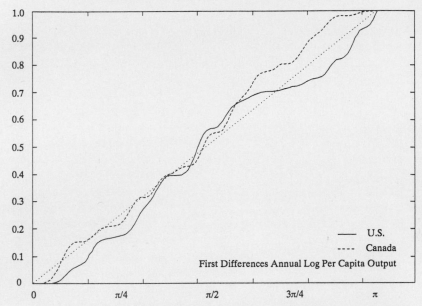

FIGURE 5.1 Spectral Distribution Functions: U.S. vs. Canada

This statistic tests whether a time series is white noise. The integral of the periodogram estimate of the spectral density permits the computation of the deviations of the sample spectral distribution function from a diagonal. The CVM statistic which equals the sum of the squared deviations in turn measures the magnitude of the deviations. The properties of this statistic are discussed in Durlauf 1990a. Column 1 of Table 5.4 presents the CVM statistics for the six countries. The white noise null is accepted for the United Kingdom, Canada, and the United States. The null is very strongly rejected for Japan, France, and Germany. These tests confirm the visual evidence that output movements in the latter three countries are dominated by slow moving components.

Notice that the small contribution of the low frequency to total variance did not cause a rejection for the United Kingdom. This should be interpreted as saying that the average deviation of the spectral distribution from a diagonal line for the United Kingdom is insufficiently large to distinguish the data from white noise. In this sense, the test is neutral with respect to which alternative hypothesis is explored. It does not focus on behavior at a specific frequency and is immune to data

and the CVM statistic as $CVM = \int_0^1 U_T(t)^2 dt$.

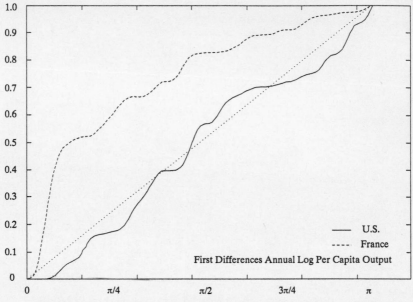

FIGURE 5.2 Spectral Distribution Functions: U.S. vs. France

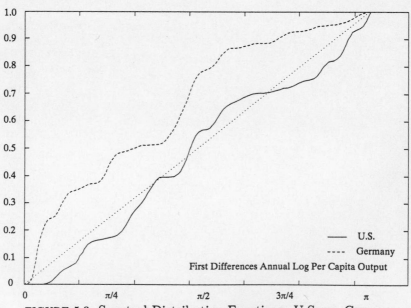

FIGURE 5.3 Spectral Distribution Functions: U.S. vs. Germany

TABLE 5.4
Time Series Behavior of Aggregate Output

Country	CVM	α_1	α_2
Canada	.05	.08 (.17)	−.17 (.17)
France	1.18*	.5* (.17)	.10 (.17)
West Germany	.53*	.37* (.17)	.10 (.17)
Japan	.84*	.45* (.17)	.02 (.17)
United Kingdom	.12	−.11 (.17)	−.28 (.17)
United States	.05	−.02 (.18)	−.06 (.17)

For regression $\Delta Y_{1-t} = C + \alpha_1 \Delta Y_{1.t-1} + \alpha_2 \Delta Y_{1.t-2}$.

CVM equals the Cramér-von Mises statistic for null hypothesis that $\Delta Y_{1.t}$ is white noise.

* Denotes significant at 5% level.

mining criticisms. By the same reasoning, the rejection of the white noise null for Japan, France, and Germany did not occur by focusing on the biggest deviation from the null hypothesis.

A second test of the properties of national output changes was performed through direct estimates of univariate autoregressions. Second order autoregressions are reported for the different countries in the last two columns of Table 5.4. Again, the AR coefficients are insignificant for the United Kingdom, Canada and the United States whereas the AR(1) not coefficients are large and significant for Japan, France, and Germany. Table 5.4 indicates that the six economies seem to fall into two distinct groups in terms of dynamic behavior.

Our reduced form time series analysis confirms the more structural approach of Taylor (1988a,b). This research not only demonstrated that the United States and Japanese economies have substantially different reactions to supply and demand shocks but also showed that these differences could be at least partially explained by the structure of wage contracts in the two countries. Consequently, there is substantial evidence in support of the position that some component of OECD output fluctuations are due to country-specific rather than common international factors. This result is consistent with other studies such as Campbell and Mankiw 1988, and Durlauf 1989.

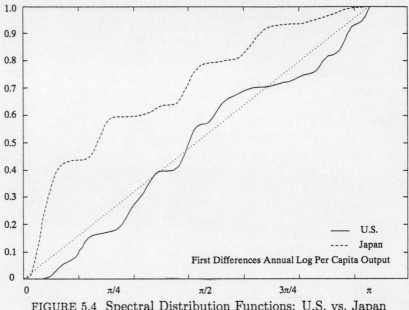

FIGURE 5.4 Spectral Distribution Functions: U.S. vs. Japan

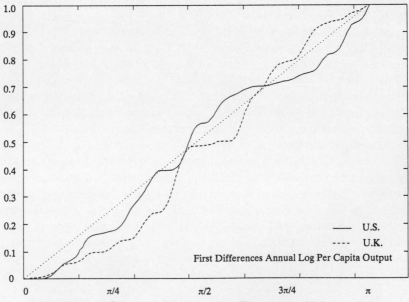

FIGURE 5.5 Spectral Distribution Functions: U.S. vs. U.K.

5 Modeling Multiple Macroeconomic Equilibria

In this section, we offer a framework for explaining the empirical discrepancies in cross-country output behavior. The approach we pursue to understanding international differences in cycles and growth centers on the difficulties of organizing economic activity in advanced industrial economies. Following the vast literature on convergence, we assume that different countries possess essentially identical preferences and technologies.[5] The source of distinct national experiences will stem from the ways in which societies deal with organizing production where there are positive complementarities to high levels of activity. From this perspective, national economies are capable of generating different production histories from the same technologies and preferences. These distinct production histories will be differentiated by both the mean and the volatility of output.

Specifically, we consider models of production complementarities which lead to multiple equilibria in aggregate output. Multiple equilibria occur because high and low levels of activity are mutually reinforcing. This idea has received increasing prominence in macroeconomics since the work of Diamond (1982), which illustrates how the social increasing returns to production and search for trading partners made multiple equilibria possible. Cooper and John (1988) have united much of this literature by showing how many models of complementarities will lead to multiple Pareto rankable equilibria. In a seminal contribution, the microeconomic foundations of complementarities has recently been explored in a game theoretic framework by Milgrom and Roberts (1989).

In the Appendix to this paper, we outline a formal model of multiple macroeconomic equilibria based upon the role of technological complementarities. The basic idea is straightforward. Each industry i at time t chooses between a high production level, which requires a fixed cost to achieve and a low production level which requires no such fixed cost. We define $\omega_{i,t}$ which equals one if the high production level is produced by industry i at t, 0 otherwise. Firms choose the high production level when intertemporal profit maximization justifies the payment of the fixed cost.

The dynamics in such an economy occur through the intertemporal complementarities which link the production level decisions of industries

---

[5]The migration of technology across borders implies that production side heterogeneity cannot explain long-run output differences. Sufficiently large heterogeneity in preferences can, of course, explain any discrepancies in output behavior. Such an approach renders all cross-country analyses uninterpretable.

over time. Specifically, we assume that each industry's productivity is affected by its own production decision along with the production decisions of similar industries the previous period. Letting $\Im_t$ denote the history of the economy, and $\Delta_{k,l}\{i - k \ldots i \ldots i + l\}$ represent an index of the industries which affect the productivity of firms in industry i, the conditional probability that industry i chooses high production at t equals

$$(5.7) \qquad Prob(\omega_{i,t} = 1 \mid \Im_{t-1}) = Prob(\omega_{i,t} \mid \omega_{j,t-1} \ \forall j \in \Delta_{k,l}).$$

The randomness in the industry production decisions comes from industry-specific productivity shocks. This structure means that high current production decisions will spill over and increase the probability of future high production.

As shown in the Appendix, when these productivity spillover effects are strong enough, the economy will exhibit multiple equilibria. In particular, we show the economy will exhibit one equilibrium where all industries are producing at high levels and another equilibrium where each period some industries produce at each level. This second equilibrium will exhibit a lower mean and a higher variance of output than the former. Intuitively, when all industries are producing at a high level, the average productivity of each industry is sufficiently high to render the industry insensitive to individual productivity shocks, whereas when few industries are producing at a high level, the impact of these shocks on individual behavior is greater. The selection of a particular equilibrium is determined by the interactions of the initial conditions in the economy with the particular conditional probability structure for production as defined by equation (5.7).

Suppose that we think of each time period t as introducing a new set of production opportunities through technical innovation. If each innovation produces a new good which is equally valued, then average or per capita output in the economy will evolve according to the sequence of resolutions of the production complementarities for each technology. The net impact of technical change on economic growth will depend both on the evolution of the production possibility frontier as well as the way that production patterns emerge across time.

Many patterns of growth are possible in this economy. Letting γ_j denote the average level of activity using the technology for the low production equilibrium and $\bar{\gamma}_j$ denote the average level of activity under the high production equilibrium, we consider three important cases. If the low equilibrium always occurs for each technical innovation, then aggregate per capita output Y_t will follow

(5.8) $Y_t = \underline{\gamma} t$

If the high level equilibrium always occurs, then

(5.9) $Y_t = \bar{\gamma} t$

Finally, consider the case that there is some probability between zero and one that the high level equilibrium occurs for each technical innovation. If we correspondingly define Ξ_j as a binary random variable associated with the jth technology which takes on the values $\underline{\gamma}$ or $\bar{\gamma}$ according to which equilibrium is selected, then

(5.10) $$Y_t = \sum_{j=0}^{t} \xi_j$$

This is a unit root process. If the Ξ_j's are independent, the aggregate output is a random walk with drift.

As these examples illustrate, a given set of technologies is compatible with many different patterns of long-term growth. The percentage of total output variance which is attributable to persistent shocks can vary across countries as well. In particular, notice that countries can have different deterministic or stochastic trends, and that the degree of persistence need not correlate with the average growth rate. Further, by choosing different degrees of dependence across the Ξ_t's, one can generate different time series properties as well.

In a model of this degree of abstraction, it is difficult to draw concrete policy conclusions. However, one result is both clear and robust to generalizations of the framework. The government can play a crucial role in the long-run growth rate of the economy through its encouragement of high production. A successful stabilization policy ensures that the sequence of high efficiency equilibria are chosen, which means both that there are no fluctuations and that the growth rate is high.

Government policies which affect microeconomic incentives may have powerful consequences for long-run growth. If government policies can affect the value of $\Theta_{k,l}^{\min}$, then stabilization policies will have powerful long-term consequences. These conditional production probabilities may be linked to macroeconomic policy along several dimensions. For example, in this volume, Bernheim and Shoven have demonstrated that risk premia are a very important element in the cost of capital. Federal Reserve policies which increase the risk premium can have adverse long-term consequences. Taylor (1989) shows how different exchange rate policies can affect interest rate, price and output volatility. Similarly, tax laws which encourage capital formation can positively affect growth rates through the many spillover effects we have discussed.

In fact, the potential for policy to affect long-term growth will introduce a bias in optimal policy formulation. Suppose that the government has to make a decision on a tax law where the consequences in terms of encouraging investment and innovation are unclear. The potential payoffs to changes which affect long-term growth are so high that in general a policymaker should err on the side of encouraging expansion. The reason is that the payoffs to the policy if efficacious will far outweigh the costs if ineffective, for a wide range of probabilities attached to the two outcomes. See Durlauf 1989 for a formalization of this idea in a decision theoretic framework.

6 Summary and Conclusions

This paper has explored a number of issues in the determination of long-run output in advanced industrialized economies. On the empirical side, we have explored the behavior of long-run output fluctuations in several OECD economies. We found little evidence of cointegration across OECD economies, nor was there strong evidence of interactions across short term fluctuations. Together, these results imply that long-run output movements in these economies are at least partially determined by country-specific factors.

On the theoretical side, we have outlined the elements of a theory of multiple equilibria in economic growth which is capable of explaining the stylized facts of national autonomy. We outline a model of local technological complementarities among industries. When these complementarities are strong enough, multiple long-run growth paths exist for the economy. The choice of equilibrium is determined by the interaction of initial conditions with the microeconomically determined probabilities of high production for each industry given the economy's history. This model suggests that domestic institutions can play a critical role in determining both short and long-run fluctuations. The model implies that successful stabilization policies can improve long-run economic growth by successfully guiding an economy to an efficient equilibrium.

The critical next step in research on these questions is the identification of specific microeconomic complementarities which seem fundamental to aggregate equilibrium. Such research not only will help to more clearly differentiate the real business cycle and multiple equilibrium paradigms, but also permit the analysis of growth implications of specific policy proposals.

TECHNICAL APPENDIX

A. Time Series Methodology

A.1 Properties of Stochastic Trends

Let $L_{Y_i}(t)$ denote the linear space spanned by $Y_{i,t}, Y_{i,t-1}, Y_{i,t-2} \cdots$. We define the deterministic trend $\beta_i t$ as

$$(5.11) \quad A\beta_i t = E(Y_{i,t} \mid L_{Y_i}(-\infty))$$

whereas the stochastic trend $S_{i,t}$ may be expressed as part of current activity which can be expected to persist into the indefinite future, subtracting off the deterministic trend.

$$(5.12) \quad S_{i,t} = \lim_{k \Rightarrow \infty} E(Y_{i,k} \mid L_{Y_i}(t)) - \lim_{k \Rightarrow \infty} E(Y_{i,k} \mid L_{Y_i}(-\infty))$$

By the law of iterated expectations, this definition implies that

$$(5.13) \quad E(S_{i,t} \mid L_{Y_i}(t-1)) = S_{i,t-1}$$

which shows that the stochastic trend in output is a random walk.

A.2 Spectral Distribution Function

Much of the discussion in Section 2 is based on the normalized spectral distribution function SDF. To see how to derive the SDF, it is useful to recall that if ΔY_t is a stationary L^2 process, then by the Cramér Representation Theorem

$$(5.14) \quad \Delta Y_t = \int_0^\pi \cos \omega t \cdot du(\omega) + \int_0^\pi \sin \omega t \cdot dv(\omega).$$

In these integrals, the terms $du(\omega)$ and $dv(\omega)$ are zero mean random variables with the properties that

  (i) $Edu(\omega)^2 = Edv(\omega)^2$,
 (ii) if $\omega_1 \neq \omega_2$, then $Edu(\omega_1)du(\omega_2) = Edv(\omega_1)dv(\omega_2) = 0$, and
(iii) $Edu(\omega_i)dv(\omega_j) = 0 \forall \omega_i, \omega_j$.

In other words, output changes can be represented as the sum of a continuum of orthogonal random variables. Recalling that the period of $\cos \omega t$ and $\sin \omega t$ is $\lambda = \frac{2\pi}{\omega}$, this means that one can equivalently think of the time series as composed of randomly weighted sines and cosines of different periods. When ω is small, then $\cos \omega t$ and $\cos \omega(t+1)$ are essentially the same, which means that low frequencies represent the contribution of the slowly changing components of the time series.

Orthogonality of $du(\omega)$ and $dv(\omega)$ when evaluated at different ω's means that one can measure the contribution of a particular frequency $\bar{\omega} \in [0, \pi]$ to the total variance of output changes

(5.15) $\cos^2(\bar{\omega})Edu(\bar{\omega})^2 + \sin^2(\bar{\omega})Edv(\bar{\omega})^2 = Edu(\bar{\omega})^2$

Further, the SDF gives a precise measure of how different intervals of cycles contribute to the volatility of output movements. The contribution of frequencies $[0,\lambda]$ to total variance equals the spectral distribution function. By construction,

(5.16) $\dfrac{\int_0^\lambda Edu(\omega)^2 d\omega}{\sigma_{\Delta Y}(0)} = \dfrac{2\int_0^\lambda f_{\Delta Y}(\omega)d\omega}{\sigma_{\Delta Y}(0)}$

White noise is a process where all frequencies contribute equally to the variance, that is, $Edu(\bar{\omega})^2$ is constant; which implies that the SDF for white noise is a diagonal line. If the SDF of a time series is concentrated below a diagonal line, this is evidence that output changes are relatively dominated by quickly reverting components, in the sense that the low frequency contribution to total variance is small relative to white noise.

B: A Model of Multiple Aggregate Equilibria

In this appendix, we outline a model of multiple equilibria. This model, based upon Durlauf 1990b and 1991, provides a framework for formally modeling the ideas in Section 2 above. We consider an infinite sequence of industries that must make a decision each period on whether or not to produce at a high or low level of production. Each industry is made up of a large number of small firms, which means that a firm takes the behavior of the industry as given when making decisions. A typical firm in industry i maximizes

(5.17) $\displaystyle\prod_{i,t} = \mathrm{E}\left(\sum_{j=0}^{\infty} \beta^j (Y_{i,t+j} - K_{i,t+j}) \mid \Im_t\right).$

$Y_{i,t}$ equals the output of the ith industry's representative firm at t; β equals a constant one-period ahead discount factor; $\Im_t$ denotes all information available to the economy at the beginning of t. Firms possess initial endowments $Y_{i,0}$ to begin investment.

The technology in the economy is very simple. At time $t-1$, a firm chooses both a technique of production and a level of capital to determine output at t.

(5.18) $Y_{1,i,t} = \begin{cases} Y_1 & \text{if } K_{i,t-1} \geq \bar{K} - \zeta_{i,t-1} \\ 0 & \text{otherwise} \end{cases}$

or

(5.19) $Y_{2,i,t} = \begin{cases} Y_2 & \text{if } K_{i,t-1} \geq \bar{K} - \eta_{i,t-1} \\ 0 & \text{otherwise} \end{cases}$

We assume that $Y_1 > Y_2$ and that $\bar{K} - \zeta_{i,t-1} > \bar{K} - \eta_{i,t-1}$. This technology says that a firm can produce at one of two levels of output, each of which requires a fixed capital input. Technology 1 produces more output in exchange for more capital. $\zeta_{i,t}$ and $\eta_{i,t}$ are industry-specific productivity shocks. High draws of either variable lead to low capital requirements for production. $\zeta_{i,t}$ and $\eta_{i,t}$ are elements of $\Im_t$. Recalling that firms within an industry are identical, we define $\omega_{i,t}$ which equals 1 if technique 1 is used by industry i at t, 0 otherwise; $\omega_t = \{\ldots \omega_{i-1,t}, \omega_{i,t}, \omega_{i+1,t} \ldots\}$ which equals the joint set of techniques employed at t; and $\Omega_t = \{\ldots \omega_{t-1}, \omega_t, \omega_{t+1} \ldots\}$ which equals the history of technique choices up to t. The entire history of technique choices can be indexed by Z^2, the two-dimensional lattice of integers.

We model technological complementarities through the dependence of $\zeta_{i,t}$ and $\eta_{i,t}$. $Prob(x \mid y)$ denotes the conditional probability measure of x given information y; $x(y)$ denotes the random variable associated with this measure. $\Delta_{k,l} = \{i - k \ldots i \ldots i + l\}$ indexes the industries which affect the productivity of firms in industry i.

Assumption B.1   Conditional Probability Structure of Productivity Shocks

  A. $Prob(\zeta_{i,t} \mid \Im_{t-1}) = Prob(\zeta_{i,t} \mid \omega_{j,t-1} \; \forall j \in \Delta_{k,l})$.
  B. $Prob(\eta_{i,t} \mid \Im_{t-1}) = Prob(\eta_{i,t} \mid \omega_{j,t-1} \; \forall j \in \Delta_{k,l})$.
  C. The random pairs $(\zeta_{i,t} - \zeta_{i,t}(\Im_{t-1}), \eta_{i,t} - \eta_{i,t}(\Im_{t-1}))$ are mutually independent of each other $\forall i$.

The economic intuition behind this assumption is straightforward. Industries are productive to a great extent because they are elements of a vast set of interactions which occur across sectors and across time. High activity in one sector improves the capacity of another to produce successfully. This is the basic idea of the social increasing returns to scale work of Romer (1986) and Lucas (1988). Romer's work is closest in spirit to ours, as his model postulates that the aggregate capital stock will affect the productivity of individual firms. In the economy, learning-by-doing spills over sectors and increases across time within an active sector. Some empirical evidence on behalf of this view has recently been developed by Caballero and Lyons (1989). Alternatively, if one sector provides capital to another, then high efficiency, low price production in one sector will make the decision to capitalize in another easier. This could be accommodated in the model by making the fixed cost K depend upon the behavior of other industries.

In equilibrium, the representative firm makes a choice of the high or low efficiency technology based upon the level of activity of the complementary industries in the previous period. From our assumptions on technologies, profit maximization follows from a rule whereby a firm chooses technique 1 if

$$(5.20) \quad Y_1 - k + \zeta_{i,t} \geq Y_2 - k + \eta_{i,t}$$

Further, given our assumptions on the productivity shocks, it is clear that

Theorem B.1 Structure of Technique Choice Conditional Probability Measures

The equilibrium technique choice conditional probability measures for each industry obey

$$(5.21) \quad Prob(\omega_{i,t} \mid \Im_{t-1}) = Prob(\omega_{i,t} \mid \omega_{j,t-1} \; \forall j \in \Delta_{k,l}).$$

We now restrict the conditional probabilities in order to discuss multiplicity and dynamics. Past choices of technique 1 are assumed to improve the current relative productivity of the technique. As a result, technique 1 choices will propagate over time. Further, we assume that $\omega_t = \underline{1}$ is a steady state, which means that when all productivity spillovers are active, the effects are so strong that high production is always optimal.

Assumption B.2 Impact of Past Technique Choices on Current Technique Probabilities

Let $\underline{\omega}$ and $\underline{\omega}'$ denote two realizations of $\underline{\omega}_{t-1}$. If $\omega_j \geq \omega_j' \; \forall j \in \Delta_{k,l}$, then

A. $Prob(\omega_{i,t} = 1 \mid \omega_{j,t-1} = \omega_j \; \forall j \in \Delta_{k,l}) \geq Prob(\omega_{i,t} = 1 \mid \omega_{j,t-1} = \omega_j' \; \forall j \in \Delta_{k,l}$.

B. $Prob(\omega_{i,t} = 1 \mid \omega_{j,t-1} = 1 \; \forall j \in \Delta_{k,l}) = 1$.

Whenever some industry fails to choose technique 1, a positive productivity spillover is lost. The loss of these spillovers reduces the probability of high production by several industries in the future. We can bound the various probabilities of high production by the values $\Theta_{k,l}^{\min}$ and $\Theta_{k,l}^{\max}$.

$$(5.22) \quad \Theta_{k,l}^{\min} \leq Prob(\omega_{i,t} = 1 \mid \omega_{j,t-1} = 0 \text{ for some } j \in \Delta_{k,l}) \leq \Theta_{k,l}^{\max}$$

The bounds $\Theta_{k,l}^{\min}$ and $\Theta_{k,l}^{\max}$ measure the strength of production complementarities in the economy. When $\Theta_{k,l}^{\max}$ is small, complementarities are powerful in that high production is unlikely unless all possible productivity spillovers are active. Conversely, if $\Theta_{k,l}^{\min}$ is large, complementarities are weak in the sense that high production occurs relatively frequently even without productivity spillovers.

We are now in a position to understand whether this economy exhibits multiple equilibria. From our assumptions, $\omega_t = \underline{l}$ is an equilibrium, since once technique 1 is chosen for all industries, high production will continue into the future. On the other hand, it is not clear that an economy starting from arbitrary initial conditions will converge to this equilibrium. In particular, if $\omega_0 = \underline{0}$, the economy may fail to achieve $\omega_t = \underline{1}$ even asymptotically. The reason for this is straightforward. If firms fail to choose the high technique with sufficient frequency, productivity spillovers may to build up over time. As a result, the economy will fail to build up sufficient momentum to converge to $\omega_t = \underline{1}$ for sufficiently high t.

In fact, the bounds $\Theta_{k,l}^{\min}$ and $\Theta_{k,l}^{\max}$ determine the number of equilibria in the economy. When $\Theta_{k,l}^{\min}$ is sufficiently large, the economy will converge to $\omega_t = \underline{1}$ for any initial conditions, whereas if $\Theta_{k,l}^{max}$ is sufficiently small, the economy will never converge to the high production equilibrium, if all industries start at the low production technique. Formally, Durlauf (1990c) proves

Theorem B.2 Existence of Multiple Equilibria
For each index set $\Delta_{k,l}$ with at least one of k or l nonzero, there exists a number $\Theta_{k,l}$, $0 < \Theta_{k,l}$, such that if $\Theta_{k,l}^{\max} \leq \Theta_{k,l}$, then

A. $\lim\limits_{t \Rightarrow \infty} Prob(\omega_{i,t} = 1 \mid \omega_0 = \underline{0}) < 1.$

B. $\lim\limits_{t \Rightarrow \infty} Prob(\omega_t = \underline{1} \mid \omega_0 = \underline{0}) = 0.$

C. $\lim\limits_{t \Rightarrow \infty} Prob(\omega_{i,t} = 1 \mid \omega_0 = \underline{1}) = 1.$

In an economy starting with all the low production technique industries, the probability that any individual industry achieves high production will be strictly bounded below one. Further, the economy will almost surely fail to converge to the high production equilibrium. If the economy ever achieves the high production equilibrium each firm will produce at the high level in all subsequent periods.

In this economy, initial conditions can play a critical role. If high production is relatively unproductive in absence of productivity spillovers, then the economy will be unable to take off. On the other hand, if production is sufficiently profitable in the early period of the economy, high production will always emerge. Observe that the time series properties of output will differ across the equilibria.

References
Arrow, K. J. 1962. The Economic Implications of Learning By Doing. *Review of Economic Studies*, 29, pp. 155–73.

Bernheim, B. D. and J. B. Shoven. 1989. Comparison of the Cost of Capital in the U.S. and Japan: The Roles of Risk and Taxes. Working Paper, Stanford University.

Blanchard, O. and D. Quah. 1988. The Dynamic Effects of Aggregate Demand and Supply Disturbances. NBER Working Paper No. 2737.

Bryant, J. 1985. A Simple Rational Expectations Keynes-Type Model. *Quarterly Journal of Economics*, 98, pp. 525–28.

Caballero, R. and R. Lyons. 1989. The Role of External Economies in U.S. Manufacturing. Working Paper, Columbia University.

Campbell, J. Y. and N. G. Mankiw. 1987. Are Output Fluctuations Transitory? *Quarterly Journal of Economics*, CII, pp. 857–80.

Campbell, J. Y. and N. G. Mankiw. 1988. International Evidence on the Persistence of Economic Fluctuations. NBER Working Paper No. 2498.

Cochrane, J. 1988. How Big is the Random Walk in GNP?, *Journal of Political Economy*, 96, pp. 893–920.

Cooper, R. and A. John. 1988. Coordinating Coordination Failures in Keynesian Models. *Quarterly Journal of Economics*, CIII, pp. 441–65.

DeLong, J. B. and L. H. Summers. 1988. How Does Macroeconomic Policy Affect Output? Brookings Papers on Economic Activity, 2, 1988, pp. 433–80.

Diamond, P. 1982. Aggregate Demand In Search Equilibrium. *Journal of Political Economy*, XC, pp. 881–94.

Dobrushin, R. L. 1968. Description of a Random Field by Means of Conditional Probabilities and Conditions for Its Regularity. *Theory of Probability and Its Applications*, 13, pp. 197–224.

Durlauf, S. N. 1989. Output Persistence, Economic Structure and the Choice of Stabilization Policy. Brookings Papers on Economic Activity, 2, pp. 69–116.

Durlauf, S. N. 1990a. Spectral Based Tests of the Martingale Hypothesis. *Journal of Econometrics*, forthcoming.

Durlauf, S. N. 1990b. Locally Interacting Systems, Coordination Failure and the Behavior of Aggregate Activity. Working Paper, Stanford University.

Durlauf, S. N. 1990c. Nonergodic Economic Growth. Working Paper, Stanford University.

Durlauf, S. N. 1991. Path Dependence in Aggregate Output. Working Paper, Stanford University.

Engle, R. and C. W. J. Granger. 1987. Cointegration and Error Correction: Representation, Estimation and Testing. *Econometrica*, 55, pp. 251–76.

King, R., C. Plosser, J. Stock, and M. Watson. 1987. Stochastic Trends and Economic Fluctuations. Working Paper, University of Rochester.

Lucas, R. E. 1988. On the Mechanics of Economic Development. *Journal of Monetary Economics*, 22, pp. 3–42.

Milgrom, P. and D. J. Roberts. 1989. Rationalizability, Learning and Equilibrium in Games with Strategic Complementarities. Working Paper, Stanford University.

Nelson, C. and C. I. Plosser. 1982. Trends and Random Walks in Macroeconomic Time Series. *Journal of Monetary Economics*, 10, pp. 139–62.

Romer, P. 1986. Increasing Returns and Long-Run Growth. *Journal of Political Economy*, 94, pp. 1002–37.

Shapiro, M. and M. Watson. 1988. Sources of Business Cycle Fluctuations, in *Macroeconomics Annual 1988*, S. Fischer, ed. Cambridge: National Bureau of Economic Research.

Summers, R. and A. Heston. 1988. A New Set of International Comparisons of Real Product and Price Level Estimates for 130 Countries, 1950–1985. *Review of Income and Wealth*, 34, pp. 1–25.

Taylor, J. B. 1988a. Synchronized Wage Determination and Macroeconomic Performance in Japan. Working Paper, Stanford University.

Taylor, J. B. 1988b. Differences in Economic Fluctuations in Japan and the U.S.: The Role of Nominal Rigidities. Center for Economic Policy Research Technical Paper No. 153.

Taylor, J. B. 1989. Policy Analysis in a Multicountry Model. To appear in *Macroeconomic Policies in an Interdependent World*, R. Bryant, D. Currie, J. Frenkel, P. Masson, and R. Portes, eds. Washington D.C.: International Monetary Fund.

Weitzman, M. 1982. Increasing Returns and the Foundations of Unemployment Theory. *Economic Journal*, 92, pp. 787–804.

Part II

The "Non-R&D" Influences on Technology Commercialization

6

Comparing the Cost of Capital in the United States and Japan

B. DOUGLAS BERNHEIM AND J. B. SHOVEN

As Michael Boskin and Lawrence Lau have noted in this volume, almost two-thirds of technical progress requires new physical investment. For a satisfactory rate of productivity growth, an economy must be competitive with respect to its net and gross rates of national investment. This situation does not augur well for the United States in its competitive race with one of its main trading partners, Japan. The United States badly trails Japan in the fraction of its GNP devoted to investment and in its rate of technical progress. The failure to keep up is often attributed to the high cost of capital in the United States and the tax system's role in creating that high cost. One remarkable aspect of this argument, however, is that it is made without a widely agreed upon definition of the cost of capital. Thus, this paper seeks to address the question of how to measure the cost of capital. We also propose to ascertain whether it is indeed true that the United States suffers from a higher cost of capital than Japan.

Our examination of literature on the cost of capital readily revealed that there are no regularly reported statistics for it in government or private publications. To confuse matters further, at least three separate definitions of the concept were found. Some papers used the inflation-adjusted interest rate as the cost of capital (see, for example, Blair and Litan 1990). Others used the Hall-Jorgenson (1967) formulation, which essentially adjusts the real interest rate for a host of tax factors (see King and Fullerton 1984). Textbooks on corporate finance, on the other hand, use something called the "weighted average cost of capital" (see, for example, Brealey and Myers 1984). The weighted average cost of capital simply takes the required market rate of return on each of the firm's securities (bonds, notes, preferred stock, common stock, etc.) and determines the before corporate tax rate of return which would be necessary for the firm to meet the market requirements on all of its securities. Our contention is that all three definitions found in the literature are seriously flawed. Therefore, our first task is to develop a better measure of the concept. The flaws in the existing approaches will become apparent as we present our preferred approach.

The cost of capital is the hurdle rate or the required rate of return which must be met by new investments if they are to be in the interest of the firm's owner. Thus, the definition of the cost of capital hinges on what determines this hurdle rate. The straightforward answer to that question revolves around the idea of opportunity cost. A new investment to be prudent must earn at least as much as other readily available uses of the same funds. A natural alternative investment to consider is a financial investment in the stock and bond markets. The idea is that a physical investment at the level of the firm should be considered as a type of financial investment in that both involve the expenditure of funds now for a random distribution of returns in the future. These physical investments made at the firm level using the owner's money must offer terms at least as attractive as those available at any brokerage house. However, if the firm considering the investment is a corporation and is contemplating using equity financing (including, perhaps, retained earnings), it must also take the corporate income tax into account in determining whether the investment in question offers a competitive rate of return.

So far, we have argued that the cost of capital for a new equity financed investment inside a corporation should be the expected or average rate of return which must be earned by the investment in order for it to offer terms which are competitive with those in the financial market, even after taking corporate level taxes into account. The two logical steps to follow in determining the cost of capital are: one, determine the terms available in financial markets and, two, calculate what the project would have to earn before taxes in order to provide those terms after corporate taxes.

As is widely known, financial markets exhibit a substantial degree of risk aversion. This risk aversion manifests itself in the much higher average returns earned on risky investments than on safe ones. We refer to the extra average or expected returns on a risky investment as a "risk premium." How large these risk premia can be is indicated by noting that the average real pre-personal-tax rate of return on the Standard and Poors 500 since 1926 has been 8.9 percent. Over the same 65 year period, the average inflation-adjusted rate of return on Treasury Bills has been 0.4 percent. The difference in average rates of return of 8.5 percent appears to reflect considerable risk aversion on the part of investors. What does this mean with regard to the cost of capital? Since financial markets offer much higher average rewards to risky investments, firms making investment decisions must assess the riskiness of their projects. If a project under consideration is extraordinarily safe, then perhaps it only has to offer a return which is competitive with Treasury bills. On the other hand, if it involves an amount of risk comparable to the S&P 500, it must then offer a much higher expected return, one which at least matches the S&P 500 after the payment of corporate taxes.

The above discussion implies that there are three important factors in determining the cost of capital: interest rates, risk premia, and taxes. It also suggests that the cost of capital cannot be conveyed by a single number. Rather, it needs to be represented as a schedule since investments with a higher risk have a corresponding higher cost of capital. Let us return briefly to the three definitions in the literature and their shortcomings. First, the real interest rate fails to reflect both taxes and risk premia. It would only be an appropriate cost of capital or hurdle rate in a theoretical world without taxes and risk. The Hall-Jorgenson approach rectifies the matter with respect to taxes, but still fails to capture the enormous risk premia imposed by financial markets. Even if one attempts to put risk premia into their framework, it cannot deal with the interaction between risk premia and tax considerations. Finally, the weighted average cost of capital does reflect all three of our factors: interest rates, risk premia, and taxes. Its shortcoming, however, is distilled in the very word "average." It is the appropriate cost of capital for the *existing* assets of the firm, but unless the new investment under consideration is *exactly* like a miniature version of the existing firm's assets, the weighted average cost of capital will be inappropriate. What is needed, and what we will try to construct here, could be termed the "marginal cost of capital." In our approach, the riskiness of each new investment must be assessed and the hurdle rate set at an appropriate level for an investment of that riskiness.

1 Constructing a Measure of the Cost of Capital

The financial markets' historical evidence of rates of return clearly indicates that risk premia are extremely important elements in the cost of capital for risky real investments. The strategy we follow in comput-

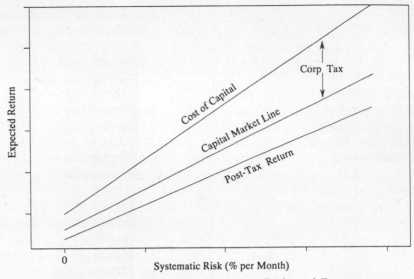

FIGURE 6.1 Relationships Between Risk and Return

ing the cost of capital actually involves identifying three schedules of expected rates of return for investments of different degrees of riskiness. One is the "capital market line," which shows the expected return in financial markets as a function of the standard deviation of systematic (undiversifiable) risk. That is, the capital market line shows the expected return on assets of different riskiness after corporate or business level taxation (i.e., at the brokerage houses). From it one can compute (using details of the tax law and certain equilibrium conditions) the cost of capital line and the post-personal-tax return to investors. The three lines are shown in Figure 6.1.

The concept of systematic risk deserves further explanation. Basically, the variability in an asset's rate of return results in two kinds of riskiness: independent risk and systematic risk. To understand the difference, consider a large number of assets, each with a highly variable but independent rate of return. In such a situation, a large number of these assets can be grouped to form a portfolio with a safe return. The riskiness of each security is eliminated in the portfolio through diversification. This type of risk is referred to as "independent risk" and should not carry a risk premium. On the other hand, if a large number of assets have perfectly correlated rates of return, no amount of diversification will eliminate or reduce the risk inherent in each asset. The systematic risk of an asset depends on the correlation of its return with a broad market portfolio. Such systematic risk, which must be borne by investors, will usually require a premium expected rate of return relative to the return on a safe asset.

The first step in our approach is to determine the position of the post-corporate tax capital market line, the brokerage house terms. This turns out to be quite difficult in its own right. Once we have done that, we make the assumption that all assets must offer equal "certainty equivalent" rates of return. We use the slope of the capital market line to determine risk premia. This permits us to compute certainty equivalent yields for risky investments. The basic arbitrage condition is that all assets must be competitive with Treasury bills on an after-tax certainty equivalence basis.

The complete mathematical presentation of our approach is contained in Bernheim and Shoven 1989. We assert there that new physical investments are subject to two separate kinds of risk. First, there is uncertainty as to the contemporaneous marginal product (i.e., the rents or net cash flow that the project will generate). Second, there is uncertainty about the actual depreciation that will be realized on the asset over the next accounting period. In equation form, the investment's net cash flow in any period is given by

(6.1) $f(k)(1 + \epsilon_f)$

where k is the physical asset and ϵ_f is a zero mean random variable. In any particular period, actual depreciation is

(6.2) $k(\delta - \epsilon_f)$

where δ is the expected exponential rate of depreciation and ϵ_f is another independent mean zero random variable. This bifurcation of risk into two components has been suggested elsewhere in the literature (Bulow and Summers 1984) and is particularly appropriate here since the corporation income tax applies to shifts in earnings due to income (ϵ_f) fluctuations but not to unanticipated depreciation (ϵ_δ).

Without taxes, the cost of capital would be equal to the real interest rate adjusted for the risk premia associated with these two types of risk. The cost of capital would exactly correspond to the capital market line shown in Figure 6.1. The precise cost of capital for any particular project could be looked up on this line or schedule once the riskiness of the project had been determined. This cost of capital, in a world without taxes, would be exactly the same regardless of whether the incremental investment was financed by debt or equity. This brings us to one of our analysis' major points, that is, the risk that is included in the cost of capital computation is associated with the physical investment and not the financial instruments used to raise the funds. Some companies (General Motors, for example) could finance a risky project with safe debt. In order to decide whether this project adds to the value of the firm, the manager should compare its expected return to the cost of capital associated with the riskiness of the investment itself and not the riskiness of the bonds.

Taxes influence the cost of capital in a variety of ways. The corporate tax rate, the pattern of depreciation, the personal tax on interest and dividends, the taxation of realized capital gains, and the lack of inflation indexation of the income definition are some of the important considerations. The taxation of an investment also depends on a firm's behavior with respect to dividend policy and the financial instruments used.

We calculate separately the cost of capital for equity financed and debt financed projects. Unlike the situation in a world without taxes, there is a difference in any given project's cost of capital depending on whether it is debt or equity financed. This is due to the different taxation of the returns (interest payments are deductible from the corporation's income tax base but fully taxable at the household level; equity return is taxable at both the corporate and personal level, with special treatment for capital gains).

Even before presenting some estimates, we can relate one of our findings from the theoretical model. All of the cost of capital formulae in Bernheim and Shoven 1989 (like that of others) have a multiplicative term of $(1 - A)/(1 - \tau)$ where A is the present value of the tax savings due to depreciation deductions and credits (per dollar of investment) and τ is the corporate tax rate. For new equity issues, this multiplicative term is the only term involving the corporate level tax, and it affects both the intercept and slope of the cost of capital line. Thus, if A were equal to τ, as with immediate expensing, the corporate tax would not be a distortionary influence on the cost of equity capital. It also means that the extent to which the corporate tax is distortionary depends crucially on the nominal interest rate. With a zero nominal interest rate, depreciation deductions, even if spread out over thirty years, are equivalent to expensing and therefore, the corporate tax does not distort investments financed with new equity. With a ten percent interest rate, however, deductions occurring more than, say, fifteen years in the future are worth twenty cents on the dollar or less. This means that A, the present value of the taxes saved with the depreciation deductions, is substantially less than the corporate tax rate, and $(1 - A)/(1 - \tau)$ is significantly greater than unity, thus raising the cost of capital. This crucial importance of the nominal interest rate (and hence, the rate of inflation) may be what reconciles the conflicting stories as to whether the corporate income tax is more or less distortionary in Japan than in the United States. The corporate tax rate is unambiguously higher in Japan. Depreciation schedules are roughly similar. However, their corporate tax system may be less distortionary due to the lower nominal interest rates and therefore, the more adequate present value of depreciation deductions. That is, in Japan A is much closer to τ.

In order to implement our measure of the cost of capital, we must first determine the position of the capital market line. What is the expected return on the market portfolio, and what is its expected riskiness? What about the expected return on an absolutely safe investment? If these two points can be determined, then they define the capital market line. If one makes a set of assumptions which basically rule out the importance of higher moments of the distribution of returns, then all assets must compete with linear combinations of the safe asset and the market portfolio. This means that all assets must offer terms on the line defined by those two points.

Unfortunately, determining the expected returns on a market portfolio (such as the S&P 500) cannot be done completely satisfactorily. If the monthly real total (dividends plus real capital gains) rates of return on the market were independently drawn from an identical distribution, then the average of a large number of realizations would give an accurate measure of the expected future returns. Similarly, the standard deviation of realizations would give an accurate guide to the standard deviation of the constant underlying real rate of return distribution. However, the real returns in U.S. and Japanese equity markets do not seem to conform to the independent draws from an identical distribution model. The long-term average realization on the S&P 500 portfolio is extraordinarily sensitive to the precise period covered. An extreme example is that the ten-year average of real monthly returns (120 draws) for the S&P 500 equities looking backwards from September 1984 was 0.704 percent per month whereas the 120-month backwards-looking average calculated from January 1984 was 0.301 per month. The two averages have 112 common entries and yet the eight substitutions cause the average realization to more than double. While this example is one of the sharpest changes observed over recent years, it is not unique. The monthly real return series for the S&P 500 appears to reflect "regime changes" or shifts in the underlying distribution.

Similarly, the return series based on the Tokyo Stock Exchange securities does not seem to conform to the independent draws from an identical distribution model. An example of the lack of stability in

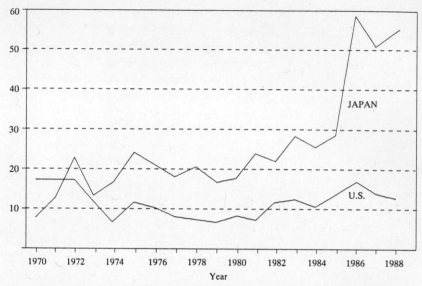

FIGURE 6.2 Price/Earnings for the United States and Japan

the return distribution of Japanese securities is that the annual total inflation-adjusted rate of return realized by the Tokyo Stock Exchange Index I companies was 61.02 percent per year for 1971 and 1972, −1.83 percent per year for 1973 to 1982, and 24.84 percent per year from 1983 through 1988.

The return series of both countries exhibit a high coefficient of variation, so a small set of observations cannot hope to indicate the expected outcome. On the other hand, extending the averaging over very long periods of time (e.g., twenty years or more) increases the likelihood that the distant observations were drawn from a different regime. Unless the movements in the distribution itself can be described, there is no easy way out of this dilemma. Despite these significant problems, most finance economists and practitioners find it useful to construct and evaluate capital market lines based on past realizations over five or ten years. We have examined the consequences of following this practice in Bernheim and Shoven 1989, but are wary of its appropriateness given the apparent lack of stability of the underlying distributions.

An alternative measure of the expected return on equity shares can be taken from the price-earnings ratio, or more precisely from the earnings themselves. The pattern of price-earnings ratios in the United States and Japan from 1970 to 1988 is shown in Figure 6.2. Basically, the P-E of the Japanese stock market, which started the period at half the

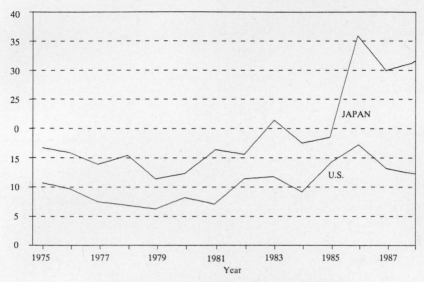

FIGURE 6.3   Adjusted Price-Earnings Ratios

level of the U.S. market, more than doubled between 1970 and 1973, remained relatively flat between 1973 and 1982, and then more than doubled again between 1982 and 1988. Clearly, most of the appreciation of the Japanese market between 1982 and 1988 was due to this rise in the P-E ratio rather than to an increase in earnings themselves. The P-E ratio of the U.S. market sagged in the mid 1970s to early 1980s. However, it does not show the dramatic ratcheting up displayed by the P-E ratio of the Japanese market.

International comparisons of price-earnings ratios are dangerous due to several factors including different accounting practices and tax rules. A recent paper by French and Poterba (1989) attempts to correct the P-E ratios of the United States and Japan for accounting differences and for double counting due to intercorporate ownership. The adjusted P-E results (based on estimating Japanese earnings with U.S. accounting practices) are shown in Figure 6.3 for the 1973–88 period. The bottom line is that accounting differences explain some of the difference in the P-E ratios between the two countries, but even after the adjustments, the P-E ratio in Japan in 1988 was 2.74 times the P-E value in the United States.

The reciprocal of the P-E ratio, or the earnings-price ratio, is an alternative candidate for the expected return of holding a stock portfolio. Abstracting from the potentially serious accounting problems, a

TABLE 6.1

Expected Short-Term Rates of Return and the
Earnings-Price Ratios for the United States

(percent per month)

Year	December Expected Short Term Return	Monthly Earnings-Price Ratio
1980	0.59097	0.95785
1981	0.29259	1.09649
1982	0.56591	0.75075
1983	0.44561	0.70028
1984	0.42511	0.88652
1985	0.32707	0.58685
1986	0.29428	0.47619
1987	0.20870	0.64599
1988	0.32512	0.71225

firm's earnings reflect equity profits after corporate taxes and after depreciation allowances sufficient to maintain the physical capital of the firm. If these earnings were fully paid out to shareholders, the firm's capital stock and hence "earnings power" would remain intact. As long as the ratio of the stock market valuation of the firm to the value of the underlying capital stock remains constant, the real rate of return of the shareholder would be given by the earnings-price ratio. Using the earnings-price ratio as the expected rate of return involves several assumptions. The two most important of these are that the expected future earnings rate on physical capital is given by the current earnings rate, and the expected real change in the market price of a constant capital stock is zero.

For some individual firms, the E-P ratio understates expected future returns because of anticipated growth in earnings. However, one should note that growth in earnings due to retentions is already reflected in today's E-P ratio. The only growth in earnings which would cause the expected rate of return to exceed the E-P ratio would be an anticipated increase in earnings from a constant capital stock. While there are some firms which may be in such a position (biotechnology companies, for example), there are others which find themselves in declining industries. It is our feeling that anticipated growth in real earnings per unit of capital is not likely to be significant at the aggregate level (i.e., for the market as a whole), even though it may be an important consideration for individual securities.

In this paper, we report the results using the adjusted earnings-price

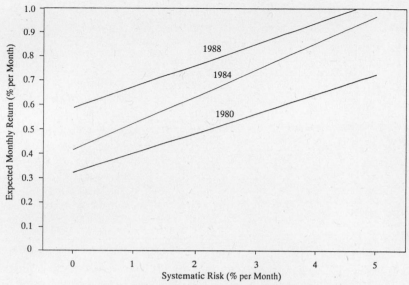

FIGURE 6.4 U.S. Capital Market Lines Using E/P

ratio as a proxy for the expected return on the market. One should be aware that there are important consequences of proceeding with this proxy. First, there may be some remaining serious accounting problems. Second, there is a dramatic divergence between the measure of expectations with this approach and those using a past realizations based measure (particularly for Japan). Third, the earnings and price figures refer not only to reproducible capital such as plant and equipment, but also to goodwill, research and development, and land. Despite these many warnings, we believe the adjusted earnings-price ratio is the best available proxy for the expected after corporate tax returns that must be offered equity financiers.

Fortunately, establishing the expected return on a safe investment is a little easier to determine, since short term nominal interest rates are readily observable. However, converting these nominal rates to real rates involves unobserved inflation expectations. For simplicity, we take inflation expectations to be average monthly rate of inflation over the previous six months, and we subtract this from the monthly rate on T bills to get our measure of the expected real safe interest rate. While we are aware that T bills are not perfectly safe in real terms, we have followed the usual practice of ignoring this inflation induced risk.

The market lines, determined using the adjusted E/P proxy for expectations, are shown in Table 6.1 and Figure 6.4 for the United States

TABLE 6.2

Expected Short Term Rates of Return
and the Earnings Price Ratios for Japan

(percent per month)

Year	December Expected Short Term Return	Monthly Earnings-Price Ratio
1980	0.58653	0.66138
1981	0.41560	0.48733
1982	0.43624	0.51125
1983	0.30574	0.39494
1984	0.30574	0.47619
1985	0.56956	0.45788
1986	0.46694	0.23343
1987	0.33611	0.27964
1988	0.24211	0.25961

and Table 6.2 and Figure 6.5 for Japan. What we see in those figures is that the capital market line is lower in Japan in each of the years displayed and that the slope of the line is much smaller (i.e., the price of risk or the degree of risk aversion displayed in financial markets is lower).

Translating these results on the capital market line to those on the cost of capital requires the application of our modeling of the tax system and the conditions of asset equilibrium. The translation involves a large number of parameters; the keys ones are shown in Table 6.3 with their assumed values for the United States and Japan. The corporate and personal tax rates for the United States for 1980 are taken from King and Fullerton 1984, while the 1980 tax rates for Japan come from Shoven and Tachibanaki 1988 and the 1988 Japanese figures from Shoven 1989. In all cases, the corporate and personal tax rates are meant to reflect sub-national levels of taxation as well as national taxation. The a and a_E parameters reflect the tremendous amount of dividend smoothing which takes place in both countries.

We examine two types of physical investments: one is an example of relatively short lived equipment, an automobile, while the other, much longer lived is a plant. We assume that cars actually depreciate exponentially at a rate of 1/7 per year in both countries, whereas plants lose productivity at a common exponential rate of 1/31.5. Table 6.3 shows the shortest available depreciation lifetimes for these assets. We assume that firms use the most accelerated form of depreciation permitted by the tax authorities. In the United States, of course, the 1986 tax

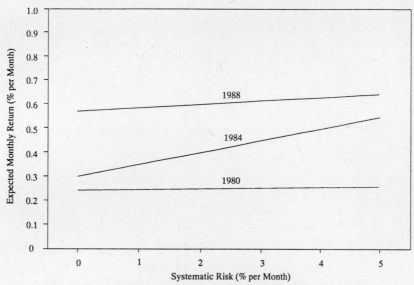

FIGURE 6.5   Japanese Capital Market Lines Using E/P

reform involved a deceleration of depreciation deductions. Previously, firms could use 150 percent declining balance methods for plants, but now plant depreciation must be straight line over 31.5 years. The depreciation deductions on autos were also decelerated, primarily because the minimum lifetime was increased from three years to five years. The investment tax credit on equipment investments (including autos) which is included in A, also was eliminated in 1986.

The results of using these parameter values in our Bernheim-Shoven (1989) model of the cost of capital cannot be summarized easily with simple numbers, but only schedules or lines. Graph sets 1 and 2 (pp. 166–169) show the cost of debt and equity financed investments in the United States and Japan. In each graph, the higher of the two essentially parallel lines is the cost of equity financed projects, whereas the lower line is for debt financing. The less steep line is the capital market line. The fact that the cost of capital line is always dramatically steeper than the capital market line in the United States indicates that the U.S. tax system discriminates against risky investments. That is, the difference between the cost of capital and the market's expected return (the government's wedge) is much higher for more risky investments. In the United States in 1980, the cost of capital for both debt and equity financed autos was less than the return to financiers represented by the capital market line. This is the result of rapid three-year depreciation, the investment tax

TABLE 6.3

Parameter Values Used in the Calculation
of the Cost of Capital

	United States		Japan	
	1980	1988	1980	1988
Corporate Tax Rate $= \tau$	.495	.380	.526	.499
Average Marginal Personal Tax Rate on Interest & Dividend $= m$	.475	.300	.124	.200
Effective Average Marginal Personal Tax Rate on Retained Earnings $= z$	.25	.21	.06	.08
Effective Average Marginal Tax Rate on Purely Nominal Capital Gains $= z_n$	.14	.13	0.0	.02
Fraction of Long-Term Earnings Paid as Dividends $= \alpha$	.5	.5	.33	.33
Fraction of Transitory Earnings Paid as Dividends $= \alpha_\epsilon$	.02	.02	.02	.02
Fraction of Total Risk Attributable to Capital Risk $= \nu$	.9	.9	.9	.9
Short-Term Nominal Interest Rate $= i$	.1566	.0809	.09756	.04092
Expected Rate of Inflation $= \pi$	.0858	.0418	.027204	.011904
Exponential Rate of Depreciation for Autos & Plants $= \delta$	1/7,1/31.5	1/7,1/31.5	1/7,1/31.5	1/7,1/31.5
Tax Depreciation Lifetimes for Autos & Plants $= L$	3/25	5/31.5	4/26	4/26
Present Value of the Tax Value of Depreciation Deduction and Tax Credits for Autos & Plants $= A$	.5342 .1659	.3325 .1418	.465 .250	.473 .344
Expected Real Rate of Return on Market Portfolio from Past Realizations $= r_1^e$	.0964	.1142	.13495	.1549
Expected Real Rate of Return on Market Portfolio from Adjusted Earnings-Price Ratio $= r_2^e$	.1149	.0855	.0794	.0312

TABLE 6.4

Cost of Capital Estimates for Autos
and Plants with Standardized Riskiness

(Standard Deviation = 4.5% per month)

Earnings/Price Basis	United States Autos	United States Plant	Japan Autos	Japan Plant
Equity Financed				
1980	0.0887	0.2073	0.1086	0.1453
1988	0.1037	0.1258	0.0408	0.0512
Debt Financed				
1980	0.0695	0.1674	0.0582	0.0739
1988	0.0829	0.0973	0.0246	0.0308

credit, relatively light capital gains tax treatment on some of equity's return, and interest deductibility from the corporate income tax. The comparison of the cost of capital in the United States and Japan is shown most clearly in Graph Set 3 (pp. 170–171). That graph set shows that the United States has a much higher price of risk for both years and both assets. However, in 1980, the cost of capital was comparable in the two countries, particularly for equity financed investments in automobiles. In 1988, the U.S. cost of capital was unambiguously higher as well as the price of risk being much greater. The lower cost of equity capital in Japan was largely due to the much higher values for a unit of earnings in Japanese securities markets. A project with a given earnings potential will command a much higher market value in Japan. This leads rather directly to the lower cost of capital and the higher rate of investment in Japan. The price of risk is lower in Japan for the same reason, but also because the lower nominal interest rates make depreciation deductions more adequate.

Despite the fact that our approach requires that cost of capital information be transmitted as a schedule rather than a number, there seems to be a substantial demand for single number summaries of the results. Because of that, we present Table 6.4, which shows the annual cost of capital for autos and plants whose riskiness is a standardized 4.5 percent per month. This amount of risk is comparable to the long-run risk of a diversified portfolio of common stocks in either country. The numbers again show Japan to have a dramatically lower cost of capital than the United States in 1988. For example, the cost of capital for an equity financed plant is approximately 12.6 percent in the United States, but only 5.1 percent in Japan. The contrast is equally sharp for autos and for debt financed investments.

Cost of Capital in the U.S. in 1980
for Equity and Debt Financed Autos

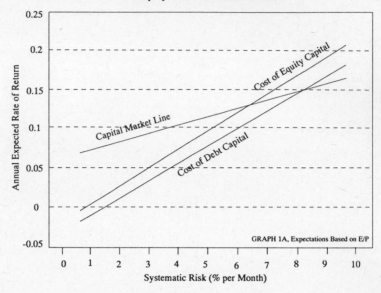

GRAPH 1A, Expectations Based on E/P

Cost of Capital in the U.S. in 1988
for Equity and Debt Financed Autos

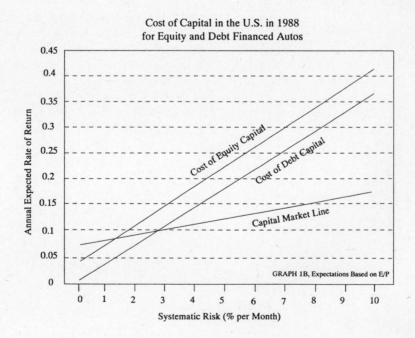

GRAPH 1B, Expectations Based on E/P

Cost of Capital in the U.S. in 1980
for Equity and Debt Financed Plants

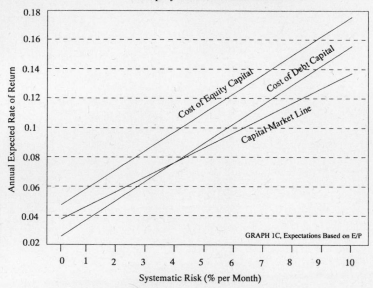

GRAPH 1C, Expectations Based on E/P

Cost of Capital in the U.S. in 1988
for Equity and Debt Financed Plants

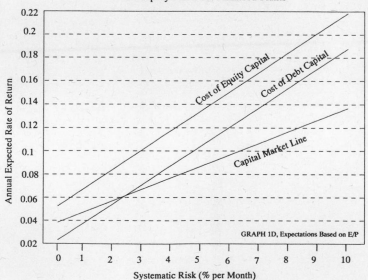

GRAPH 1D, Expectations Based on E/P

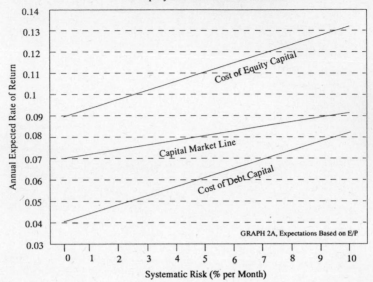

**Cost of Capital in Japan in 1980
for Equity and Debt Financed Autos**

Cost of Equity Capital

Capital Market Line

Cost of Debt Capital

GRAPH 2A, Expectations Based on E/P

Annual Expected Rate of Return

Systematic Risk (% per Month)

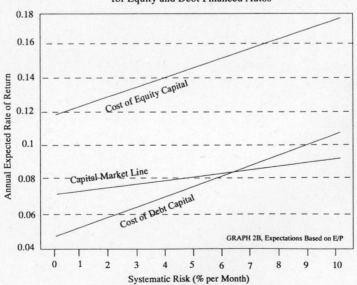

**Cost of Capital in Japan in 1988
for Equity and Debt Financed Autos**

Cost of Equity Capital

Capital Market Line

Cost of Debt Capital

GRAPH 2B, Expectations Based on E/P

Annual Expected Rate of Return

Systematic Risk (% per Month)

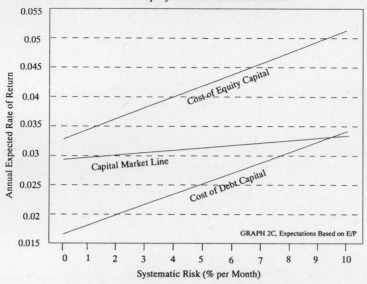

**Cost of Capital in Japan in 1980
for Equity and Debt Financed Plants**

Annual Expected Rate of Return

Cost of Equity Capital

Capital Market Line

Cost of Debt Capital

GRAPH 2C, Expectations Based on E/P

Systematic Risk (% per Month)

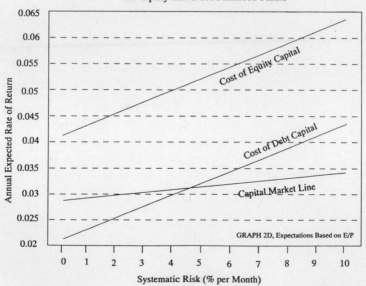

**Cost of Capital in Japan in 1988
for Equity and Debt Financed Plants**

Annual Expected Rate of Return

Cost of Equity Capital

Cost of Debt Capital

Capital Market Line

GRAPH 2D, Expectations Based on E/P

Systematic Risk (% per Month)

Cost of Capital for Equity Financed
Autos in the U.S. and Japan in 1980

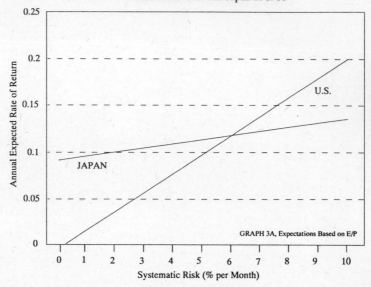

GRAPH 3A, Expectations Based on E/P

Cost of Capital for Equity Financed
Plants in the U.S. and Japan in 1980

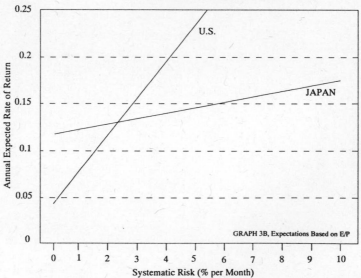

GRAPH 3B, Expectations Based on E/P

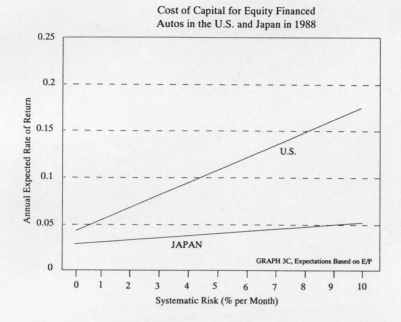

Cost of Capital for Equity Financed
Autos in the U.S. and Japan in 1988

U.S.

JAPAN

GRAPH 3C, Expectations Based on E/P

Annual Expected Rate of Return

Systematic Risk (% per Month)

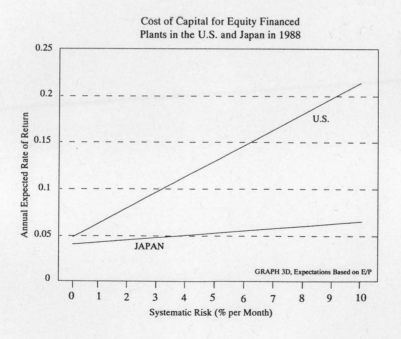

Cost of Capital for Equity Financed
Plants in the U.S. and Japan in 1988

U.S.

JAPAN

GRAPH 3D, Expectations Based on E/P

Annual Expected Rate of Return

Systematic Risk (% per Month)

In 1980, the U.S. cost of capital was more comparable to that in Japan, particularly for autos and equipment. The combination of accelerated depreciation and the investment tax credit kept the U.S. cost-competitive. The cost of equity capital numbers in Table 6.4, ranging from 8.9 to 20.7 percent in the United States and from 4.1 to 14.5 percent for Japan, agree much more closely to the statements of businessmen regarding hurdle rates than do the 3.0 to 6.0 percent figures generated by approaches which fail to factor in risk premia.

2 Conclusions Regarding the Cost of Capital and Investment

Our research seems to confirm the oft-made allegation that the United States suffers from a cost of capital disadvantage relative to Japan. Somehow journalists and business economists came to this realization even though the analytical framework to support their conclusion was lacking. Given this finding regarding the relative cost of capital, it is not surprising that the United States lags in investment relative to GNP.

Our analysis has highlighted several other findings with respect to the cost of capital issue. First, and perhaps most important, we find that risk premia are extremely important in determining the hurdle rates on new investments. Any analysis which is fundamentally based on safe interest rates will give misleading cost of capital answers unless it explicitly incorporates risk. Second, the U.S. tax system interacts with risk premia in a non-neutral manner. In general, riskier projects face higher effective tax rates than safer projects. In the diagrams in this paper, this is reflected by the always steeper slope of the cost of capital line than the capital market line. This tax discrimination against riskier projects could be reduced or eliminated with some tax reforms such as expensing investments or indexing depreciation deductions for inflation.

Our analysis also shows that debt financed investments always face a lower cost of capital than equity financed ones. This discrimination could be eliminated by removing the double taxation of equity or by removing the deductibility of interest expenses.

Our general conclusion is that the cost of capital handicap faced by the United States relative to Japan should be a serious public policy concern. Further, there are policies such as the adoption of a corporate cash flow tax which would dramatically improve the problem. We urge that the appropriate government agencies and our fellow academic economists pay more attention to this problem and its possible remedies. The growth rate and even the long-run vitality of our economy depends on rectifying this situation.

References

Auerbach, A. J. and L. Kotlikoff. 1983. National Savings, Economic Welfare, and the Structure of Taxation. In *Behavioral Simulation Methods in Tax Policy Analysis*, Martin Feldstein, ed. Chicago: University of Chicago Press, pp. 459–98.

Bernheim, B. D., J. K. Scholz, and J. B. Shoven. 1989. Consumption Taxation in a General Equilibrium Model: How Reliable Are Simulation Results? NBER Conference on Saving, January 6–7, 1989, Maui, Hawaii.

Bernheim, B. D. and J. B. Shoven. 1987. Taxation and the Cost of Capital: An International Comparison. In *The Consumption Tax: A Better Alternative?* Charles E. Walker and Mark A. Bloomfield, eds. Cambridge, MA: Ballinger Publishing Company, pp. 61–86.

Bernheim, B. D. and J. B. Shoven. 1989. Comparison of the Cost of Capital in the U.S. and Japan: The Roles of Risk and Taxes, CEPR Working Paper, No. 179.

Blair, Margaret M., and Robert E. Litan. 1990. Corporate Leverage and Leveraged Buyouts in the Eighties. In *Debt, Taxes, and Corporate Restructuring*, John B. Shoven and Joel Waldfogel, eds. Washington D.C.: The Brookings Institution.

Bradford, D. F. *Untangling the Income Tax*. Cambridge, MA: Harvard University Press, April 1986.

Brealy, R. and S. Myers. 1984. *Principles of Corporate Finance*, 2nd edition. New York: McGraw-Hill.

Bulow, J. I. and Summers, L. H. 1984. The Taxation of Risky Assets. *Journal of Political Economy*, 92(1) (February), pp. 20–39.

Feldstein, M. S. and J. Slemrod. 1978. How Inflation Distorts the Taxation of Capital Gains. *Harvard Business Review*, September/October, pp. 20–22.

French, K. R. and Poterba, J. N. 1989. Are Japanese Stock Prices Too High? Paper presented at the Center for Research in Security Prices Seminar on the Analysis of Security Prices. Chicago, IL, May 4, 1989.

Fullerton, D., R. Gillette, and J. Mackie. 1987. Investment Incentives Under the Tax Reform Act of 1986, in *Compendium of Tax Research 1987*. Office of Tax Analysis, Department of the Treasury, Washington D.C.

Fullerton, D., Y. K. Henderson, and J. Mackie. Investment Allocation and Growth Under the Tax Reform Act of 1986. In *Compendium of Tax Research 1987*, Office of Tax Analysis, Department of the Treasury, Washington D.C.

Fullerton, D., J. B. Shoven, and J. Whalley. 1983. Replacing the U.S. Income Tax With a Progressive Consumption Tax. *Journal of Public Economics*, 20 (February), pp. 3–23.

Gravelle, J. G. 1989. Income, Consumption, and Wage Taxation in a Life Cycle Model: Separating Efficiency from Redistribution. Congressional Research Service mimeograph, June 1989.

Hall, R. E. and D. W. Jorgenson. 1967. Tax Policy and Investment Behavior, *American Economic Review* 57: 391–414.

Hatsopoulos, G. N. and Brooks, S. H. The Cost of Capital in the United States and Japan. Paper presented at the International Conference on the Cost of Capital, Kennedy School of Government, Harvard University, Cambridge, MA, November 19–21, 1987.

Hendershott, P. H. 1988. The Tax Reform Act of 1986 and Economic Growth, National Bureau of Economic Research Working Paper No. 2553, Cambridge, MA.

King, M. A. and D. Fullerton. 1984. The Taxation of Income From Capital. Chicago: National Bureau of Economic Research/University of Chicago Press.

McLure, C. E., Jr. The 1986 Act: Tax Reform's Finest Hour or Death Throes of the Income Tax? *National Tax Journal*, vol. 41, no. 3 (September), pp. 303–15.

Mossin, J. 1973. *Theory of Financial Markets*. Englewood Cliffs, NJ: Prentice-Hall.

Poterba, J. M. 1989. Dividends, Capital Gains, and the Corporate Veil: Evidence from Britain, Canada, and the United States. *National Saving and Economic Performance*, B. Douglas Bernheim and John B. Shoven, eds. Chicago: NBER/University of Chicago Press.

Poterba, J. M. and L. H. Summers. 1988. Mean Reversion in Stock Prices, *Journal of Financial Economics*, 22, pp. 27–59.

Shoven, J. B. 1989. The Japanese Tax Reform and the Effective Rate of Tax on Japanese Corporate Investments, National Bureau of Economic Research Working Paper No. 2791.

Shoven, J. B. and Tachibanaki, T. 1988. The Taxation of Income from Capital in Japan. In *Government Policy Towards Industry in the United States and Japan*, John B. Shoven, ed. Cambridge: Cambridge University Press, pp. 51–96.

Summers, L. H. 1981. Capital Taxation and Accumulation in a Life Cycle Growth Model. *American Economic Review*, 71 (September), pp. 533–44.

Summers, L. H. 1987. Should Tax Reform Level the Playing Field? Mimeograph from the 1987 Proceedings of the National Tax Association.

Summers, L. H. 1987. Reforming Tax Reform for Growth and Competitiveness. Mimeograph, Harvard University.

Venti, S. F. and D. A. Wise. 1988. The Determinants of IRA Contributions and the Effect of Limit Changes. In *Pensions in the U.S. Economy*, Zvi Bodie, John B. Shoven, and David A. Wise, eds. Chicago: University of Chicago Press, pp. 9–52.

Vogt, S. 1988. Taxing Consumption as a Fiscal Policy Alternative: A Review of the Debate. Center for the Study of American Business Working Paper No. 120, Washington University, St. Louis, MO, July 1988.

7

Strategies for Capturing the Financial Benefits from Technological Innovation

DAVID J. TEECE

Despite the supposed existence of "first mover" advantages, first-to-market innovators with new products and processes do not always win. Situations where firms were first to commercialize a new product, but did not participate in the profits that were subsequently generated from the innovation, are increasingly common. For instance, transistors, integrated circuits, computer tomography, magnetic resonance imaging, video recorders, and color television are among the many advances made in the West and first embodied in new products in the West. "Yet in each of these areas the Japanese have subsequently gained a strong and sometimes dominant position" (Gomory and Schmitt 1988, p. 113).

The archetypical example of the phenomenon was the experience of the British company EMI with the body scanner.[1] In the early 1970s, the U.K. firm Electrical Musical Industries (EMI) Ltd. was in a variety of product lines including phonographic records, movies, and advanced electronics. EMI had developed high resolution TVs in the 1930s, pioneered airborne radar during World War II, and developed the United Kingdom's first all solid-state computers in 1952. In the late 1960s, Godfrey Hounsfield, an EMI senior research engineer, engaged in pattern

---

This article is based on an earlier article of mine titled "Profiting from Technological Innovation," published in *Research Policy*, December 1986.

[1] The EMI story is summarized in Michael Martin 1984.

recognition research which resulted in his displaying a scan of an animal's brain. Subsequent clinical work established that computerized axial tomography (CAT) was viable for generating cross-sectional "views" of the human body, the greatest advance in radiology since the discovery of x-rays in 1895. While EMI was initially successful with its CAT scanner, within six years of its introduction into the United States in 1973 the company had lost market leadership, and two years later dropped out of the CAT scanner business. Other companies, though late entrants, dominate the market today.

Other examples include Bowmar, which introduced the pocket calculator, but was unable to withstand competition from Texas Instruments, Hewlett Packard and others, and went out of business. Xerox failed to succeed with its entry into the office computer business with its "Star" system, even though Apple succeeded with the MacIntosh which contained many of Xerox's key product ideas, such as the "mouse" and "windows."[2] The de Havilland Comet saga has some of the same features. The Comet I jet was introduced into the commercial airline business two years or so before Boeing introduced the 707, but de Havilland was unable, for reasons discussed later, to capitalize on its substantial early advantage.

If there are innovators who lose, there must be followers and/or imitators who win. A classic example is IBM with its PC line, a great success when it was introduced in 1981. Neither the architecture nor components embedded in the original IBM PC were commonly considered advanced when introduced; nor was the way technology was packaged a significant departure from then-current practice. Yet the IBM PC was fabulously successful and established MS-DOS as the leading operating system for sixteen-bit PCs. By the end of 1984, IBM had shipped over 500,000 PCs, and many considered that it had irreversibly eclipsed Apple in the PC industry.[3]

The body of this article describes an analytic approach innovators can use to analyze new market opportunities. The approach employs a mix of old and new concepts including appropriability conditions, the stage in the design cycle, the complementarity of assets, and contracting versus integration strategies. Using these concepts, it is possible to identify strategies most likely to win; to identify strategies that can help

---

[2] See "The Lab that Ran Away from Xerox," *Fortune*, September 5, 1983. The Alto computer developed at PARC preceded the 'Star' and contained many important features subsequently adopted in the PC industry.

[3] MITS introduced the first personal computer, the Altair, experienced a burst of sales, then slid quietly into oblivion.

the innovator win big; and to explain when and why innovators will lose, and therefore should not enter markets even though their innovation has real commercial value.

1 The Role of the Appropriability Regime

The most fundamental reason why innovators with good marketable ideas fail to open up markets successfully is that they are operating in an environment where their know-how is difficult to protect. This constrains their ability to appropriate the economic benefits arising from their ideas and their product/process concepts. The two most important environmental factors conditioning this are the efficacy of legal protection mechanisms, and the nature of their technology.

1.1 Legal Protection

Technological innovation arises from the utilization of certain physical items, like machines, along with human skills in particular organizational contexts. But whereas non-human assets like machines have clear rights of ownership, the know-how and skills developed by an innovating firm cannot be protected unless they fall into one or another of the various categories of recognized intellectual property—patents, copyrights, trade secrets, trademarks (see Table 7.1)—in which case title can be established, rights of assignment recognized, and limited protection afforded.[4]

Even when intellectual property protection exists, it does not confer perfect appropriability, although patents do afford considerable protection of intellectual property in some instances, such as with new chemical products and drugs (Winter 1987). Many patents can be "invented around" at modest costs.[5] Often patents provide little protection because the legal and financial requirements for upholding their validity or for proving their infringement are high. And copyright can often be worked around as it is the expression of ideas and not the ideas themselves that the law protects (see Table 7.1).

In some industries, particularly where the innovation is embedded in

---

[4]For a fuller treatment of intellectual property in the strategic management context, see Teece 1989a.

[5]Mansfield, Schwartz, and Wagner (1981) found that about 60 percent of the patented innovations in their sample were imitated within four years. In a later study, Mansfield (1985) found that information concerning product and process development decisions was generally in the hands of at least several rivals within twelve to eighteen months, on average, after the decision was made to acquire it. Process development decisions tend to leak out more slowly than product development decisions in practically all industries, but the difference on average was found to be less than six months.

TABLE 7.1

Some Comparative Characteristics of Four Types of Intellectual Property

(United States only)

	Patent	Copyright	Trade Secret	Trademark
Term	17 years from date of grant	Life of the author plus 50 years from date of creation of a work, or in the case of a "Work for Hire," 75 years from date of creation	Perpetuity	Perpetuity so long as the mark does not become generic
Matter that is protected	Invention or discovery must be a new and useful process, machine, manufacture or composition of matter or a new and useful improvement thereof	Original works of authorship fixed in any tangible medium from which the work can be perceived, reproduced or otherwise communicated either directly or with the aid of a machine or device	Information used in one's business that supports a competitive position	Words or symbols
Condition for protection	The invention must not (1) have been known or used by others in the U.S., (2) have been patented or described in a printed publication in the U.S. or any foreign country, (3) have been in public use or on sale in the US for more than one year prior to the date of application, or (4) have been abandoned	The work must be original	Confidentiality	Registration

processes, trade secrets are a viable alternative to patents. Trade secret protection is possible, however, only if a firm can put its product before the public and still keep the underlying technology secret. Usually only chemical formulas and industrial-commercial processes can be protected as trade secrets after they are "out."

1.2   Nature of the Technology

The degree to which knowledge about an innovation is tacit, or easily codified, also affects the ease of imitation. Tacit knowledge is, by definition, difficult to articulate and so is hard to imitate or to pass on unless those who possess the know-how can demonstrate it to others (Polanyi 1962, Teece 1981). Codified knowledge is easier to transmit and receive and is more exposed to industrial espionage.

Simplistically then, we can divide appropriability regimes into those that are "weak" (innovations are difficult to protect because they can be easily codified and legal protection of intellectual property is ineffective) and those that are "strong" (innovations are easy to protect because knowledge about them is tacit and/or they are well protected legally).[6] Despite recent strengthening of the protection of intellectual property,[7] strong appropriability is the exception rather than the rule. Only in a few industries, most notably drugs and chemicals, is intellectual property thought of as the main mechanism to isolate innovators from followers (Winter 1987). However, the number of such industries may be expanding because of recent court decisions.

As we shall see, innovators who try to enter markets where intellectual property protection is good have a reasonable chance of winning commercially; in weak regimes they often lose. Since strong appropriability is still the exception rather than the rule, innovators in most industries must adopt clever market entry strategies if they are to keep imitators and other followers at bay.

---

[6]Of course, even if a strong patent exists, it will eventually expire. Hence, even regimes that start out strong because of an enforceable patent will eventually become weak when the patent expires.

[7]At least in the United States, the 1980s have witnessed a marked strengthening of intellectual property. For instance, the Copyright Act has been rewritten four times to include new forms of original composition in the list of protected works. In 1980, the Copyright Act was extended to computer programs. Perhaps of greater significance has been the vesting in October 1, 1982, of a single court with subject matter jurisdiction over all appeals from the U.S. Patent and Trademark Office in connection with pending patent applications and patent infringement cases from the U.S. District Courts. This measure has brought both uniformity and greater strength to the U.S. patent system.

2 Market Entry Strategies Before the Emergence of a "Dominant Design"

The best market entry strategies for innovators with good, marketable ideas depend not only on the appropriability regime, but also on where the industry is in its development cycle (Abernathy and Utterback 1978). In the early stages of development of many industries and product families, designs are fluid, manufacturing processes are loosely and adaptively organized, and generalized capital is used in production. Competition among firms manifests itself in competition among designs, which are markedly different from each other.

At some point in time, and after considerable trial and error in the marketplace, one design or a narrow class of designs begins to emerge, either naturally or through sponsoring, as the more promising. Such a design must be able to meet a whole set of user needs in a relatively complete fashion.[8] The Model T Ford, the Douglas DC-3, the IBM 360, and the VHS are examples of dominant designs in the automobile, aircraft, computer and VCR industries, respectively.[9]

The existence of a dominant design watershed is of great significance to the strategic positions of the innovator and its competitors.[10] Once a dominant design emerges, competition migrates to price and away from design. Competitive success then shifts to a whole new set of variables. Scale and learning become much more important, and specialized capital is developed and deployed as incumbents seek to lower unit costs through exploiting economies of scale and learning. Reduced uncertainty over product design provides an opportunity to amortize specialized long-lived investments.

Product innovation is not necessarily halted once the dominant design emerges; as Clark (1985) points out, it can occur lower down in the design hierarchy. While improvement is likely to be incremental, cumulatively it is of great importance. For instance, a "V" cylinder configuration emerged in automobile engine blocks during the 1930s with the emergence of the Ford V-8 engine. Niches were quickly found for

---

[8]Sometimes the market can lock in, through small chance events, on the design that is not the best of available alternatives. See Arthur 1988.

[9]A dominant design, almost by definition, can only be identified ex post. In certain cases a number of indicators, such as rapid sales growth, can be observed that can help one predict dominant designs.

[10]Needless to say, dominant design and standards can be sponsored. This is sometimes a viable strategy, particularly in network industries. The rapidly expanding literature on standards and standards promotion will not be surveyed here. An excellent summary can be found in Tirole 1989, Section 10.6.

it. Moreover, once the product design stabilizes, there is likely to be a surge of process innovation as producers attempt to lower production costs for the new product.

The product/process innovation cycle does not characterize all industries and product families. It seems more suited to mass markets where consumer tastes are relatively homogeneous. It would appear to be less characteristic of small niche markets where the absence of scale and learning economies attaches much less of a penalty to multiple designs. In these instances, generalized equipment will be employed in production.[11]

2.1 Strong Appropriability

In any case, none of these threats need unduly worry an innovator who is preparing to enter a market with good intellectual property protection, because the innovator knows that its ideas must form some part of the eventual dominant design. Even if the innovator comes to market in this early phase with a sound product concept but the wrong design, good intellectual property protection will afford the time needed to perform the trials needed to get the design right. As suggested earlier, the best initial design concepts often turn out to be hopelessly wrong, but if the innovator possesses an impenetrable thicket of patents, or has technology that is simply difficult to copy, then the market may well afford the innovator the necessary time to ascertain the right design without being eclipsed by imitators who fix the initial design flaws. The innovator can thus proceed through this early phase of industry development confidently, concentrating the bulk of his strategic thinking on how to win after this dominant design emerges, as discussed later on.

2.2 Weak Appropriability

In the early design development of an industry with weak appropriability, the innovator has to recognize that although it may have been responsible for the fundamental scientific breakthroughs as well as the basic design of the new product, as with Ampex and the VCR (Rosenbloom and Abernathy 1982), imitators may enter the fray, modifying the product in important ways, yet relying on the fundamental designs pioneered by the innovator. When the game of musical chairs stops and

---

[11]In terms of this framework, industries that are not characterized by a dominant design are likely to remain fragmented. The chances of winning big on a single product are thus low. However, considerable profit opportunities will exist. Capturing these profits is likely to require the use of complementary assets dedicated to the support of narrow classes of customers.

a dominant design emerges, the innovator might well end up positioned disadvantageously relative to a follower. When imitation is easy and is quickly coupled with design modification before the emergence of a dominant design, followers have a good chance of having their modified product anointed as the industry standard, often to the great disadvantage of the innovator.

In this situation the innovator must be careful to let the basic design "float" until sufficient evidence has accumulated that a design has been delivered that is likely to become the industry standard. In some industries there may be little opportunity for product modification. In other industries it may be quite considerable. The reason for letting the design float is that once a heavy capital investment is committed to a particular design, the innovator will have created certain irreversibilities that will prove to be a severe handicap if the initial guesses with respect to what the market wants turn out to be wrong.

The early history of the automobile industry exemplifies exceedingly well the importance for subsequent success of selecting the right design. None of the early producers of steam cars survived the early shakeout when the closed body internal combustion engine automobile emerged as the dominant design. The steam car, nevertheless, had numerous early virtues, such as reliability, which the internal combustion engine autos could not deliver.

The British fiasco with the Comet I is also instructive. De Havilland had picked an early design with significant flaws. Stresses at the right angular corners of the windows were higher than predicted; hence the metal fatigue that led to fatal crashes. Significant irreversibilities flowing from loss of reputation seemed to prevent the innovator from successfully converting to what subsequently emerged as the dominant design. But for the accidents, de Havilland might have retained their lead in jet airliner design and construction. As a general principle, it appears that innovators in weak appropriability regimes need to be especially intimately coupled to the market so that user needs can fully impact designs.

When multiple, parallel, and sequential prototyping is feasible, it has clear advantages. Unfortunately, such an approach is generally prohibitively costly. When development costs for a large commercial aircraft exceed one billion dollars, variations on a theme are all that is possible.

In summary, the probability may be low that an innovator in a market with weak appropriability will enter possessing the dominant design. The probabilities will be higher the lower the relative cost of prototyping, and the more tightly coupled the firm is to the market.

The latter is a function of organizational design, and can be influenced by managerial choice. The former is embedded in the technology, and cannot be influenced, except in minor ways, by managerial decisions. Hence, in industries with large developmental and prototyping costs—and hence significant irreversibilities—and where imitation of the product concept is easy, then the probability that the innovator will be among the winners when the dominant design emerges may be very low indeed, unless the innovator has other strengths. In this situation, it may not make sense for the innovator to compete at all.

3 Market Entry Strategies after the Emergence of the Dominant Design

3.1 Role of Complementary Assets

Suppose the innovator is still in the game when the dominant design has emerged. It then has to face a new problem—that of securing access to complementary assets and possibly complementary products. An innovation consists of certain technical knowledge about how to do things better than the existing state of the art. In order for a new device to deliver value to the consumer, it must be sold or utilized in some fashion in the market, usually in conjunction with other devices. The bundling of devices into systems or subsystems is usually necessary to create a product. Thus the Intel 8080 microprocessor was more than a chip. "It comprised the chip to be sure, but also application notes, ads, microprocessor development systems and emulators, software, field applications engineering, and single board computers, customer education programs, and Bob Noyce's ability to capture the public imagination" (Davidow 1986, p. 26). One might also add that it required manufacturing, marketing, and sale-capacity at Intel to bring the chip to the market successfully.

In short, the successful commercialization of an innovation requires that the know-how in question be utilized in conjunction with the services of other assets (see Figure 7.1). Two classes of complementarities can be distinguished. First, there are buyer complementarities. A "product" can be thought of as the totality of what a customer buys. It is not only the physical device or service from which the customer gets direct utility but includes a number of other factors, products, and services that makes the innovation desirable, useful, and convenient. Thus computer hardware must have software available, autos must have gasoline, textbooks must have examination copies, and copiers need copier paper

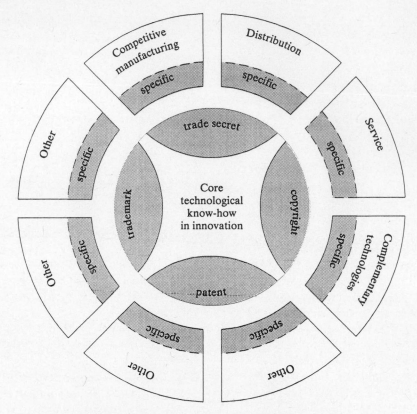

Shaded area represents the less imitatable portion of the value chain.
Outer segments represent complementary assets; inner circle segments represent know-how.

FIGURE 7.1 Representative Complementary Assets
Needed to Commercialize an Innovation

and service.[12] If any item is not properly augmented in this fashion, it might be thought of as a device rather than a product.[13] Innovating firms frequently make the mistake of underestimating the complementarities that must be available for a product to sell well.

The second, though partly overlapping, class of complementarities

---

[12]The invention of the disposable cartridge containing toner and ink and the parts most likely to wear out is a striking example of how the cost of providing these complementarities can be reduced through clever design which integrates technological and business considerations.

[13]When a device is properly augmented with the relevant complementarities, it becomes a product (Davidow 1986, p. 27).

exist on the supplier side. These are the other parts of the "value chain" that the innovator must assemble or otherwise access in order to ensure that the product is produced and delivered to the customer. Obvious examples are manufacturing, distribution, and after sales support.[14] The delivery of value to the customer will often require that the complementary assets utilized will be highly particularized to the innovation. For example, the manufacture of VCRs in volume requires special precision manufacturing equipment to make the magnetic heads.

Before the dominant design emerges, production volumes are low, and there is little to be gained in deploying specialized assets, as scale economies are unavailable and price is not a principal competitive factor. However, as the leading design or designs begin to be revealed by the market, volumes increase and opportunities for economies of scale will induce firms to begin gearing up for mass production by acquiring specialized tooling and equipment, and possibly specialized distribution as well. Since these investments involve significant irreversibilities, producers are likely to proceed with caution. Islands of asset specificity will thus begin to form in a sea of generalized assets.

Whatever the reason for needing specific assets—whether it is the emergence of a dominant design requiring scaleup or simply the particularities of the technology—the impact of their ownership on competition and the distribution of profits among industry participants is of great significance. This is because specific assets almost always have, or quickly develop, firm-specific attributes. Their idiosyncrasy may be embedded in physical capital, specialized knowledge, or locational or relational specificities (Williamson 1985). These are often difficult to imitate, which makes them valuable to hold and an important source of competitive advantage. They provide the foundation for what has been termed "isolating mechanisms" (Rumelt 1987, pp. 145–46).

Specific complementary assets can thus represent the second line of defense for new products and processes. If intellectual property law does not provide the requisite isolating mechanisms for the innovator, ownership of complementary assets can often do so. If the required complementary assets are generic, then this will not be possible; but if they are specific, or if specific assets can be used to enhance the value

---

[14]The "value chain" is used by Porter (1985) and others to divide a firm into discrete activities. However, the assets and capabilities required to commercialize new technology often lie outside the firm's existing boundaries. The value chain does not provide guidance as to which activities are to be done "in-house" and which are to be left to others.

of the innovation to the ultimate end user, then ownership or control of specific assets can be the source of a sustainable competitive advantage.

This is because specific assets often arise from a firm's prior history. In a sense, they represent the firm's particular assemblage of physical assets and prior learning. Specific assets often have a high tacit component and are thus not readily imitatable, absent the exit of a number of key individuals, an outcome that the innovating firm needs to prevent if it is to benefit from innovation that it has funded.[15]

3.2   Strategies for Incumbents

And here lie the advantages to incumbency.[16] Firms that are simply "in the business" will have built or acquired many of the necessary complementary capacities, unless the innovation is so radical that it sweeps them away. Market entry strategies for them are relatively straightforward if the innovation requires support from complementary capacities already in place and controlled by the innovator. The incumbent is in a position to move forward quickly and confidently. And even if it is not innovating, but others are, it may find the value of its complementary assets, like distribution and manufacturing, increasing if the innovation generated by others can in fact utilize the assets it already has in place. If this situation is a permanent one—that is, if a stream of competing enhancing innovations is expected from external sources while little is expected from internal sources—the incumbent could find itself an attractive acquisition candidate. Thus, even if an incumbent firm losses it innovative capacities, it is still in a position to prosper if (1) it owns or otherwise controls complementary capacities that are not easy to imitate, and (2) innovation continues from external sources and is competency enhancing rather than competency destroying. In this regard one can see that it is not necessary to be "the innovator" in order to profit from innovation. If the technological contribution embedded

---

[15]This is not the place to go into the various mechanisms for reducing the probability of this outcome.

[16]Readers may note a close relationship to the prescriptive theory in this section and the descriptive theory of von Hippel (1988, Chapter 4) which takes the boundaries of the firm as given and asks how the existing configuration of firms and industries impact the locus of innovation. Von Hippel implicitly and sometimes explicitly assumes that complementary assets are completely immobile and that intellectual property protection is weak, and that licensing is ineffective. Given these parameters, von Hippel predicts the locus of innovation according to the firm's ability to internalize the benefits. This section looks at the issue from the opposite perspective and attempts to predict the distribution of economic rents according to the firm's prior positioning in complementary assets. The two frameworks are mutually supportive, though I believe the one developed here is more general.

in an innovation is easy to copy, incumbents who do not innovate are often in the position to channel rents in their direction, and away from innovators attempting entry.

3.3   Strategies for New Entrants

Suppose the innovating firm is a new entrant. It has two pure strategies to select from with respect to accessing complementary assets. At one extreme it could integrate into (i.e., build or acquire) all of the necessary complementary assets. This is likely to be unnecessary as well as prohibitively expensive. It is well to recognize that the variety of assets and competences that need to be accessed is likely to be quite large, even to commercialize only modestly complex technologies. To produce a personal computer, for instance, a company needs to access expertise in semiconductor technology, disk drive technology, networking technology, keyboard technology, and several others. No company has kept pace in all of these areas by itself.

At the other extreme, the innovator could attempt to access these assets through straightforward contractual relationships (e.g., component supply contracts, fabrication contracts, distribution contracts, etc.). In many instances contracts may suffice, although they do expose the innovator to various hazards and dependencies that it may well wish to avoid. An analysis of the properties of the two extreme forms ought to be instructive. A brief synopsis of hybrid arrangements then follows.

3.3.1   Contractual Modes

The advantages of contractual agreements—whereby the innovator contracts with independent suppliers, manufacturers or distributors—are obvious. The new entrant will not have to make the up-front capital expenditures needed to build or buy the assets in question. This reduces capital risks as well as cash requirements.

Contractual relationships do have a number of major limitations, however. First, by contracting out the firm may deny itself valuable learning. Second, contractual problems could well arise because the recontracting hazards that exist when the contract is supporting investment in specific assets. Quite simply, one or possibly both of the parties could be stranded with no options if highly specific investments are needed to support exchange (Williamson 1985).

Instances of both parties making irreversible capital investments are nevertheless frequent. A good example of this can be found in the development of two prominent laser printers available today: the Hewlett Packard (HP) LaserJet and the Apple Laserwriter. Both these firms individually cooperated with, as it happens, the same copier company (Canon) to develop laser printer engines and cases for their different types of micro/personal computers. In both instances, all three companies made substantial capital investments and exposed themselves to financial risk. In addition to the bilateral risk involved, both Apple and HP were exposed to an additional type of risk: the risk that Canon would enter the market as a competitor with a considerable cost advantage because its scale and learning advantages on the engine. Canon did sign limited non-compete agreements with both companies, but as these agreements elapsed, it is evident that Canon entered the market with a low cost laser printer.

In short, the obvious benefits from contracting need to be balanced against the risks. Briefly, there is (1) the risk that the partner will not perform according to the innovator's perception of what the contract requires; and (2) there is the added danger that the partner may imitate or build on the innovator's technology and attempt to compete with the innovator. This latter possibility is particularly acute if the provider of the complementary asset is uniquely situated with respect to the specialized assets in question, and has the capacity to absorb and improve the technology.

3.3.2 Integration Modes

Integration modes are distinguished from pure contractual modes in that they typically facilitate greater control, greater access to commercial information, and the internalization of learning (Teece 1976). Owning rather than renting the requisite specialized assets has clear advantages when the complementary assets are in fixed supply over the relevant time period. It is critical, however, that ownership be obtained before the requirements of the innovation become publicly known; otherwise the price of the assets in question is likely to be raised. The incumbent seller, realizing the value of the asset to the innovator, may well be able to extract a portion, if not all, of the profits which the innovation has the potential to generate by charging a price that reflects the value of the asset to the innovator. Such "bottleneck" situations are not uncommon, particularly in distribution. In this regard, an innovator is clearly better off if it has a high performing product/process that requires a different set of complementary capacities to that already possessed by incumbents.

In instances where there is no incumbent with the requisite complementary capacities the innovator may be forced to integrate or to find a partner willing to make the necessary de novo investment, and in the process may destroy the value of the incumbents' assets. As a practical matter, however, the new entrant may not have the time to acquire or build all of the complementary assets needed to commercialize the innovation. Accordingly, new entrant-innovators need to rank complementary, specialized assets as to their importance. If the assets are critical, ownership is warranted; if the firm is cash constrained, a minority position may well represent a sensible compromise.[17] Needless to say, when imitation is easy, strategic moves to build or buy complementary assets that are specialized must occur with due reference to the moves of competitors. There is no point moving to build a specialized asset, for instance, if one's imitators can do it faster and cheaper.

3.3.3 Hybrid Strategies: Interfirm Agreements and Alliances

Reality rarely provides extreme or pure cases. Decisions to integrate or license involve tradeoffs, compromises, and mixed approaches. It is not surprising, therefore, that the business world is characterized by mixed modes of organization, involving judicious blends of contracting and integration. Moreover, longstanding contractual relationships can be functionally akin to integration; internalization can be so decentralized as to be akin to contracts. A particularly commonplace entry strategy that emerged in the 1970s and 1980s was the strategic alliance. Strategic alliances involve more than a simple transactional relationship.

More precisely, interfirm agreements can be classified as unilateral (where A sells X to B) or bilateral (whereby A agrees to buy Y from B as a condition for making the sale of X, and both parties understand that the transaction will be continued only if reciprocity is observed).[18] A strategic alliance can be defined as a bilateral relationship characterized by the commitment of two or more partner firms to reach a common goal,[19] and which entails the pooling of specialized assets and capabilities. Thus a strategic alliance might include one or more of the following: technology swaps, joint R&D or co-development, and the sharing of complementary assets (for example, when one party does manufacturing, the other distribution, for a co-developed product).

Strategic alliances can be differentiated from exchange transactions, such as a simple licensing agreement with specified royalties, because in an exchange transaction the object of the transaction is supplied by the

---

[17]For a more complete treatment of the impact of governance structure on innovation, see Teece 1989b.
[18]See Williamson 1985, and Jorde and Teece 1989.
[19]If the common goal was price-fixing, or naked market share division without any offsetting efficiency effects, such an agreement would constitute a cartel, and not an alliance.

TABLE 7.2

Taxonomy of Technological Development
and Commercialization Arrangements

	Nonequity	Equity
Contract	short/medium-term cash-based contracts, e.g., patent and know-how licenses	passive stock holdings portfolio diversification
Alliance	mid/long-term bilateral contracts (non cash-based)	joint venture (operating or non-operating), consortium, minority equity holdings, and cross holdings
Integration	-0-	wholly owned affiliate or subsidiary

selling firm to the buying firm in exchange for cash. Exchange transactions are unilateral and not bilateral. A strategic alliance by definition can never have one side receiving cash alone. Nor do strategic alliances include mergers, because by definition alliances cannot involve acquisition of another firm's assets or controlling interest in another firm's stock. Alliances need not involve equity swaps or equity investments, though they often do. Strategic alliances without equity typically consist of contracts between or among partner firms that are nonaffiliated. Equity alliances can take many forms, including minority equity holdings, consortia, and joint ventures.

Strategic alliances have increased in frequency in recent years and are particularly characteristic of high-technology industries.[20] Joint R&D, know-how, manufacturing, and marketing agreements go well beyond exchange agreements because they can be used to access complementary technologies and complementary assets. The object of the transaction, such as the development or launch of a new product, usually does not exist at the time contracts are inked.

Alternative governance structures are summarized in Table 7.2. Strategic alliances, including consortia and joint ventures, are often an effective and efficient way to commercialize innovation, particularly when an industry is fragmented. As compared to mergers, strategic alliances and other forms of interfirm agreements are of course far less restrictive.

Alliances are also conceptually quite different from cartels. First, they do not involve output restrictions, or price hikes. Second, the interfirm agreements that underpin them are usually for only a limited

---

[20]For a review and compendium of industry studies, see David Mowery, ed., 1988.

time period, and are assembled and disassembled as circumstances warrant. Third, they may be less comprehensive, as typically only a limited range of the firm's activities are enveloped in such agreements. Alliances are efficiency enhancing; cartels are not.[21]

Alliances are often superior to license agreements, particularly when commercialization of the technology in question is less well developed, leakage is a significant problem, and future learning is critical. The joint venture structure in particular permits future contingencies to be handled flexibly through the joint venture's governing board. Unilateral contractual relationships, as represented by licenses, offer less flexibility. Unexpected changes must be accommodated through renegotiation, and this exposes one or more parties to significant business risks.

An example of how changing technology can impact corporate boundaries, and determine the identity of the party that sits at the nexus of contracts needed to develop and manufacture complex products, can be found in the jet fighter business. Avionics now constitutes about one-third the cost of a fighter, up from fifteen percent a decade ago. It is expected to be even more important in the future, not just in terms of cost, but also in terms of performance. Given the fairly widespread diffusion of airframe and propulsion technology, the superiority of fighters today and in the future will depend primarily upon the sophistication and capability of the aircraft's electronics. Indeed, computer manufacturers like AT&T and IBM may well show up in the ranks of future prime contractors for advanced weapons systems, including fighters.

Of particular relevance here is the USAF's Advanced Tactical Fighter (ATF). While it is still too early to discuss electronics definitively, a number of technological areas can be identified that will be required in the ATF, including the integration of flight/propulsion control systems. Fly-by-wire will also be necessary, as will integration of cockpit displays, probably with a voice command system. Advanced heads-up display and VHSIC technology will also be critical.

These requirements mean that, in order to compete in the advanced fighter market in the future, it is imperative that prime contractors be on the leading edge with respect to avionics technology. If a manufacturer of fighters fails to develop or acquire such technology, it must expect to be closed out of a large and growing portion of the market.

It is unlikely that airframe companies will be able to easily contract with electronics companies for the requisite subsystems. Because avionics is becoming the core technology that will dictate other ele-

---

[21]For a more detailed discussion, see Jorde and Teece 1992.

ments of design, it will not be enough for airframe companies to contract with both avionics and propulsion companies. Indeed, the leading fighter manufacturers, such as General Dynamics and McDonnell Douglas, have developed in-house avionics capabilities. There are simply some technologies that are so central to the competitive performance of the system—in this case fighter aircraft—that an innovator cannot simply "sub it out." These technologies will be those that are the "pacing" technologies—that is, those where learning proceeds the fastest. The firms that design and build these systems are likely to be at the apex of the design hierarchy (Clark 1985). Eventually, they will be at the apex of the contracting hierarchy. Thus, unless the current airframe manufacturers can develop and promote these technologies, the chances of them managing the rest of the system (as prime contractor) in the long run is remote.

In this regard, a note of caution may be in order for those industries that turn automatically to outsourcing to solve their cost problems, a practice engaged in by U.S. auto companies in the 1980s and by U.S. consumer electronics companies in the 1970s. If outsourcing is taken to the point where the locus of systems learning migrates to key suppliers, the tables could easily be turned, and the subcontractor could become the prime contractor, and vice versa. Thus it is conceivable that if electronics becomes critical in automobile technology and the present auto assemblers do not keep up, the design hierarchy could become inverted, with companies like GM, Ford, and Chrysler performing contract assembly for the electronics giants of tomorrow, who would then assume the role of systems integrator for automobiles. A similar scenario could unfold in commercial aircraft.

4 Predicting the Outcomes of Market Entry Strategies

4.1 Success is Uncommon

As we have seen, the key decision the innovator has to make after the emergence of a dominant design is what to do about complementary assets: contract for their use, integrate to control them, or something in-between. The best strategy and chance of success will depend on these factors: (1) intellectual property protection: weak or strong; (2) interdependence between the innovator (and imitators) and the owners of the complementary assets; (3) the innovator's competitive position vis-à-vis potential imitators with respect to his ability to access complementary assets. The strategies that flow from this analysis are summarized in

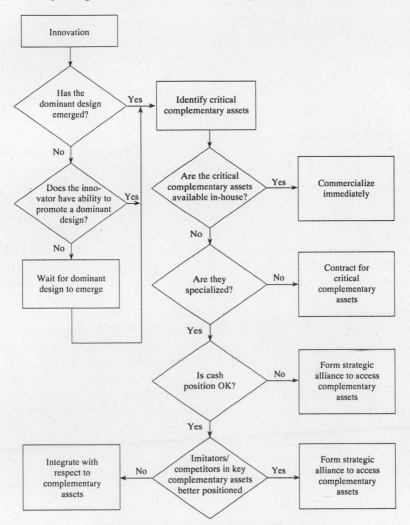

FIGURE 7.2 Market Entry Strategies:
Case of Weak Appropriability Regime

Figure 7.2. Embedded in this figure are heuristics for ascertaining the best strategy for the innovator.

The analysis helps one understand some stylized observations about success and failure. For instance, it is commonly observed that many small entrepreneurial firms that generate new, commercially valuable technology fail, while large multinational firms, often with a less meritorious record with respect to innovation, survive and prosper. One set of reasons for this phenomenon is now clear. Large firms are more likely to already possess the relevant specialized complementary assets within their boundaries, so that extraction of profits from innovation is likely to be more complete, and survival more assured. Startup firms are less likely to have the relevant specialized assets within their boundaries and so will have to incur the expense either of trying to build them or of trying to develop alliances with competitors/owners of the specialized assets. This is one way to interpret the symbiotic relationship between small firms and large firms with respect to innovation. However, it ought to be apparent that the symbiotic relationship between big and small does not stem from size itself, but from the portfolio of specialized and cospecialized assets that the big firm is likely to control.

Figure 7.3 summarizes an innovator's best strategy when complementary assets are specialized, indicates the innovator's chances of winning, and predicts how much of the benefits will be retained by the innovator if it wins. The best situation for the innovator is, of course, where appropriability is strong and where the innovator is strongly placed with regard to the suppliers of complementary assets. In this case (cell 1), the innovator can (metaphorically) sit back and contract for the use of the necessary assets in the confident expectation of winning the lion's share of the benefits generated by the innovation. Provided appropriability is strong, a contracting strategy can be a winner, even when the complementary assets are critical.[22]

Only if the complementary assets that the innovation uniquely requires are already tied up must the innovator contemplate integration (cell 4) when appropriability is strong. Even then, the difficulty of imitation affords the innovator the opportunity to do so on favorable terms. In extreme cases, where an innovator with watertight legal and technical protection confronts a pure monopoly seller of a critical asset, the innovator might have to share the benefits with the asset owner. These situations are rather rare, but they do occur, as when an innovator needs a component from a sole source supplier and where self supply is prohibitively costly.

Contractual strategies are appropriate wherever imitation is difficult, the innovator is well placed with regard to the suppliers of complementary assets, and learning is not firm specific. However, the innovator's

---

[22]For simplicity, this analysis does make assumptions about the locus of future learning. To the extent that learning is connected to the complementary assets, and the learning is firm specific, contracting should be abandoned in favor of integration.

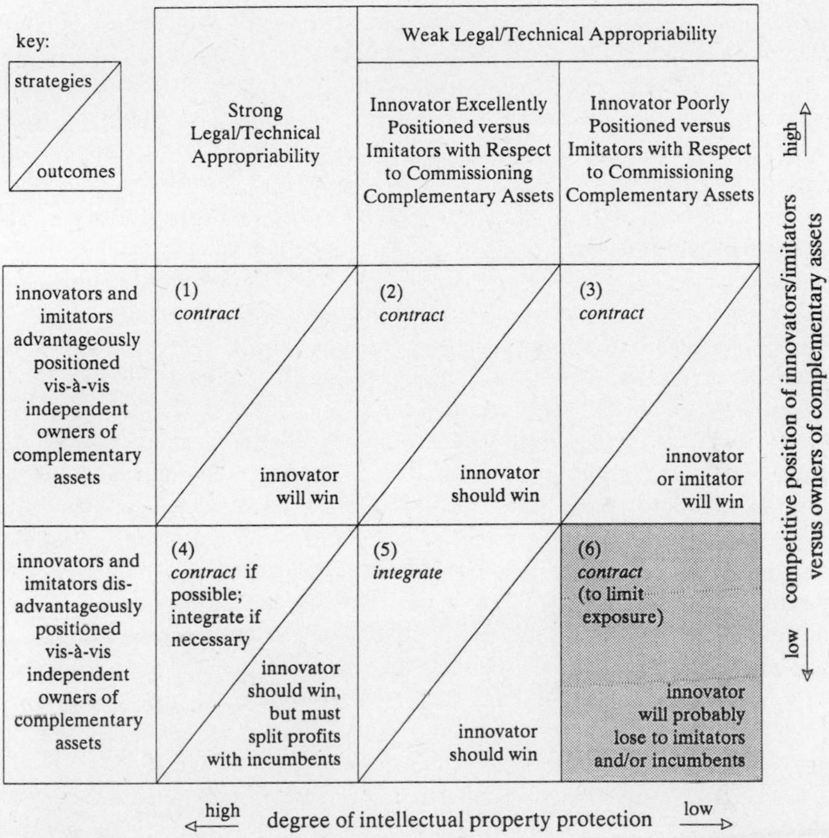

key: strategies ⟋ outcomes	Strong Legal/Technical Appropriability	Weak Legal/Technical Appropriability	
		Innovator Excellently Positioned versus Imitators with Respect to Commissioning Complementary Assets	Innovator Poorly Positioned versus Imitators with Respect to Commissioning Complementary Assets
innovators and imitators advantageously positioned vis-à-vis independent owners of complementary assets	(1) *contract* innovator will win	(2) *contract* innovator should win	(3) *contract* innovator or imitator will win
innovators and imitators disadvantageously positioned vis-à-vis independent owners of complementary assets	(4) *contract* if possible; integrate if necessary innovator should win, but must split profits with incumbents	(5) *integrate* innovator should win	(6) *contract* (to limit exposure) innovator will probably lose to imitators and/or incumbents

high — competitive position of innovators/imitators versus owners of complementary assets — low

⟵ high   degree of intellectual property protection   low ⟶

SOURCE: Based on Teece 1987

FIGURE 7.3 Market Entry Strategies and Outcomes for Innovators:
Specialized Asset-New Entrant Case

chances of winning diminish as appropriability weakens and imitators are better positioned with respect to commissioning the complementary assets.

The most difficult decisions for innovators occurs when the owners of complementary assets are strongly placed (e.g., because there are few suppliers, the asset in question being a critical bottleneck in the market entry program) and the innovation can only be exploited in conjunction with the complementary asset but not vice versa. If the innovator is well placed with respect to its potential imitators (cell 5), it should press ahead and integrate to control the assets, recognizing, however, that it will have to pay a high price for access and is still by no means certain of winning.

In the worst case (cell 6, which is shaded in Figure 7.3) when imitators are better placed (e.g., because they own the relevant assets) and imitation is easy, the innovator cannot afford to take the risk of committing substantial resources to integrating. The innovator should contract to limit exposure, but this will further reduce its already low chances of winning; and if it wins, the asset owner will be ideally placed in the future to extract maximum benefits from the contract. In fact, in this worst of all possible worlds, the innovator should probably not have entered the game in the first place. Unless the innovator is very lucky, it will lose, no matter how terrific the breakthrough it has commercialized.

Unfortunately for the innovator, all too often in the world of international commerce, legal protection of intellectual property is weak, complementary assets (like specialized marketing and selling systems) are more difficult to replicate than the innovation itself, and they are in the hands of actual or potential competitors. Since these situations can be foreseen to a great extent, this analysis provides a valuable tool for innovating firms to plan their R&D programs. After all, there is no advantage to the firm striving to produce a valuable technological breakthrough when success will most likely cost money while benefitting better positioned competitors.

4.2   The Cost of Adopting the Wrong Market Entry Strategy

As we have seen, even optimal strategies may not guarantee success. However, wrong strategies can make things much worse. On the one hand, if an innovator chooses to integrate when it should have contracted (or not even have entered the game), it will certainly commit more resources than necessary for no additional strategic benefit. Furthermore, if integration takes the form of building extra capacity in specialized assets, it may lead to industry overcapacity and, hence, may depress overall industry profitability unnecessarily. Conversely, if the innovator contracts when it should integrate (e.g., with a monopoly supplier), it may never get into the game because an imitator (or the monopoly supplier) may cut the innovator out. Even if this does not happen, the asset supplier will always be in a position to extort maximum benefits from the innovators wherever the contract fails to take account of new situations that may arise.

Access to capital often plays a role in causing innovators to fall into these traps. Large established firms with adequate cash positions have a tendency to integrate where there is no strategic advantage to doing so, while undercapitalized startups are notorious for attempting to use collaborative arrangements riddled with strategic hazards. A common defense in the latter instance is to point out that undercapitalized firms have few options, but there is always the option not to enter at all if the new entrant's strategic base is too weak. Nevertheless, some firms occasionally succeed with non-optimal strategies. When they do, it is primarily because of luck. Similarly, well-positioned firms may have a run of hard luck that puts them out of the game.

4.3 First Mover Advantages and Development Speed

The above framework can help managers assess the importance of timing and speed. As a general rule, the framework suggests that in product areas where patents work, it is important to secure a blocking patent if one is potentially available. Hence, in these instances the speed that matters relates to qualifying for the patent. Such circumstances are rare however.

Weak appropriability is far the more common circumstance.[23] What matters here is both (1) lining up the complementary capabilities ahead of competition and (2) completing the development manufacturing cycle ahead of competition. Clearly, these two considerations are interrelated.

With respect to the former, incumbent firms do not have to be the first movers in order to win. As Mitchell's (1989) study of diagnostic imaging has so clearly demonstrated, the optimal strategy for the incumbent (with complementary assets in place) is often to wait until new entrants have proved the feasibility of the technology and the existence of an attractive market.

Once incumbents and new entrants are equalized with respect to complementary assets, and a firm develops and launches a new product or product enhancements that are exposed to reverse engineering, then in the next round of innovation/product improvement the firm that can complete the development manufacturing cycle most rapidly is the firm that will be perceived as the technological leader most likely to win market share. As Gomory (1987) has succinctly explained:

If one company has a three-year cycle and one has a two, the one with the two-year cycle will have its process and design into production and in the market one year before the other. The one with the shorter will appear to have newer products with new technologies etc. In fact, both companies will be working from the same storehouse of technology. It is the speed of the [man-

---

[23]Even when a patent exists, it will at some point expire, causing the technology to fall into the weak appropriability scenario.

ufacturing and development] cycle that appears as technological innovation and leadership. It only takes a few turns of the cycle to build a commanding product lead. (p. 73) -1z

Needless to say, in order to accomplish such "quick cycle" innovation, firms must maintain fluid but close ties between development and manufacturing. This is not at all easy to accomplish. Despite its importance, this topic is outside the scope of this paper.[24]

5 Conclusions

We can now see very clearly why EMI failed, despite EMI having created an order-of-magnitude technological breakthrough. The scanner that EMI developed was of a technical sophistication much higher than would normally be found in a hospital, requiring a high level of training, support, and servicing. EMI had none of these capabilities, could not easily contract for them, and appears to have been slow to realize their importance. It most probably could have formed a partnership with a company like Siemens to access the requisite capabilities. Its failure to do so was a strategic error compounded by the very limited protection that intellectual property law afforded the scanner. Although subsequent court decisions have upheld some of EMI's patent claims, once the product was in the market it could be reverse engineered and its essential features copied. Two competitors, GE and Technicare, already possessed the complementary capabilities that the scanner required, and they were also technologically capable. In addition, both were experienced marketers of medical equipment, and had reputations for quality, reliability, and service. GE and Technicare were thus able to commit their R&D resources to developing a competitive scanner, borrowing ideas from EMI's scanner, which they undoubtedly had access to through cooperative hospitals, and improving on it where they could while they rushed to market. GE began taking orders in 1976 and soon after made inroads on EMI.

By 1978 EMI had lost market share leadership to Technicare, which was in turn quickly overtaken by GE.[25] In October 1979, Godfrey Hounsfield of EMI shared the Nobel prize for invention of the CT scanner. Despite this honor, and the public recognition of its role in bringing this

---

[24]It has received attention elsewhere; for example, Teece 1989c.

[25]In 1977, concern for rising health care costs caused the Carter Administration to introduce "certificate of need" regulation requiring HEW's approval on expenditures for big ticket items like CAT scanners. This severely cut the size of the available market.

medical breakthrough to the world, the collapse of its scanner business forced EMI in the same year to merge with Thorn Electrical Industries, Ltd. GE subsequently acquired what was EMI's scanner business from Thorn for what amounted to a pittance.[26] Though royalties continued to flow to EMI, the company had failed to capture the lion's share of the profits generated by the innovation it had pioneered and successfully commercialized.

If EMI illustrates how a company with outstanding technology and an excellent product can fail to profit from innovation while the imitators succeeded, the story of the IBM PC indicates how a new product representing a very modest technological advance can yield remarkable returns to the developer. As noted earlier, the IBM PC, introduced in 1981, was a success despite the fact that the architecture was rather ordinary. Philip Estridge's design team in Boca Raton, Florida, decided to use existing technology to produce a solid, reliable computer rather than one which was technologically way out in front. With a one-year mandate to develop a PC, Estridge's team could do little else. However, the IBM PC did use what at the time was a new sixteen-bit microprocessor (the Intel 8088) and a new disk operating system (DOS) adapted for IBM by Microsoft. Other than the microprocessor and the operating system, the IBM PC incorporated existing micro "standards" and used off-the-shelf parts from outside vendors.

The key to the PC's success was not the technology. It was the set of complementary assets that IBM either had or quickly and skillfully assembled around the PC. In order to expand the market for PCs, there was a clear need for an expandable, flexible microcomputer system with extensive applications software. IBM could have based its PC system on its own patented hardware and copyrighted software. Such an approach would cause complementary products to be cospecialized, forcing IBM to develop peripherals and a comprehensive library of software in a very short time.

Instead, IBM adopted what might be called an "induced contractual" approach. By adopting an open system architecture and by making the operating system information publicly available, a spectacular output of third party software was induced. IBM estimated that by mid-1983 at least 3,000 hardware and software products were available for the PC.[27] Put differently, IBM pulled together the complementary assets, particularly software, which success required, without even using contracts, let

---

[26]See "GE Gobbles a Rival in CT Scanners," *Business Week*, May 19, 1980.
[27]Gens and Christiansen 1983.

alone integration. This was despite the fact that the software developers were creating assets that were in part cospecialized to the IBM PC, at least in the first instance before clones emerged.

A number of special factors made this seem a reasonable risk to the software writers. A critical one was IBM's name and commitment to the project. Its brand image was a tremendous asset, as it implied that the product would be marketed and serviced in the IBM tradition (Davidow 1986, p. 31). It guaranteed the PC-DOS would become an industry standard, so that the software business would not be solely dependent on IBM, because emulators were sure to enter. It guaranteed access to retailer distribution outlets on competitive terms. The consequence was that IBM, with excellent management and strategic vision,[28] was able to take a product which represented at best a modest technological accomplishment and turn it into a fabulous commercial success. The case demonstrates the role that complementary assets and clever market entry strategies play in determining outcomes.

The lessons for managers from this framework can be summarized as follows:

o The probability of successful commercialization can be analyzed rationally in advance—focus your R&D program on delivering innovations that will have a good chance of benefitting you rather than your competitors.

o In the early design phase, stay particularly close to the market and be flexible. Try to get your design accepted as the industry standard, but be flexible enough to adapt if some other standard is adopted.

o Identify the relevant specialized complementary assets early, and decide in advance whether you can contract for their use, or build or acquire them. You cannot win trying to commercialize a new "device" no matter how advanced; you must have a "product."

o Monitor competitors and recognize that the toughest competitor may be an imitator who owns important complementary assets.

o Foreign rivals may use their government to deny you access to the complementary assets that are needed to successfully enter the foreign market. This will be damaging, particularly if the foreign firms are free to enter your domestic market.

o If, like EMI, you develop an order-of-magnitude innovation that

---

[28]IBM may have come to subsequently regret the tremendous assistance it gave Microsoft by anointing MS-DOS, which IBM did not own, as the industry standard.

will require significant investment in complementary assets, if you don't have the cash, find a strategic partner or sell the idea out.

o Strength in an area of technology will not necessarily compensate for weaknesses elsewhere. As technology becomes more public and less proprietary through easier imitation, then strength in manufacturing and other capabilities are necessary to derive advantage from whatever technology an innovator may possess.

These lessons are particularly pertinent to small, science-based companies. Founders with superb technical backgrounds are often slow to learn that generating advanced products and processes that meet a clear market need is not enough to guarantee success. Even with a terrific new device, the innovator is likely to lose unless his intellectual property is extremely well protected and/or he is strategically well positioned with respect to key complementary assets. Many technology driven companies fall into the trap of believing the best price-performance item on the market will necessarily win. Needless to say, many science-based firms, after a shaky start, soon recognize these principles and adjust their strategies accordingly.

There are attractive implications for firms that are not good at developing new technologies; the framework advanced here suggests that in many cases the lion's share of the available profits can be captured by the owners of (specialized) complementary assets. If a technology is easy to copy, but the needed specialized assets are hard to copy, the rents from the innovation will be directed away from the innovators and toward the owners of complementary assets. That is why excellence in manufacturing and distribution often turn out to be of great advantage. The identification of entry strategies for imitations follows naturally from this framework.[29]

There are powerful implications for nations that perceive their comparative advantage to be in innovation. Unless one's trading partners are backward, it will never be enough to have the best science and engineering establishment and the most creative engineers and designers.[30] Since the fruits of scientific effort are increasingly open to all with the capacity to receive, extracting value from a nation's science and engineering prowess will require its firms to have competitive capacities in certain of the key complementary assets, such as manufacturing. In many cases this will need to be onshore in order to fashion defensible competitive strategies. Public policies that do not recognize that trans-

---

[29]The author plans to draw these out more specifically in a subsequent paper.
[30]For an alternative view, see Miles and Snow 1986.

lating scientific and technological leadership into commercial leadership
in most cases requires parallel excellence in capacities complementary to
the innovation process will doom a nation to economic decline—possibly
tempered only by a bounty of Nobel prizes. What *Business Week* calls
"hollow corporations" will in general bring hollow success if the intended
focus of their activities is innovation.

In short, turning on both the private and public R&D spigots will not
be enough for pioneering nations to regain their international competitiveness. It is the complementary capacities that must be built if a nation
with the technological lead is also to lay claim to the commercial lead.

To summarize, the lessons for governments of innovating nations include the following:

o Pay attention to intellectual property protection domestically and
  internationally. Tighten where it is sensible to do so.

o Recognize that science and technology alone will not provide sufficient foundation to guarantee economic growth. Targeting the
  innovation process requires attention to the capacities complementary to innovation. In particular, manufacturing matters. So does
  infrastructure.

o Access to foreign markets is critical to innovators, particularly if
  protected foreign firms are imitators or followers. The importance
  of access transcends the incremental profits in the foreign market,
  as access can serve to check imitators, especially when imitators
  own critical complementary assets not currently owned by the innovator.

o Strategic alliances confer benefits when it comes to developing and
  capturing value from innovation. Strength in capacities complementary to R&D will promote economic welfare where imitation is
  easy. Antitrust policy needs to be revised accordingly.

o Certain free market dogma is misleading. Unassisted markets will
  not necessarily reward the efficient. Trade protection is not always tantamount to "shooting oneself in the foot," especially if
  one's firms have the capacity to imitate and improve. National
  policies must be crafted which are sensitive to what it takes to win
  when intellectual property protection is spotty, and when foreign
  governments deny access to complementary capacities.

o Shrewd private sector strategies, coupled with sensible trade policy and infrastructure development, can ensure that an innovating
  nation will enjoy the fruits of its technological prowess.

These insights come from a framework which explicitly introduces the intellectual property regime and specialized complementary assets as strategic variables. Further research on these factors can enrich both the strategy and industrial organization literatures while revealing heuristics that can be useful managerial guides in an area of critical importance to the firm and the nation.[31]

References

Abernathy, W. J. and J. W. Utterback. 1978. Patterns of Industrial Innovation, *Technology Review*, vol. 80, no. 7 (January/July), pp. 40–47.

Arthur, B. 1988. Competing Technologies: An Overview. In G. Dosi, C. Freeman, R. Nelson, G. Silverberg, and L. Soete, eds., *Technical Change and Economic Theory*. London: Pinter.

Clark, Kim B. 1985. The Interaction of Design Hierarchies and Market Concepts in Technological Evolution, *Research Policy*, 14, pp. 235–51.

Davidow, William H. 1986. *Marketing High Technology*. New York: The Free Press.

Dosi, G. 1982. Technological Paradigms and Technological Trajectories, *Research Policy*, 11, no. 3, pp. 147–62.

Dosi, G., C. Freeman, R. Nelson, G. Silverberg, and L. Soete, eds. 1988. *Technical Change and Economic Theory*. London: Pinter.

Gens, F. and C. Christiansen. 1983. Could 1,000,000 IBM PC Users Be Wrong? *Byte*, November, p. 88.

Gomory, Ralph E. 1987. Dominant Science Does Not Mean Dominant Product, *Research and Development*, November, pp. 72–74.

Gomory, Ralph E. and Roland W. Schmitt. 1988. Science and Product, *Science*, May 27, 1988, pp. 113–20.

Hippel, Eric von. 1988. *The Sources of Innovation*. New York: Oxford.

Jorde, T. M., and D. J. Teece. 1989. Competition and Cooperation: Striking the Right Balance, *California Management Review*, vol. 31, no. 3 (Spring).

Jorde, T. M., and D. J. Teece. 1992. *Antitrust, Innovation, and Competitiveness*. Oxford: Oxford University Press.

Lieberman, M., and D. Montgomery. 1988. First Mover Advantages, *Strategic Management Journal*, 9, pp. 41–58.

---

[31]Neither the strategy literature nor the industrial organization literature has systematically explored the role of intellectual property for corporate strategy and the organization of industry. Nor have complementary assets been an explicit part of the analysis. Indirectly, however, aspects of the complementary assets story have crept in under discussion of lead time advantages (which enter here through the concept of incumbency) and learning curve advantages (which explain how specialized assets are sometimes created). I assert that first mover advantages (Lieberman and Montgomery 1988; Porter 1985, Chapter 5) are best understood as a special case of the theory developed here (see section First Mover Advantages and Development Speed), and that both literatures need to place much more emphasis on how to create specialized assets, and how their presence or absence should influence R&D resource allocation.

Mansfield, E. 1985. How Rapidly Does New Industrial Technology Leak Out? *Journal of Industrial Economics*, vol. 34, no. 2 (December), pp. 217–23.

Mansfield, E., M. Schwartz, and S. Wagner. 1981. Imitation Costs and Patents: An Empirical Study, *Economic Journal*, 91 (December), pp. 907–18.

Martin, M. 1984. *Managing Technological Innovation and Entrepreneurship*. Reston, VA: Reston Publishing Company.

Miles, R. E., and C. C. Snow. 1986. Network Organizations: New Concepts for New Forms, *California Management Review*, vol. 28, no. 3 (Spring), pp. 66–73.

Mitchell, W. 1988. Dynamic Commercialization: An Organizational Economic Analysis of Innovation in the Medical Diagnostic Imaging Industry. Unpublished Ph.D. thesis, University of California at Berkeley. (Selected portions published in *Administrative Science Quarterly*.)

Mowery, D., ed. 1988. *International Collaborative Ventures in U.S. Manufacturing*. Cambridge, MA: Ballinger.

Polanyi, M. 1962. *Personal Knowledge: Toward a Post Critical Philosophy*. New York: Harper & Row.

Porter, M. 1985. *Competitive Advantage*. New York: The Free Press.

Rosenbloom, R. and W. Abernathy. 1982. The Climate for Innovation in Industry, *Research Policy*, 11, pp. 209–25.

Rumelt, R. 1987. Theory, Strategy, and Entrepreneurship. In D. Teece, ed., *The Competitive Challenge*. Cambridge, MA: Ballinger.

Sykes, H. 1986. Lessons from a New Venture Program, *Harvard Business Review*, 3 (May/June), pp. 69–74.

Teece, D. J. 1976. *Vertical Integration and Vertical Divestiture in the U.S. Oil Industry*. Stanford, CA: Stanford University Institute for Energy Studies

Teece, D. J. 1981. The Market for Know-How and the Efficient International Transfer of Technology, *Annals of the American Academy of Political and Social Science*, November.

Teece, D. J. 1986. Profiting from Technological Innovation, *Research Policy*, December.

Teece, D. J. 1987. *The Competitive Challenge: Strategies for Industrial Innovation and Renewal*. Cambridge, MA: Ballinger.

Teece, D. J. 1988. Capturing Value from Technological Innovation, *Interfaces*, vol. 18, no. 3 (May/June), pp. 46–61.

Teece, D. J. 1989a. Intellectual Property and Strategic Management. Unpublished manuscript, Center for Research in Management, University of California at Berkeley, July.

Teece, D. J. 1989b. Innovation and the Organization of Industry. Unpublished manuscript, Center for Research, University of California at Berkeley.

Teece, D. J. 1989c. Interorganizational Requirements of the Innovation Process, *Managerial and Decision Economics*.

Tirole, J. 1989. *The Theory of Industrial Organization*. Cambridge, MA: M.I.T. Press.

Williamson, O. E. 1975. *Markets and Hierarchies*. New York: The Free Press.

Williamson, O. E. 1985. *The Economic Institutions of Capitalism*. New York: The Free Press.

Winter, S. G. 1987. Know-How and Competence as Strategic Assets. In D. Teece, ed., *The Competitive Challenge*. Cambridge, MA: Ballinger.

8

Liability and Insurance Problems in the Commercialization of New Products: A Perspective from the United States and England

PETER HUBER

Who most needs insurance? The explorer, the venturer, and the distant trader. The modern insurance industry traces its roots to seventeenth-century England, where underwriters signed up at Lloyd's coffeehouse to insure the boldest entrepreneurs of that day, seafaring traders.

Today's explorers are as likely to be sailing the uncharted waters of biotechnology or toxic waste cleanup, AIDS drugs, asbestos substitutes, new vaccines, or unorthodox aircraft designs. The commercialization of such goods requires accident insurance. Without insurance, the products will not come to market at all, or they will be brought to market late, by thinly capitalized firms, which rely on the corporate veil and bankruptcy law instead of insurance to limit liability.[1]

There seem to be limits to the total uncertainty that any undertaking can bear, or so business experience suggests. Sellers of old products with established markets can sometimes shoulder the risk that attends

---

Parts of this paper are adapted from P. Huber, *Liability: The Legal Revolution and its Consequences* (1988, Basic Books). I am most grateful to Sir Maurice Hodgson, Chairman of the Lord Chancellor's Civil Justice Review Advisory Committee, for his input on the differences between U.S. and English law, and for his comments on this article.

[1] See generally R. Weaver, Impact of Product Liability on the Development of New Medical Technologies (American Medical Association, June 26, 1988).

uninsured operation. It is the new venture with the unfamiliar product that can least tolerate the extra instability of going without insurance. This is most particularly true, of course, when the risks of casualty—ships succumbing to unexpected storms, or patients succumbing to the side effects of powerful drugs—are significant.

In the past three decades, U.S. liability law has changed in many ways that have undermined insurance markets. That effect was unintended. The architects of the modern U.S. liability system fully expected that insurance would continue to provide a broad financial umbrella over the expanding new tort system. But changes in liability law were implemented with little understanding of how insurance markets operate. The result has been a sharp increase in demand for liability insurance, but a marked decline in supply. The most dramatic adjustment came in the early 1980s, with an insurance crisis eerily reminiscent of the endless gas lines during the Arab oil embargo.

A prudent insurer must know, first of all, just *who* it is insuring. It will write one contract for a driver with a history of drinking at the wheel, another (at a quite different price) for a well-run municipality, and yet another for a car manufacturer. But U.S. courts have steadily expanded principles of "joint" liability, which sweep together dozens, sometimes hundreds, of defendants in a single courtroom. "Several" liability then allows the full costs of an accident to be channeled to the wealthiest—inevitably the best insured—defendant.

A second essential ingredient of intelligent insurance is accurate timing. How much risk an insurer is covering depends not only on when the policy begins and ends—which the insurer itself can control—but also on when legal claims are born and expire. In pricing a policy, the insurer counts on earning some investment income between the time when premiums are collected and the time when claims are paid. A reasonably predictable and quick clock on insurance claims also allows the insurer to identify better and worse risks. It will not do to sell liability insurance at a flat price to all comers for twenty years, and only then discover that some in the group are careful and others (as finally judged by the tort system) are grossly careless. But U.S. courts have been steadily dismantling principles of ripeness and limitation that once kept litigation on a fairly strict and predictable timetable. Lawsuits today can look forward and backward for decades, or even generations, with no one knowing which of the dozens of policies a customer might purchase during that period will then be called into play.

An insurer also needs a reliable yardstick for pricing injury. A policy speaks of accidents and such, but the final accounting is always in cash. A stable rate of exchange between injuries and dollars is therefore essential. The conversion is not hard to make when the injury is a broken leg that must be set or lost wages that must be replaced—which is why you can easily buy first-party health or disability insurance to cover such contingencies. But the rate of exchange is quite indeterminate for pain and suffering, loss of society, or criminal fines and penalties. For precisely this reason, first-party insurance never covers such things. In U.S. courts, as elsewhere, it used to be that tangible economic losses—like medical costs and lost wages—always accounted for most damages ultimately paid. In recent decades, however, there has been great expansion in U.S. awards for such things as punishment, pain, suffering, cancer phobia, and loss of society.

A last essential ingredient of rational insurance is knowing just what risk is being covered. If an insurer sells a policy to a vaccine manufacturer, it must plan to cover the risks of vaccines, not the general health problems faced by young children. A policy priced to cover injuries caused by a spermicide cannot also cover birth defects originating in quite independent genetic accidents or a mother's drinking habit. Here too, in U.S. courts as elsewhere, the legal rules used to be fairly accommodating: claims of cause and effect were usually tested skeptically, and the courts declined to speculate about causes remote in time and place. But U.S. courts have gradually come to accept ignorance, or at least substantial uncertainty, about the risks of such things as toxic wastes as sufficient reason for setting the judicial machinery in motion.

These changes in U.S. law have undermined liability insurance across the board. Other changes seem to have been aimed more specifically, if unintentionally, at innovation and new technology.

Alone in the world so far, U.S. courts have abandoned the negligence standard for product liability; they ask juries to pass judgment instead on the adequacy of product design and manufacture.[2] The negligence standard inquired whether the technologist—the human actor on the scene—was careful, prudently trained, and properly supervised. A standard of "reasonable prudence" is likely to be congenial to the best and brightest—the best trained technologists working at the leading edge of their professions. It is at the frontiers of science, after all, that the best engineers, pharmacologists, doctors, and chemists typically congregate. Under a standard of "strict liability" for product "defects," however, the

---

[2]The EEC countries are very tentatively moving toward some form of strict liability for manufacturing defects; as of this writing, however, no other country comes close to enforcing the sweeping doctrine of strict liability for "design defects" applied in U.S. courts.

people themselves, and their good care, good training, and good faith, are irrelevant. The new inquest concerns the product itself and its alleged defects. Today's U.S. tort system places technology itself in the dock.

This is not a happy change for the innovator, most especially not when the liability system relies on juries as final decision makers, as is still the case in U.S. courts. Jurors can perhaps make reasonably sensible intuitive judgments about people—even about professionals—for we are all in the people-judging business every day of our lives. But jurors are not experts about technology itself, and intuition here is a poor guide. The layperson tends to associate safety and acceptability with age, familiarity, and ubiquity; this seems to be an atavistic instinct, a vestige from the days when safety and familiarity were indeed one and the same. Technologies that are novel, exotic, unfamiliar, or adventuresome are instinctively viewed as unwelcome and fraught with danger.

At the same time, U.S. courts have become exceedingly particular about safety warnings. It will not suffice to warn of a risk of death; the seller must warn specifically of the risk of stroke or serum sickness or acute encephalopathy. It will not suffice to warn the prescribing doctor of a drug's hazards; the drug company must somehow also get the information to the patient too. Grossly obvious risks must be flagged, but so must some risks that arise only from the most bizarre forms of consumer abuse. The only way to learn how to meet such demands is through long perseverance in the market and the courts. The warnings provided with oral contraceptives have been honed for thirty years, to the point where they occupy pages of densely detailed text, with substantial parts developed, one at a time, in the aftermath of adverse liability verdicts. No equivalent detail can be provided for a truly new IUD, because the risks are much less familiar—even if their general nature is known and even if the IUD is demonstrably safer, overall, than the pill it could replace.

The U.S. law's attitude to safety improvements made after an accident has also changed. Uniformly, the courts used to bar a plaintiff from offering any evidence of such conduct—the redesign or recall of a product, a post-sale or postaccident change in a process, or the addition of a new safety system. Safety improvements must be encouraged, the logic ran, and so should never be used to condemn past shortcomings. But in the 1960s, this rule too came under direct attack.[3] Evidence of

---

[3]See J. Hoffman and G. Zuckerman, "Tort Reform and the Rules of Evidence: Sav-

subsequent remedial measures was first admitted to impeach witnesses, then to prove that the defendant controlled the premises in question, then to show that conditions had changed since the time of the accident or that changes in design were feasible. The exceptions nibbled away at the rule until some courts were emboldened to sweep the tattered remains aside entirely.

Changes in U.S. rules on when suits may be filed have had a second effect, particularly hostile to the commercialization of new products. Suits can now be filed decades after the car gear box, herbicide, or strip-mining machine was designed or used. Defects in technology, like negligence in human conduct, depend critically on the context of time and place. The best-designed cars of 1950 were clearly deficient by 1980 standards, as were the best medical procedures, industrial chemicals, pesticides, or home appliances. When the sun never sets on the possibility of litigation, each improvement in method, material, or design can establish a new standard against which all of your earlier undertakings, no matter of what vintage, may then be judged. Finding a way to do better today immediately invites an indictment of what you did less well yesterday, or twenty years ago. Mature firms that sell long-lived products therefore have good reason to hesitate before commercializing a significant new improvement; doing so may very well promote successful litigation against the company's older vaccine, small plane, escalator, or lathe, still in use in the market.

There has also been a sharp rise in demands for, and awards of, punitive damages in U.S. courts. Many factors go into such awards, and some of the awards are undoubtedly justified by egregious misconduct. But here too, abandoning familiar and established technology may, for legal purposes, be the riskiest course of all. Many of the breakthrough punitive awards were assessed in the early 1970s against manufacturers of small cars, when such vehicles (and their attendant risks) were still novelties on the U.S. market. The great wave of punitive awards against contraceptive manufacturers was triggered by the Dalkon Shield IUD—undoubtedly an inferior product, but made especially vulnerable by the fact that it represented a sharp and identifiable break with past contraceptive technologies. By contrast, it can hardly be termed "reckless" or "outrageous misconduct" (the usual triggers for punitive damages) to do what the great mass of society has done for decades. This may help to explain why U.S. tobacco companies have yet to be assessed any punitive

ing the Rule Excluding Evidence of Subsequent Remedial Actions," *Tort & Ins. L. J.*, 22, p. 497 (Summer 1987).

damages, and indeed have so far lost only one verdict for compensatory damages.

It is useful for the American reader to consider at least one alternative legal regime and its possible effects on the climate for commercializing new products. A comparison with England is particularly instructive because the U.S. common law is so deeply rooted in English soil. Sir Maurice Hodgson points to the following features of English liability law that are distinctively different, today, from the law in U.S. courts.

Civil courts in England and Wales do not have juries. They were abolished years ago. The case is put to the judge who determines both liability and damages. Judges, who tend to be older and better educated than the average juror, to have better instincts about the comparative advantages of old and new technologies. And English product liability law still turns on a human-centered standard of negligence, not a technology-centered standard of product "defect." In addition, a judge, who sees many cases and many plaintiffs, is more likely to apply the objective criteria of damages than a one-time juror, who may be tempted to regard tort law as some kind of instrument of social justice.

Civil and criminal jurisdiction are sharply separated and differentiated in Britain. Punishment is the prerogative of the criminal law and there is no concept of punitive damages. This all but eliminates the possibility of a Manville- or Robins-like bankruptcy for any product manufacturer in the United Kingdom. While run-away liability of this kind is rare even in the United States, that it is possible at all has plainly discouraged new ventures with certain technologies like contraceptives and fiber insulators.

Two other differences between U.S. and English law concern lawyers' fees and the payments of costs. Plaintiff lawyers in the United States are remunerated almost without exception by contingency fee. England is now contemplating some modest changes in rules on "speculative actions," but there are no proposals as yet for adopting anything like the open-ended percentage-of-award contingency fees that are standard in the United States.

The U.S. contingency fee system powerfully encourages litigation, and the most vigorous pursuit of large damage awards. Contingency fees, combined with open-ended awards for pain, suffering, and punishment, reduce incentives for settlement and encourage lawyers to prosecute cases, even if centered on long-shot claims, to cash in handsomely on the rare but munificent run-away verdict. To the extent that new technologies are more vulnerable to large verdicts for intangible losses, as they do appear to be in U.S. courts, contingency fee arrangements indirectly add to the legal system's bias against what is new.

The treatment of court costs is also different on the two sides of the Atlantic. In the United States costs do not, as in England and Wales, follow the verdict. Each side, in other words, bears its own litigation costs, regardless of who wins. U.S. courts do reserve the power to shift costs, but almost never exercise it.

The fear of having to pay the other party's costs if a case is lost is a strong deterrent to frivolous litigation. And such a rule entirely cuts off one kind of legal strategy—the war of attrition—that has become increasingly attractive to the U.S. plaintiff's bar. By 1985, for example, one out of four U.S. obstetrician-gynecologists had been sued. Almost three-quarters won their cases, but it cost the physician or insurer an average of $20,000 per claim to do so.[4] Costs like these often make it prudent to settle claims with no merit whatsoever—especially when other facets of the law always keeps open the possibility of defeat, no matter how weak the underlying case by conventional standards. Guerrilla warfare in the U.S. courts is fought even more effectively against product manufacturers. In the space of a few years, nearly 800 lawsuits were filed against Searle's Copper-7 IUD. Some 470 were disposed of one way or another, one-third with no payment and no trial, the remaining two-thirds with very small settlements. By 1986, Searle had actually been to court for full trials in only ten cases, of which it had won eight. The cases continued to mount, however, and so did the expenses. Searle spent $1.5 million defending just four cases that it won outright; the several hundred million dollars in legal expenses that Searle faced were hundreds of times its annual profit on the device.[5] Merrill-Dow Pharmaceuticals was defeated in similar manner. For twenty-seven years the company manufactured and sold Bendectin, an antinausea morning sickness drug used, over the years, in 33 million pregnancies. As the liability rules changed, however, lawsuits against Bendectin burgeoned. Merrill-Dow won almost all the early cases, but each trial stimulated a flood of new claims. Eleven hundred claims were consolidated in a single class action. Faced with the staggering legal costs of taking such a case to trial, Merrill-Dow offered to settle for $120 million. Plaintiffs' lawyers torpedoed the offer. The case was then tried, and Merrill-Dow

---

[4]S. Johnson, "Malpractice Costs vs. Health Care for Women," *New York Times*, July 14, 1985, p. A14.

[5]E. Connell, "The Crisis in Contraception," *Technology Review*, May/June 1987, p. 47; "Birth Control: Vanishing Options," *Time*, September 1, 1986, p. 78.

won again. But hundreds of other claimants had opted out of the class action, and their cases were still pending. And here and there, lightning did strike. In 1987 a jury awarded $2 million to a six-year-old boy born with club feet. Another jury awarded $95 million for another birth defect. Throughout, the overwhelming scientific consensus, in the FDA and in all respectable scientific circles, had not moved an inch: Bendectin does not, in fact, cause birth defects. Such histories—which all but assure that no major new IUDs or morning sickness drugs will be commercialized in the United States—would never have been written under the English rule for allocating litigation costs.

A final difference, often remarked upon, is the existence, in the United Kingdom, of a comprehensive, if basic, system of remedies for those injured by accidents. It comprises the National Health Service and the social security system. Together these effectively provide a broad "no fault" scheme at the base. Tort law offers additional compensation for negligence, subject to mitigation for contributory negligence. This arrangement weakens the case (if there is one) for the use of tort as a form of social justice.

The difference in the funding of medical services in England and the United States, however, is not as sharp, or as important, as it is sometimes thought to be. There is a large, though not quite as seamless, blanket of health, accident, and disability insurance in the United States too, much of it written privately, some also provided by the government. No accident victim is left lying in the street, and most accident victims are covered by first-party health, accident, or life insurance of some kind. But except in a few cases where rules have been changed legislatively, U.S. courts still insist on applying the "collateral source" rule, which forbids the presentation of any evidence concerning outside insurance, and which denies any reduction in liability awards for compensation received from direct insurance. The principle firmly applied in U.S. courts is that liability awards are to provide first-dollar coverage; other forms of insurance provide last-dollar coverage or the outright windfall of double payment. Some substantial fraction of the growth in product liability awards, for example, is attributable to insurance companies themselves, who sue outside suppliers of products to recover payments made to employees through workers' compensation programs.

It is all but impossible to quantify the effects of the different legal rules on the commercialization of new products in England and the United States. There are many other confounding variables. If one looks to the commercialization of new drugs, for example, differences in liability law certainly have some impact, but there are also major differences in preemptive administrative regulation, which is slower in the United States than in Britain. One can identify various U.S. industries that have slowed their commercialization of products because of liability concerns—general aviation, contraceptives, medical equipment, and sporting goods, for example—but it is once again difficult to quantify the comparative advantage gained by operating abroad in a more hospitable legal climate.

Examples of the adverse impact on the United States abound, however. Burt Rutan, the pioneering designer of the *Voyager*, used to sell construction plans for novel airplanes to do-it-yourselfers. In 1985, fearful of the lawsuits that would follow if a home-built plane crashed, he took the plans off the market. Meanwhile, the U.S. general aviation industry, once the unquestioned world leader, has almost ceased operations because of liability problems. In 1977, small-plane manufacturers in the United States paid a total of $24 million in liability claims. By 1985, their payout was $210 million. Companies like Beech, Cessna, and Piper sharply curtailed or completely suspended production; they quickly discovered that the new-model planes, carrying a 50 percent surcharge for liability insurance, could no longer compete with used planes already on the market. The market is moving steadily to manufacturers in Europe and South America.

For similar reasons, Monsanto decided in 1987 not to market a promising new filler and insulator made of calcium sodium metaphosphate. The material is almost certainly safer than asbestos, which it could help replace in brakes and gaskets. But safer is not good enough in today's climate of infectious litigation.

Liability fears have likewise caused the withdrawal of exotic drugs that the FDA deems safe and effective, including some for which no close substitute is known. In the past fifteen years, most companies have ceased U.S. research on contraceptives and drugs to combat infertility and morning sickness. Carl Djerassi, developer of the oral contraceptive, notes that no new active ingredients have appeared in oral contraceptives sold in the United States since the 1960s; three new ones (desogestrel, norgestimate, and gestodene) were introduced in Europe in the 1980s. Djerassi sees reform of the U.S. liability system as the most urgent priority in resurrecting the U.S. contraceptive industry.[6]

Some of the most regressive effects have been felt precisely where aggressive commercialization of new products is most urgently needed. U.S. liability today is highly—but often indiscriminately—contagious,

---

[6]C. Djerassi, The Bitter Pill, *Science*, vol. 245, pp. 356–57 (July 28, 1989).

which means that the introduction of new products is undercut the most in the markets already swept up in a hurricane of litigation—contraceptives, vaccines, obstetrical services, and light aircraft, for example. Yet it is often in such markets that new products are most urgently needed.

The commercialization of new products in the United States is thus discouraged on two sides. The availability of insurance has declined, and prices have increased, as the courts have adopted rules that discourage rational underwriting and systematic selection and differentiation of risks. At the same time, U.S. courts are effectively demanding that all goods and services come wrapped in a special-purpose insurance contract. The availability of insurance, most especially modestly priced insurance, depends largely on an accumulation of accident experience. That is something that established technologies always have and truly innovative ones never do. Insurance is easiest to find when a good has been used by many people for many years, so that the caprices of tort liability have been as far as possible washed out and the statistics of experience speak for themselves. Innovation necessarily starts without an established market, and so is often condemned to start without insurance as well. For the prudent businessperson, a start without insurance is often worse than no start at all.

We can perhaps gain some additional insight into the relation between liability law and commercialization by digressing briefly to look at libel law and the rationales behind some recent changes.

The one branch of U.S. tort law that has sharply contracted in recent years is the law of defamation. Strict liability, once the legal norm in libel cases, has been abolished in the United States. Punitive damages have been very sharply curtailed. And while juries are still used, there is exceptionally careful and complete review of all verdicts by both trial and appellate judges. For practical purposes, U.S. defamation law has almost abolished the jury in libel cases, as England has done in product liability cases.

The cut-back in U.S. defamation law has been led by the Supreme Court, in writing what amounts to a new common law of the first amendment. That defamation verdicts can be imposed only for factually false publications is not sufficient protection for the bold, venturesome press that we desire, the Court has reasoned. Strict liability will have a chilling effect on the press—will discourage the commercialization, so to speak, of valuable public information. Even the negligence standard is too strict, the Court has concluded, at least insofar as news about "public figures" is concerned; here liability is permitted only upon a showing of knowledge or reckless disregard of the falsehood being published. Too-strict liability will do more harm than good, by discouraging the diligent and accurate journalist more than it disciplines the incompetent and the mendacious.

England apparently accepts much the same logic, but for providers of products and services, not for journalists. Defamation law is exceptional in England, too. Very roughly speaking, English defamation law is as open-ended as U.S. product law; U.S. defamation law is as restrained as English product law. Juries are still used in England in libel cases. Strict liability—liability imposed because the published fact is false, regardless of whether or not the publisher was negligent—applies.

The different rules do have some notable effects. English libel cases are common, and spectacular verdicts, much like U.S. verdicts in product cases, are quite frequent. An English jury, for example, recently awarded 600,000 pounds to the wife of the "Yorkshire Ripper" for libel by *Private Eye*. The verdict was appealed, but the dispute was then settled out of court for one tenth of the original award. "There is some regret at this settlement," says Sir Maurice Hodgson, "because it was hoped that the Appeal judges would give some guidance on what sort of sum was reasonable in the circumstances... This episode has opened up debates not only on what level of damages is reasonable in matters of this kind but also on the question of whether juries are appropriate in libel cases." Unsurprisingly, U.S. libel lawyers representing plaintiffs have strained to have U.S. courts apply English law in any libel case brought against an English publisher. This is in marked contrast to product cases, where both English and American plaintiffs' lawyers much prefer to litigate, if they can, under U.S. laws.

Do these different legal regimes have any evident impact on the "commercialization" of new products in the information business? No one who has observed the London tabloids in action can conclude that the British press has been entirely cowed. Nonetheless, many American observers believe that the U.S. press is more confident and aggressive in publishing accounts of institutional corruption, political malfeasance, and other forms of official misconduct. But once again there are other confounding factors. Britain has a far more sweeping Official Secrets law, for example. And it may be that the greater caution of the British press in the face of greater libel threats is in fact desirable, an indication of more responsible behavior. It is still puzzling to see such different assumptions about the beneficial and harmful effects of liability law accepted in two legal systems with common roots.

Is there any reason to suppose that liability's chilling effects are greater for the press than for product manufacturers? Or that juries can better identify product "defects" than published falsehood? It seems unlikely. Manufacturing defects—the manufacturing world's equivalent of the typographical error—are easy enough to spot. But most product cases today turn on alleged defects in design. The design of an airplane engine or a bioengineered vaccine is as subtle and complex as the composition of a news story. Yes, journalism requires all sorts of judgment calls, on what to include, what to paraphrase, what to omit, what to double check, what to infer or even guess at. But the design engineer engages in the same series of judgment calls, one after the next, and on matters much further from every day experience.

Does journalism, then, require greater protection from liability because of jury bias or prejudice? There is no reason to suppose that this is the case either. If juries tilt against large, powerful, institutional defendants, and research suggests that they do, Ford motor company is every bit as vulnerable as CBS.

How can a manufacturer commercialize a new but somewhat risky product without affordable insurance, in a legal environment that always holds out the possibility for large verdicts, numerous lawsuits, and substantial expenses even in defending non-meritorious claims?

One possibility, often suggested of late as a means for bringing the new French abortifacient, RU-486, into the U.S. market, is to create a thinly capitalized, single-product firm, operate it without insurance, siphon off profits quickly, and use the specter of immediate bankruptcy to deter litigation. One small-plane manufacturer is currently attempting to operate in this way, and the option appears to be increasingly attractive for other small startup companies. This is not much of a solution, however. Most new products of any consequence require large, up-front investments to create a chain of distribution, publicize the product, and for such things as drugs, to shepherd the product through the very expensive and lengthy regulatory process.

A second possibility is to build a case-hardened litigation team and adopt a scorched earth litigation policy to deter attack. Every large manufacturer does this to some extent, though the smart money seems to have concluded that this strategy is simply infeasible with certain classes of products used, for example, by pregnant women and young children. Tobacco companies are the only ones that have used this approach with almost complete success, and they have little in the way of new products to commercialize.

Yet another approach for the U.S. manufacturer is to establish a foreign base for the early stages of commercialization. This is by no means a complete solution, because it is becoming easier to pursue manufacturers back to U.S. shores, wherever products may be sold or used. Foreign manufacturers do have one important advantage. They can effectively limit their total liability for their U.S. operations by the cold-blooded expedient of keeping few of their assets in this country. Transnational recognition of judgment rules are still cumbersome enough to provide an effective shield against the wholesale export of U.S. liability verdicts to collect against the foreign-based assets of a foreign firm, no matter where its products may have been marketed.

How can policy be changed to make liability a less serious obstacle to the commercialization of new products? One approach is to increase the supply of insurance. The other, to reduce the need for it.

Government has granted immunity, indemnification guarantees, or alternative insurance coverage to providers of various niche products and services, ranging from childhood vaccines to nuclear weapons. Similar approaches have been considered for "orphan drugs," which may be kept from market because the costs of insurance may exceed any potential profit. In other areas, however, immunities from liability have been cut back rather than expanded. Charities and the government itself used to enjoy blanket immunities from liability, but no longer do. In waiving its own absolute immunity from suit, however, the U.S. government carefully insisted on trial by judge, not jury, forbade punitive damages, and denied liability for any "discretionary function," the government equivalent of the judgment calls that a private company routinely makes in designing a new product.

More recently, and much less fruitfully, have been attempts to expand supply or curtail the cost of insurance by direct legislative decree. When insurers have canceled policies of their highest-risk customers, many states have passed laws limiting the insurer's freedom to cancel or refuse renewal. When coverage for particular lines has become entirely unavailable, some states have created joint underwriting associations, which force all private insurers doing any business in the state to offer coverage collectively, through the pool, for the otherwise uninsurable. When insurance prices have then risen still faster, several other states have turned to price controls. When insurers have attempted to flee, a few states have even challenged their right to depart. It seems likely, however, that such efforts will quite quickly reduce, rather than increase, the supply of insurance.

The second possible approach is to curtail demand, the need for insurance. This means changing the legal rules, so that cases are less frequent and judgments more modest. One of the blunter and more controversial approaches is to establish dollar caps on awards of such things as pain-and-suffering, or to limit punitive awards to some fixed multiple of the compensatory judgment. Many states have placed limits of one sort or another on joint liability, to reduce the increasingly magnetic properties of insurance, which otherwise cause the best insured to attract the most lawsuits. There have been various modest efforts to improve the quality of scientific testimony in the courts, so that factually unfounded claims cannot be used for their speculative value before a jury that may not understand the science in question. Other sensible guidelines for how the law can be changed to limit the chilling effects of open-ended liability can be found in the Federal Tort Claims Act, and in modern U.S. libel law.

In the end, the search must be for rules that allow society to say yes to new and better products, with the same conviction and force that an open-ended liability system can say no to old and inferior ones. In many areas of policy, the answers given depend largely on the question asked. For several decades, U.S. policy makers in the courts and elsewhere, have asked: What is unduly risky? And how can risk be deterred? But an equally important pair of questions is: What is acceptably safe? And how can safety be embraced?

What indeed are our legal vehicles for getting to yes?

The first is private contract, which restores to individual buyers and sellers the power to make binding deals—deals addressing safety and risk along with other matters of risk—that will then be enforced, as written, in the courts. Current legal policy generally refuses to enforce such contracts, at least insofar as they address safety matters, on the assumption that consumers have unequal bargaining power, inadequate information, or simply too little patience to consider these matters.

But such rules could be changed. When we deal with the essentially private risks, of transportation or recreation for example, fair warning and conscious choice by the consumer must be made to count for much more than it does today. Not because individual choices will always be wise—they surely won't be—but because such a system at least permits positive choice and the acceptance of change.

A second vehicle for getting to yes is through the positive choice made by surrogates. Some safety matters will always remain too complex or far-reaching to be collapsed into the world of purely private agreement. Informed consent by the individual is never going to take care of such things as chemical waste disposal, mass vaccination, or central power generation. These are and obviously must remain matters to be delegated to expert agencies acting for the collective good. But if they are to be useful at all, agents must be able to buy as well as to sell. For safety agencies, this means not just rejecting bad safety choices, but also embracing good ones. Yet the long-standing rule, still rigidly applied in U.S. courts, is that even the most complete conformity to applicable regulations is no shield against liability.

The courts should be strongly encouraged to respect the risk and safety choices made by expert agencies. It is politically unrealistic to propose that liability be cut off entirely for any activity expressly approved by a qualified regulatory agency. But liability could at least be firmly curtailed in such circumstances. Some states, like New Jersey, have already adopted reform along these lines for pharmaceuticals and other products comprehensively regulated by federal agencies. At the very minimum, complete compliance with a comprehensive licensing order should provide liability protection against punitive damages—awards that have a strong, depressive effect on innovative experiment and change. It has always been true that ignorance of the law is no excuse. At present, knowledge of the law is no excuse either. It should be.

We may conclude, however, where we began, with the issue of insurance. Insurance remains the best and most robust market instrument for making the random predictable. It is possible to insure against misadventure in the courts, just as it is possible to insure against wind and wave on the high seas. But only up to a point. When the legal rules are such that insurance itself attracts litigation, when the rules change too quickly for actuarial assessment, and when systematic biases creep in against new technologies and innovative undertakings, insurance will rise in price and narrow in coverage. No other country in the world has gone through liability insurance shocks like those experienced in the United States. The reason is that no other country has transformed its liability system so completely and abruptly. It is within the direct power of U.S. courts to restore a climate that is healthy for private insurance markets. The courts thus have a very direct role to play in shaping policies to renew the spirit of enterprise in the United States, and recreate a social climate that welcomes, indeed encourages, the commercialization of new products.

Part III

International Contrasts in Technology Commercialization

9

The Japanese Pattern of Innovation and Its Evolution

KEN-ICHI IMAI

The purpose of this paper is to explore the basic reasons for Japan's success in the commercialization of new technologies. A coherent framework is presented which analyzes the evolution of Japanese innovations from process innovation to systemic innovation, and the associated organizational change from a mechanic system to an organic network.

In Section 1, the characteristics of Japanese innovation are discussed in relation to the product-process dichotomy of innovation and a concept of systemic innovation.

In Section 2, the commercialization process of new technologies in Japan is investigated as a sequential learning process. Emphasis is placed on user-producer interaction, human resource creation, organizational adaptation, and the role of R&D. A new concept, the "micro-macro information loops," is also presented and explained.

Finally in Section 3, the above discussions are summarized to provide an overall perspective on Japan's national system of innovation.

1 Characteristics of Japanese Innovation: From Process-Product Innovation to Systemic Innovation

Recent discussions of the nature of Japanese innovation have characterized it as chiefly consisting of process innovation rather than product innovation and as focused on incremental innovation rather than radical innovation. Mansfield, for example, wrote that "the American firms in my sample devote about two-thirds of their R&D expenditures to improved product technology (new products and product changes) and about one-third to improved process technology (new processes and process changes). Among the Japanese firms, on the other hand, the proportions are reversed, two thirds going for improved process technology and one-third going for improved product technology," in his often-cited paper, "Industrial R&D in Japan and the United States: A Comparative Study" (Mansfield 1988, p. 326).

This will not be contradicted by Japanese historical statistical data. It is true that Japanese R&D funds have largely been devoted to incremental process innovations during the past twenty years. But, we propose that today the situation is very different. During much of the past two decades, simple learning by doing and gradual improvements and cost reduction of the manufacturing process indeed constituted the basic method of innovation. More recently, however, innovation has moved away from simple cost reduction and towards product differentiation. Process improvements and product differentiations are now being fused, and the former substantially contributes to the latter. R&D which is oriented to process improvements also is fused into product innovation.

Also, although Japanese innovation has been largely incremental in nature, its contents have gradually changed. As bit-by-bit improvements have accumulated and as the number of linkages among components of the Japanese "innovation system" have increased, a critical mass is appearing which necessitates the global reconception or crystallization of the system. The realized new system includes a qualitative jump that transcends simple incremental innovation. Starting from piece-by-piece improvement, Japanese innovation has reached this stage. This jump is not yet radical innovation, but it is creating essentially a qualitatively different system. The emphasis in this paper is on the systemic innovation in this sense.

In sum, then, we think that Japanese innovation today should be investigated in terms of "systemic innovation" rather than in terms of the product-process innovation dichotomy (Imai and Baba 1989). Our definition of systemic innovation underlines a new form of interlinkages between component elements and the new logic of institutional adaptation to it which will be fully explained in the latter sections.

2 A Sequential Learning Process

In order to discuss Japanese innovation from the above perspective, it is crucial to treat market and entrepreneurial coordination as a "process" and hence it is essential to include a time dimension. We follow the neo-Schumpeterian approach in this respect. Innovation dynamics are a sequential process of chain reactions, including proliferation of new divisions of labor, accumulation of knowledge and capital, and the creation of resources. A trajectory of this chain reaction is necessarily path-dependent, and the time enters in an irreversible way. This process accompanies organizational innovations involving interpenetration of the market and the organization (Imai and Itami 1984).

We emphasize the importance of resource creation in this sequential dynamic. The reason for Japanese success in the commercialization of innovation will eventually boil down to the creation of human resources in the sequential learning process.

2.1   An Outline of the Sequence

The Japanese industrial system after the Second World War has been created as a combination of an old technological paradigm and a new technological paradigm. The old paradigm is an extension of the mechanization process familiar since the second industrial revolution. The new paradigm is an information technology-centered one which has emerged in the last fifteen years.

The old paradigm was active in mature mass production technologies like steel, chemicals, synthetic fibers, etc. The large-scale plant-based industries producing these basic materials were the strategic industries of the initial phase of rapid growth. This phase was followed by a second "old paradigm" phase in which the automobile industry, the household electric appliance industry, and other so-called assembly industries became the core of technological and industrial development.

The new paradigm emerged after the 1973 oil crisis which involved and was enabled by the widespread introduction of microelectronics and information technologies into the Japanese industrial system. This not only created a clustering of information industries but also rejuvenated the old paradigm industries.

This introduction of the phase distinction of technological development has been done not for the purpose of analyzing each phase in detail, but for just explaining the background which will be required to analyze the *sequential* learning process in Japanese firms and industries.

Let us now explore the sequential development of the Japanese innovation system. We begin by describing events that took place during Japan's period of especially rapid growth (1955–64). Then, we turn to events stimulated by the oil crisis of 1973.

During the "rapid growth period" of 1955 to 1964, Japanese firms laid the foundation of inter-firm cooperation which creates a positive-sum game theoretic situation among participating firms. Japanese industries have often developed cooperative vertical (and sometimes horizontal) relationships among firms. The *Keiretsu*, a group of cooperative (and often subcontracting) firms, is an example (Imai 1988). A long-term semi-fixed relationship between user and suppliers, or between affiliated firms, subcontractors, vendors, and others enables the participants to share information about the nature of technology and the product involved. The long-run transaction involved in such relationships includes not only an economic component, but also a social one in which trust, loyalty, and power are transacted.

The emergence of these behaviors in Japan may have been in part a function of culture. But, in my view, the most crucial factor which contributed to creating them was the rapid industrial growth in Japan in the 1960s. Continuation of fast economic growth fostered optimistic long-run expectations, and expected growth of the firm's sales was crucial for creating mutual trust between a parent company and subcontractors, between a user firm and vendors, and between related firms. The prototype of the Japanese industrial organization was formulated under these special conditions of very rapid economic growth.

Next came the oil crisis of 1973—which may have been a blessing in disguise for Japanese industry, giving it the motivation to explore the age of new technical innovation. With the stimulus of the oil crisis, Japanese technologies and industries entered into the new technological paradigm mentioned above. The depletion of oil and other natural resources drove Japan, a country with scarcely any natural resources, to make an overall effort to conserve energy by utilizing all available technologies. Fortunately, this occurred concurrently with the beginning of the industrial application of microelectronics. Each company studied the possibility of reducing energy consumption in its manufacturing and distribution process by utilizing microelectronics and information technologies. This occurred as a "piece-by-piece" and a "bit-by-bit" approach, which incorporated many incremental process improvements. Thus, the oil shock destroyed the optimistic long-run expectation of Japanese firms, but at the same time created a foundation of incremental innovations and reshuffled the boundaries of firms, promoting the new interaction between firms for the purpose of utilizing microelectronics.

Incremental innovation of the type induced by the oil crisis was firm-specific and factory-specific because it was not an application of a general discipline or general inputs but rather a problem solving heuristic contingent on each location. It required a dense information exchange between users and capital goods suppliers. As is well known now, user-producer interactions are crucial for recent innovations (e.g., Lundvall 1988), and this new approach firmly took root as the energy saving improvement gained momentum.

At the same time, two additional important things were happening. First, thanks to the widespread introduction of electronics, the price and quality of capital goods were substantially improved. Second, human resources adaptable to the new paradigm were being created. Workers had to learn electronics and they were re-educated chiefly in-house. (The cost of this education became a sunk cost which was to be recovered from future revenues.)

The expectation of declining prices of capital goods on the one hand and increasing wage costs on the other strongly pushed Japanese firms to promote further rationalization of their manufacturing and distribution process utilizing more and more new capital goods. New capital goods embodied more information technologies and the systematization of them required more and more information and related technologies. Consequently, this process of rationalization was the major vehicle which Japanese industries utilized to dash into the midst of the new paradigm.

The sequential learning process described above is the essence of what characterizes the evolution of Japan's industrial system. What should be emphasized here is that during this learning process, the range of technological choice of Japanese firms was substantially widened, and their capability for adapting to turbulent economic environments increased to a large extent.

The learning-by-doing in the initial stage was a method of reducing costs and energy consumption. But its subsequent development by interacting with "learning-by-using" and "learning-by-interaction" changed its major role to a method of dynamic product differentiation. This process is in line with that described by Lachmann, a representative of the neo-Austrian school who said "once a new machine has been introduced, different people will use it in different ways in order to produce different products, or different varieties of the same product, which have to compete with each other for the same customers. It is the divergence of interpretations of the range of potentialities of the new machine which here lends shape and direction to the market process" (Lachmann 1986, p. 57).

In addition to the above, the pattern of innovation progressed towards a more systemic type of innovation, transcending the simple product-process dichotomy of innovation. Japanese industrial groups and network type organizations were well fitted for systemic innovation. A variety of differentiated and diversified goods and services were produced and combined into system products like OA (office automation) equipment.

Under these circumstances, recent structural changes to cope with the yen appreciation became possible. Japanese firms strongly believed in the inevitability of structural transformation needed to increase domestic demand, and have taken action to make new product developments for such domestic demand. Not only are corporate development expenditures increasing rapidly but research personnel are also actively consulting the marketplace to obtain on-the-spot information related to new products. This leads to a new conceptualization of products which have a potential to create new demands.

So much for a brief summary of the sequential learning process in the Japanese industrial system. Now, let us investigate the above points more analytically to obtain a deeper understanding of how each contributed to a successful commercialization process.

2.2   User-Producer Interaction

One major source of Japanese innovations and their commercial success lies in the nature of user-producer interaction. This includes a sequential and dense communication between a demanding user and a producer. This is especially true between a user and a capital goods supplier. Recently, the user-producer interaction hypothesis seems to have been gradually accepted, and "users as innovators" is perhaps the most carefully studied case (Lundvall 1984, von Hippel 1976). The Japanese case includes wider applications, and the user-producer interactions have created capital goods which have subsequently upgraded their quality through learning by diffusion across industrial sectors.

In the early stage of industrial development in the older technological regime, the machine tool industry played a crucially important role in the widespread diffusion of technologies. As Rosenberg noted "the machine tool industry may be regarded as a center for the acquisition and diffusion of new skills and techniques in a machino-facture type of economy. Its chief importance, therefore, lay in its strategic role in the learning process associated with industrialization" (Rosenberg 1963, pp. 425–26).

Similarly, "mechatronics," a Japanese combination of the machine and electronics industries, has become a center for the acquisition and diffusion of electronics in the modern Japanese economy. However, the role played by the mechatronics industry is much wider than that of the machine tool industry. In the case of the older machine tool industry, its role was chiefly related to processing technology. In the case of the mechatronics industry today, not only material processing but also assembling and especially testing are crucial arenas of application.

Furthermore, mechatronics enhances the technological coordination between processing, assembling and testing that is essential for creating a technological system with high precision.

This systemic character of mechatronics changed the meaning of process technology for production. Traditionally production was viewed as a process of transforming general inputs into finished products, but recently it has become an "activity of research and coordination" to find a best configuration of a system (Amendola and Gafford 1988).

2.3   Human Resource Creation

In parallel with the improvements of capital goods described above, the skills of labor were also being substantially upgraded. Importantly, the upgrading was of a type that made labor more flexible. Although specific skills focused on narrow jobs were developed, job rotation and simultaneous in-house education on electronics and information technologies were undertaken. As a result, Japanese workers truly became human resources (not simple labor inputs) adaptable to the new technology regime.

Of course, the process is not as simple as just now described. An important reservation is that these educational and training programs were done chiefly for the core members of big firms. The Japanese labor force in big firms is a combination of permanently employed (in principle) core members and temporary workers. In our view, the core members are educated in the long-run perspective within a firm and compensated for in part by the profits of a firm. Responding to this, they work hard and compete with each other in the long run for higher status positions. The important jobs which require deep skill and adaptation and intense coordination among workers are carried out by these core members. On the other hand, rather simple standardized jobs are done by temporary workers and also by subcontractors' workers. These relationships are depicted in Figure 9.1 (Imai and Komiya 1989).

This combination gave flexibility to the human resource capability in Japanese firms. Jobs were often reshuffled adapting to the changing needs of the day. Some of them were transferred to temporary workers and some of them were completed by a project team of core members. During these processes core members were well educated in order to foster their capability to coordinate among the diversified jobs needed for adapting and absorbing new technologies.

The educational and training programs just described created costs, even though they were performed by way of on-the-job training. Education needs a gestation period, and therefore, until its effect is fully

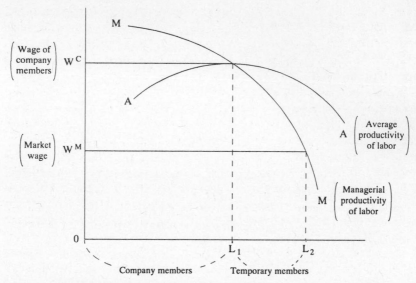

FIGURE 9.1 Behavioral Characteristics of Japanese Firms

reflected in company performance, education costs become a kind of sunk cost. How to recover this sunk cost is a problem which has not yet been discussed but it shapes the direction of sequential learning processes (Amendola and Gafford 1988). Without liquidity, firms cannot allocate their resources to the education and training of workers who do not directly contribute to current production activities. And, in a period of technological transition, firms have to expend funds for R&D and fixed investments embodying the new technologies as well. Therefore, they are forced to make a decisive choice between human resources, R&D and new fixed investments. The latter two are fixed costs, but there is a way to finance these by borrowing from outside on the security of equipment, buildings and land. However, in the case of expenditures to create human resources, such financing is impossible, and so firms have to rely solely on their own liquidity.

Japanese firms fortunately had ample liquidity after the oil crisis when they faced the technological transition to electronics because it was a turning point from the rapid economic growth to a lower rate of growth. However, it was a transitory coincidence just at that time. In the case of the sequential learning process, human resources have to be continuously educated, and therefore liquidity for that purpose is also continuously required. What made it possible for Japanese firms to provide this continuous liquidity was sales proceeds from their current operation.

Most Japanese firms tried to expand their sales even if the resultant profit rate was very low. Usually this is explained as the expansion-oriented bias of Japanese firms. However, in the context described above, such a view may be superficial. To maintain continuous sales proceeds even if profit rates were low was a good strategy for firms in order to provide the liquidity that was required to create human resources needed in the continuous, sequential learning process. Consequently, the Japanese human resources needed in order to cope with the new technologies regime have been gradually created through learning by doing and learning by interaction that have been allowed by the sales proceeds produced by such a learning process itself.

This is quite an important point because during the continuous learning process the range of adaptability to new technology in each firm gradually increased. The Japanese personnel management system in large corporations promotes a rotation of workers among different sections, and workers climb spirally through their career ladder, experiencing several different jobs. It is not unusual for core members to have a career path including production, marketing, and R&D, accompanied by an intermittent training period.

In sum, in this section we have endeavored to explain: (i) why Japanese firms have succeeded in creating their human resources, (ii) why they have had increasing adaptability to new technologies and capability to cope with new problem-solving as a group. While outside education and training, and movement of workers between firms could also create in-depth capability in workers, the Japanese way of in-house sequential learning created a group capability with wide adaptability to changing economic environments. This capability was seen in full play in recent times, as Japanese firms successfully diversified into new products in order to create new domestic demand in Japan, so as to overcome the substantial yen appreciations.

2.4   Overlapping of Development Phases

During the sequential learning process in which Japanese new product development has shifted from process to product and further to systemic innovation, associated corporate and industrial organizations as well have sequentially changed in order to create innovations in each phase. The essence of the process of Japanese new product development is explained as an overlapping phase approach which is used in many Japanese companies (Imai et al. 1985).

A simplified illustration of the overlapping nature of phase management is depicted in Figure 9.2. The sequential approach, labelled

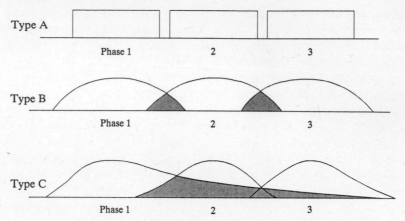

FIGURE 9.2 Sequential (A) vs. Overlapping (B) Phases
of Development

Example: Overlapping Nature of the Development Schedule at Fuji-Xerox

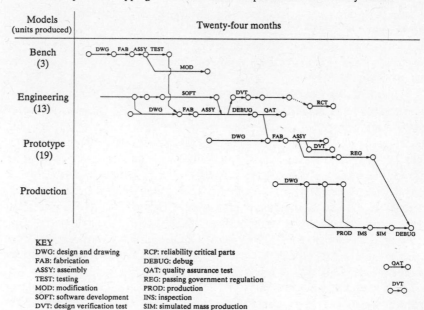

KEY
DWG: design and drawing   RCP: reliability critical parts
FAB: fabrication   DEBUG: debug
ASSY: assembly   QAT: quality assurance test
TEST: testing   REG: passing government regulation
MOD: modification   PROD: production
SOFT: software development   INS: inspection
DVT: design verification test   SIM: simulated mass production

Type A, is typified by the NASA-type phased program planning (PPP) system adopted by a number of U.S. companies. Under this system, a new product development project moves through different phases—for example, concept, feasibility, definition, design, and production—in a logical, step-by-step fashion. The project proceeds to the next phase only after all the requirements are satisfied, thereby minimizing risk. However, at the same time, a bottleneck in one phase can slow down the entire development process. The overlapping approach is represented by Type B, in which the overlap occurs only at the meeting of adjacent phases, and Type C, in which the overlap extends across several phases. The product development process at the companies in which we made in-depth case studies are all closer to Type C than to Type B.

The overlapping approach has both merits and faults. The obvious merits include: (1) faster development, (2) increased flexibility, and (3) information sharing. On the other hand, the burden of managing the process increases exponentially. By its nature, the overlapping approach amplifies ambiguity, tension, and conflict within the group. The burden to coordinate the intake and dissemination of information also increases, as does management's responsibility to carry out ad hoc and intensive on-the-job training.

Japanese companies adopted this method in order to develop new products rapidly and with maximum flexibility. Rather, it might be better to say, they were forced to do so in a severe competitive environment of product development. They found, however, as a result of learning by doing, that phase overlapping is an essential linkage mechanism of networking.

The loose coupling of phases in the development process changes the character of the division of labor. Division of labor works well in a Type A system where the tasks to be accomplished in each phase are clearly delineated. Each project member knows his or her responsibility, seeks depth of knowledge in a specialized area, and is evaluated on an individual basis. But such segmentalism works against the grain of a loosely coupled system (i.e., Types B and C) where the norm is to reach out across functional boundaries as well as across different phases. A project member is expected to interact extensively, to share everything from risk, responsibility, and information, to decision-making, and to acquire breadth of knowledge and skills. Under a loosely coupled system, then, the tasks can only be accomplished through what we termed "shared division of labor," or "network division of labor." It is an important point that this shared division of labor takes place not only within the company but also with outside suppliers. The overlapping approach requires extensive social interaction on the part of all those involved in the project, as well as the existence of a cooperative network with suppliers. A project team asks suppliers to come to their factory and start working together with them as early in the development process as possible. The project team is also welcome to visit the suppliers' plants. This kind of mutual exchange and openness about information works to enhance flexibility. Early participation on the part of the suppliers enables them to understand where they are positioned within the entire process. Furthermore, by working with the suppliers on a regular basis, project members learn how to induce suppliers to create precisely what they are looking for, even if they only show suppliers a rough sketch.

In sum, the overlapping of phases in the development process, even if it seems to be very simple at first glance, is a crucial way to create linkages between people in different functions or sections and between different firms by a commitment to joint partnership. These linkages made it possible for Japanese firms to perform dynamic adaptation when market conditions changed.

2.5 Micro-Macro Information Loops

Speed and flexibility in the companies' new product development crucially depends on the working of an intra-company network and an outside suppliers' network. The structure and behavior of Japanese suppliers' networks have recently been much discussed (Imai 1988, Asanuma 1989). What we will need to elaborate here is the character of the information exchange within the network. To do this we introduce a new concept we call "micro-macro information loops."

First of all, let us illustrate the concept of micro-macro information loops via a concrete example. Figure 9.3 refers to the structure of the supplier network for Fuji-Xerox copiers. Toritsu-Kogyo, one of the primary subcontractors for Fuji-Xerox, operates a factory of its own and utilizes 77 secondary subcontractors. A large number of these subcontractors are located within walking distance from Toritsu-Kogyo. Toritsu-Kogyo is a fairly large subcontractor by Japanese standards. With sales of about seven million dollars, Toritsu-Kogyo has 50 employees, while most of the secondary subcontractors have less than ten employees. A clear-cut division of labor emerges as each secondary subcontractor tries to differentiate itself from the rest according to two criteria. The first differentiation occurs on the basis of the seventeen skills listed in Figure 9.3. The second differentiation is based on prod-

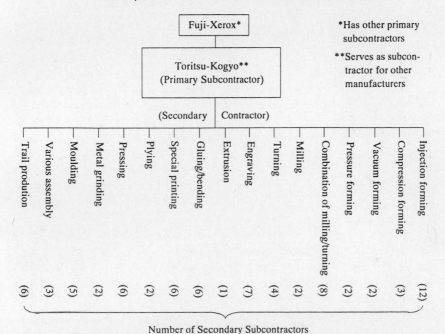

FIGURE 9.3 Inter-organizational Network for Fuji-Xerox

ucts handled. The six secondary subcontractors in pressing, for example, may each handle a different-sized product.

Vertical information exchange occurs within the three levels of the vertical network, that is, the lead manufacturer, primary subcontractor, and secondary subcontractors. The relationship between the lead manufacturer and primary subcontractor is a vertical strong tie with intense information exchange. Information flow between the lead manufacturer and the secondary subcontractor is funneled through the primary subcontractor, who thus plays an important linking role.

Through these information linkages, a "micro-macro information loop" is created within a network. This loop is defined as an indirect informational linkage between micro- and macro-economic actors which influences the behavior of each. In our case in Figure 9.3, the macro-economic actor corresponds to the lead manufacturer and micro-economic actors to the various secondary subcontractors. These subcontractors are completely independent small companies, but their behaviors are coordinated through this "micro-macro information loop" in which the primary subcontractor acts as go-between. This is in essence an invisible coordination mechanism within the network in general. The

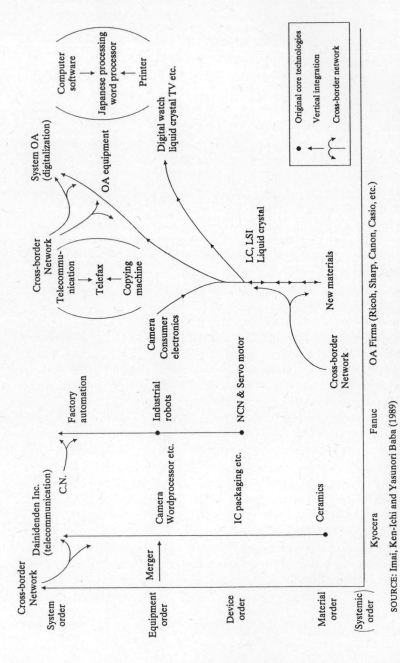

FIGURE 9.4 The Diversification and Network Pattern
of the Japanese IT Firms

SOURCE: Imai, Ken-Ichi and Yasunori Baba (1989)

implication of this concept will be further elaborated upon in the discussion of the national system of innovation in Section 3.

2.6 Relationship to R&D

Japanese industries and firms, with the flexible internal organization and elaborate inter-firm networks, have been quite successful in attaining speedy and flexible development of new products. These developments have featured a great variety of incremental improvements in component technologies, and Japanese firms were able to maintain their competitive edge by accumulating incremental innovations until the first half of the 1980s. At that point, however, they met increasing obstacles to further development. When innovations in technologies and related social and economic systems occur in a decentralized and dispersed manner, there will inevitably arise gaps or disequilibrium between subsystems. If such gaps are small, interfaces to link subsystems can be devised, and balance will be restored. This is the usual method of incremental innovation. However, it is apparent that such patchworks have a limitation; large gaps that could not be overcome by such patchworks could emerge, along with serious conflicts between subsystems. It becomes impossible to fill such gaps or to resolve such conflicts without upgrading of the system level or a new conceptualization of a whole system.

This situation has naturally oriented Japanese firms towards the systemic innovation mentioned above. Figure 9.4 depicts the development of innovation levels from material order, to device order, to equipment order, and to system order in several cases. For example, the robot was a typical example of incremental innovation. However, triggered by the advancement of information technologies in the early 1980s, the industrial robot business seems to have moved to a new phase of systemic development: the transition was set in the direction of factory automation (FA) where discrete robots and equipment are linked into total plant automation tailored to the specific needs of the customers. In this systemic automation phase, the services required by the users have become flexible, and robot manufacturers diversified to foster the software capability and consulting capability. Likewise in office automation (OA) the digitalization of telecommunications renovated the entire industrial scene. It provided the opportunity first to systematize in-house information processing, and then to extend the network to the other firms by increasing the inter-operability of the systems concerned. In this new setting, each piece of stand-alone equipment was linked with all others, and the whole collection of equipment was "born again" as part of an OA system.

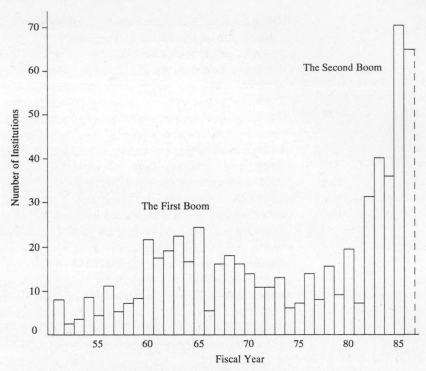

SOURCE: Science and Technology Agency (March 1987).

FIGURE 9.5 Setting up of R&D Institutions

In response to the rise of systemic innovation, Japanese firms stren-
uously began to enlarge and reshuffle their R&D activities. This created
a second boom in setting up corporate R&D institutions after 1985 as
is shown in Figure 9.5. The first boom occurred in the beginning of the
1960s when central R&D institutions for each of the big companies were
built for the purpose of creating their own technological basis, supplant-
ing the simple strategy of catching up with foreign technologies. Many
of them were chemical-oriented. On the other hand, the second boom
now is for the purpose of coping with systemic innovation, the impact
of information technologies and the coming new technology paradigm.
Usually they have a multidisciplinary character, including research into
new materials, electronics, mechatronics, and biotechnologies.

In accordance with this trend, the R&D expenditures of Japanese
firms substantially increased. Further, their emphasis shifted from development to applied research and further to basic research as R&D investment became a strategic investment for Japanese firms comparable

TABLE 9.1

A recent trend in R&D indicators in Japan

	R&D Expenditure 100 million yen		R&D Sales %		R&D Fixed investment %		Basic R&D Total R&D %	
	1976	1986	1976	1986	1976	1986	1976	1986
Steel	998	2,553	1.0	2.5	7.6	31.9	6.8	6.0
Chemicals	3,519	9,836	2.4	4.3	42.3	72.8	8.9	11.5
Non-ferrous metals	286	1,102	1.0	2.1	20.4	39.2	4.5	4.5
Ceramics	521	1,876	1.4	2.9	31.1	48.0	4.0	7.2
Machinery	1,386	3,791	1.7	2.8	61.6	87.5	4.3	6.6
Electrical machinery	4,917	19,800	3.7	5.5	104.4	94.4	4.4	4.1
Automobile	2,193	8,404	2.2	3.2	37.9	50.8	4.6	4.8
Textile	209	627	0.7	1.2	18.7	31.7	2.4	6.4
Pulp and Paper	136	339	0.5	0.8	6.1	8.0	4.4	5.0

SOURCE: MITI, *Sangyo gijutsu no doko to dakai* (Trends and Tasks of Industrial Technology in Japan), Tsusho Sangyo Chosakai, Tokyo, 1988 (ISBN4-8065-2304-6), pp. 150–55. (Abstracted from Table 2-2-2-5.)

to fixed capital investment as is shown in Table 9.1. Also, the diversification of R&D expenditures which indicates the interdisciplinary character mentioned above is shown in Table 9.2, as the ratio of R&D expenditures which was directed to the product fields that belong to other sectors.

Another noteworthy recent trend in corporate R&D institutions is their concentration into the Tokyo area. (In the case of big companies whose equity capital exceeds 100 million yen, more than sixty percent of R&D institutions are concentrated in the greater Tokyo area.) This is because the interactive process of information creation, learning, and research, which is crucial for systemic innovation, requires an environment with intense interaction, which is best provided by the very dense markets in the Tokyo area. Interaction includes three dimensions: between users and suppliers, between R&D, marketing, and manufacturing, and between physical products, software, and services. Recurrent interactions on each of these dimensions generates dynamic information which gives insight for further research. The character of the new industrial society is one of continuous iterative innovation generated by linkages across the border of specific sectors and specific scientific disciplines.

3 Discussion

Modern technological innovation is the result of a cumulative learning process within an industrial system. Technological breakthroughs take

TABLE 9.2
R&D Expenditure by Industry and Product Field

Sector	R&D for the own products field (%)		R&D for other products field (%)	
	1975	1985	1975	1985
Food	64.4	66.5	35.6	33.5
Textile mill products	39.7	28.3	60.3	71.7
Pulp and paper	79.8	71.6	20.2	28.4
Publishing and printing	69.8	70.7	30.2	29.3
Chemical products	90.2	91.2	9.8	8.8
Petroleum and coal products	57.2	56.9	42.8	43.1
Rubber products	93.5	85.6	6.5	14.4
Ceramics	65.0	54.3	34.0	45.7
Iron and steel	76.5	71.5	23.5	28.5
Non-ferrous metals and products	73.3	49.6	26.7	50.4
Fabricated metal products	51.5	40.6	48.5	59.4
Machinery, except electrical	80.4	71.3	19.6	28.7
Electrical machinery	89.8	89.2	10.2	10.8
Transportation equipment	81.9	86.0	18.1	13.0
Precision machinery	66.6	53.7	33.4	46.3

SOURCE: Bureau of Statistics, Office of Prime Minister, Japan, *Report on the Survey of Research and Development in Japan*, 1975 and 1985.

shape through feedbacks received from not only the firms that enter production on the basis of the technology in question but also from the machinery makers that develop the needed machinery from "co-specialized assets," in general. (See David Teece's paper in this volume.) Widespread information exchanges among the related firms is required. Also, as is insightfully explained in the Kline-Rosenberg chain-link model of innovation, the informational feedback between analytical design, re-design, production, and marketing, and between each of them and the accumulated knowledge of research, is crucial for innovation.

The size of innovating firms has been one of the controversial points in the discussion of technological innovation since the celebrated work of Joseph Schumpeter. Interpreted in light of the above discussion, how-ever, it seems to us that the size of the network involved in the learning process of technology development is more pertinent than is the size of an individual firm for the success of technological innovation. Also, the efficiency of the diffusion and transfer of technology within the multi-firm system is important to success.

From this perspective, we would like to discuss Japan's national system of innovation in this final section in an "information network" framework. (According to Freeman, the national system of innovation is defined as "the network of institutions in the public and private sectors whose activities band interactions initiate, import, modify, and diffuse new technologies." Freeman 1989, p. 1.)

Recall the concept of the "micro-macro information loops" that was presented previously as useful in explaining the workings of industrial networks which include a lead manufacturer, primary subcontractors, and secondary subcontractors. We now propose that the Japanese networks in many levels, including national, regional, inter-firm, and intra-firm, have an isomorphic character in their essential working mechanism (Aoki 1988). Therefore, we can safely extend the above concept to other network levels.

Japan's remarkable economic growth and innovation are based on its structural adaptability and flexibility to uncertain economic conditions. How has the Japanese capacity to cope with structural change been created? There seems to be some important invisible mechanism which has yet to be properly identified. In my view, the government's participation in creating "micro-macro information loops" has played a crucial role in this respect.

The price mechanism is an "invisible hand" which links micro and macro actors by compressing relevant information into a price signal. However, information transmitted in the market is not transmitted by price alone. Rather, technological and other information also moves between actors in the market. Especially as the strategic importance of technology increases, information which is not directly reflected in prices has become essential.

Some argue that such non-price information is reflected in prices through the direct and indirect effects of interactive economic activities. If the situation is viewed this way, however, an important point will be missed. What is important is to determine precisely what kinds of information other than prices are transmitted among economic actors. To understand the workings of today's market economy and of corporate behavior within it, it is crucial to give explicit consideration to how various economic actors are interconnected in the economy, and how information flows between them. The economic analyses developed so far, however, are often based on an unrealistic model of structure and information.

In general equilibrium theory, the existence of a fictitious "auctioneer" is assumed who is supposed to adjust demand and supply at no cost. To analyze the dynamic real world, we have to assume that each actor strives to communicate, obtain, process, interpret, and generate information through strong and weak ties with other actors at substantial costs.

Recently, certain kinds of information have been treated by economic theory as commodities to be transacted in the market. A major part of industrial information, however, is still communicated among actors outside of explicit market transactions. This is the basic reason why interorganizational linkages (such as networks), or "micro-macro information loops" in my term, are formulated and why the formation of such linkages is such an important factor in economic decision-making. My theoretical position is that the "micro-macro information loop" is an *ex ante* coordinative mechanism operating through various information exchanges, whereas the market is an *ex post* selection mechanism based on price signals (in a sense that the above *ex ante* coordination has already been done).

The Japanese system is coordinated through multilayer networks, including inter-firm, inter-industry, inter-industries, government and university and other networks, which have their own "micro-macro information loop" as a linkage and coordinating method. The supplier network explained above is an example at the inter-firm level. MITI's well-known "Industrial Structure Council" is at the inter-industry level. The Japanese industrial groups like Mitsui, Mitsubishi, and Sumitomo are also at the inter-industry level. The technopolis is an attempt to create an inter-industries, government and university network.

In every country there are personal networks of elite university graduates who will play an important role for information exchange and transmission. But Japan is perhaps unique in that MITI's elite officials with a sense of macro policy have coordinated industrial networks by interfacing micro-information in each firm and macro-information of an industry and the economy as a whole, by utilizing personal networks which consist of key persons who graduated from elite universities. In short, these MITI personnel have played a linking role in creating "micro-macro information loops" in Japan.

4 Conclusions

Japanese network institutions in the private and public sectors have been engaged in sequential learning and structural changes in the way described in this chapter. It has a process of continuous adaptation to the requirements of Japanese industrial system which were imposed by external conditions like the oil crisis, external trade imbalance, or the yen appreciation. Fortunately, the actual performance of innovation and structural adjustment has been extremely successful, and has in fact exceeded our expectations. A lot of analyses and explanations have been made in attempts to understand the reasons for this success. In this chapter, we have tried to improve these analyses by showing the *sequential* nature of the learning process engaged in by Japanese industry, and also by introducing the new concept "micro-macro information loop" and describing what we believe to be its important contribution to the success of Japanese industry.

References

Amendola, Mario and Jean-Luc Gafford. 1988. *The Innovative Choice— An Economic Analysis of the Dynamics of Technology*. New York: Basil Blackwell.

Aoki, Masahiko. 1988. *Information, Incentives, and Bargaining in the Japanese Economy*. New York: Cambridge University Press.

Asanuma, Banri. 1989. Manufacturer-Supplier Relationships in Japan and the Concept of Relation-Specific Skill, *The Journal of the Japanese and International Economies*, vol. 3, no. 1 (March), pp. 1–30.

Dosi, Giovanni, et al., eds., 1988. *Technical Change and Economic Theory*. London: Pinter Publishers.

Freeman, Christopher. 1987. *Technology Policy and Economic Performance— Lessons from Japan*. London: Pinter Publishers.

Hippel, Eric von. 1976. The Dominant Role of Users in the Scientific Instrument Innovation Process, *Research Policy*, July.

Imai, Ken-ichi. 1988. Japan's Government Policies and Structural Change: The Importance of Micro-Macro Information Loops. Paper presented at the XIVth World Congress of the International Political Science Association in Washington, D.C., August 28–September 1, 1988.

Imai, Ken-ichi. 1988. Japan's Corporate Networks, *Japanese Economic Studies*.

Imai, Ken-ichi and Yasunori Baba. 1989. Systemic Innovation and Cross-Border Networks: Transcending Markets and Hierarchies to Create a New Techno-Economic System. Paper presented at the International Seminar on the Contributions of Science and Technology to Economic Growth at the OECD, Paris, June, 1989.

Imai Ken-ichi, Ikujiro Nonaka, and Hirotaka Takeuchi. 1985. Managing the New Product Development Process: How Japanese Companies Learn and Unlearn, in Kim B. Clark, Robert H. Hayes, and Christopher Lorenz, eds., *The Uneasy Alliance: Managing the Productivity-Technology Dilemma*. Boston: Harvard Business School Press.

Imai, Ken-ichi and Hiroyuki Itami. 1984. Interpenetration of Organization and Market: Japan's Firm and Market in Comparison with the U.S., *International Journal of Industrial Organization*, 2.

Imai, Ken-ichi and Ryutaro Komiya. 1989. *The Japanese Firm*. (In Japanese.) Tokyo: The University of Tokyo Press.

Kline, Stephen J. and Nathan Rosenberg. 1986. An Overview of Innovation, in Ralph Landau and Nathan Rosenberg eds., *The Positive Sum Strategy*, Washington D.C.: National Academy Press.

Lachmann, Ludwig M. 1986. *The Market as an Economic Process*. Oxford: Basic Blackwell.

Lundvall, Bengt-Ake. 1988. Innovation as an Interactive Process: From User-Producer Interaction to the National System of Innovation, in Giovanni Dosi, et al., eds., *Technical Change and Economic Theory*. London: Pinter Publishers.

Lundvall, Bengt-Ake. 1984. User-Producer Interaction and Innovation. Paper presented at the TIP-Workshop, Stanford University, December 4, 1984.

Mansfield, Edwin. 1988. Industrial R&D in Japan and the United States: A Comparative Study, *American Economic Review*, May, 1988.

Mansfield, Edwin. 1988. The Speed and Cost of Industrial Innovation in Japan and the United States: External vs. Internal Technology. Manuscript, University of Pennsylvania.

Nelson, Richard. 1982. The Role of Knowledge in R&D Efficiency, *Quarterly Journal of Economics*, 97, pp. 453–70.

Rosenberg, Nathan. 1963. Technological Change in the Machine Tool Industry 1840–1910, *Journal of Economic History*, vol.23, no. 4 (December), pp. 425–26.

10

The Organization of the
Innovative Process

FRANCO MALERBA

This paper proposes that the characteristics of the knowledge base underpinning innovative activities, the conditions of demand, and the type of competition greatly affect the organization of the innovative process and the strategies of commercialization of new technologies at both the industry and firm levels. The knowledge base, demand, and competition affect the type and speed of learning processes, the choice of sustainable innovative approaches and behaviors of firms, the capabilities and size required for survival and growth, and the intensity of the selection process in the industry.

Differences in the organization of innovative activities and the strategies of commercialization of new technologies may emerge not only across but even within high-technology industries, if the various lines of business within an industry are characterized by a different knowledge base, condition of demand, and type of competition. For example, in lines of business characterized by costly R&D and productive investments, a standard demand and a global competition, large multinational firms will be present. On the other hand, in lines of business characterized by limited R&D, limited productive investments, and a custom or local demand, small specialized firms may predominate. Finally, in lines of business characterized by a diversified knowledge base, products which are systems and which integrate several components, as well as different types of demand, a wide heterogeneity of firms in terms of size, capability, behavior, strategy, and degree of internationalization may be present. In extreme cases it is therefore possible that two quite different types of organizations of innovative activities may emerge within the

TABLE 10.1

World Market Shares of European Firms in Semiconductors,
Semiconductor Equipment, and Computers

		Europe	United States	Japan
Semiconductors	1982	12.7	51.4	35.3
	1990	10.5	36.5	49.5
Semiconductor Equipment	1983	8.0	67.0	25.0
	1987	10.0	55.0	35.0
Computers	1989	15.8	62.7	21.5

SOURCES: For semiconductor and semiconductor equipment: Dataquest in EEC (1989). For computer technologies: Datamation. The Datamation sample is composed of the major information technologies producers.

same broad industry: one dominated by large firms, the other by an heterogeneity of firms in terms of capability, size, and specialization.

The possibility of the presence of quite different types of organizations of innovative activities and of strategies of commercializing new technologies implies also that public policy must be in tune with the specific organization of innovative activities in which it is intervening. This is not an easy task, but the range of actors and tools which can be mobilized by public policy is broad: universities as well as public and non-profit institutions (in both basic and applied research and in training), public procurement, national or regional technological infrastructures, and fiscal incentives or public subsidies to innovating firms.

Yet substantial differences may emerge across countries in both the philosophy and timing of public intervention, as well as in the types and capabilities of the institutions involved. These differences are the result of the economic and socio-political histories of the various countries.

Within this broad framework, in this paper the organization of innovative activities and the role of institutions in the semiconductor and computer industries in Europe during the 1980s (see Tables 10.1, 10.2, and 10.3) will be examined.

The computer and semiconductor industries are interesting cases for two reasons. First, computers and semiconductors have been at the center of the electronics revolution throughout the postwar period. Both industries have shown high rates of innovation, and have shaped the development of the advanced industrial economies. Accordingly, they may be considered key, strategic, or leading industries. Second, institutions have played a major role in these industries. The role of public agencies in fostering the development of these industries has influenced the evolution and competitiveness of national firms (Dosi 1984; Malerba 1985;

TABLE 10.2

The Semiconductor (SC) Industry: Leading Companies in Europe

Companies	SC Sales 1984	SC Sales 1990	Sales Growth 84/90	SC Sales/ Total Sales 1988	R&D/ Sales 1990	World Market Shares 1984	World Market Shares 1990
Philips	1325	1932	45.8%	6.2%		5.0	3.3
SGS-Thomson	636	1463	130.0%	100%	18%	2.4	2.5
Siemens	450	1221	171.3%	1.3%		1.7	2.1

SOURCES: SGS-Thomson Electronics Business, various years.

Steinmueller 1986 and 1988; and, Okimoto, Sugano and Weinstein (eds.) 1984).

The patterns that emerged in the organization of innovative activities in Europe during the 1980s cut across the semiconductor and computer industries because a major rift took place between basic components and basic systems, and system applications. In this paper it will be shown that in basic components and basic systems, high costs of R&D and productive investments, standard demand and global competition determined an organization of innovative activities and a commercialization of new technologies dominated by large firms. During the 1980s technological change pushed up the scale and cost of R&D and production and therefore generated in Europe an increase in industrial concentration through the radical reorganization and restructuring of many firms. Manufacturing capabilities, design, and quality control became key variables in international competitiveness within these industries. In system applications, by contrast, the organization of innovative activities was characterized by a wide heterogeneity of firms, because of the flexibility of technology, the systemic features of products and the variety of demand. During the 1980s technological change generated new opportunities for the growth of different types of firms, and increased the variety of possible innovative behaviors and strategies open to firms. Factors such as system integration, user-producer relationships, distribution capabilities and cooperative networks became relevant for the success of European firms in system applications.

In this paper it will be shown also that public policy in Europe during the 1980s had to conform to these different organizations of innovative activities. As a consequence, policies adopted for basic components and basic systems were quite different from those for system applications. In the first case public policies were mainly mission-oriented, while in the latter case they were concentrated on technical standard setting human capital formation, information provision and technology diffusion.

TABLE 10.3
European Information Technology (IT) Firms

Firms	Sales IT mill.$	Sales IT on Total Sales	Share IT per Segment %							Market Share World	Market Share Europe
			Main-frame	Mini	Micro	Data-comm.	Peripher.	Software	Serv./ Maint.		
1984											
Siemens	2789.5	17.3	29.0	6.3	.6	34.5	17.2	1.4	11.0	1.9	7.0
Philips	1090.3	20.0	–	23.0	9.2	27.5	23.0	7.3	10.0	.7	N.A.
Olivetti	2012.4	69.6	1.5	26.8	14.4	4.3	17.9	4.7	30.4	1.4	N.A.
Bull	1555.6	100.0	32.1	6.4	6.4	–	38.9	6.4	9.8	1.0	3.2
Nixdorf	1147.7	100.0	–	29.6	–	–	43.6	14.0	12.0	.8	3.4
Nokia	181.0	11.5	N.A.	N.A.	N.A.	N.A.	N.A.	N.A.	N.A.	.1	.6
STC-ICL	1222.7	100.0	29.6	16.4	–	3.3	36.5	6.5	7.7	.8	3.4
Cap Gemini Sogeti	706.0	100.0	N.A.	N.A.	N.A.	N.A.	N.A.	N.A.	N.A.	.1	.5
1990											
Siemens-Nixdorf	7735.1	100.0	13.2	11.9	11.2	1.6	25.1	11.9	25.1	2.5	N.A.
Philips	3283.9	10.7	–	10.2	17.5	15.5	35.4	3.2	18.2	1.1	3.0
Olivetti	6414.5	14.5	1.9	8.2	27.9	3.4	17.4	9.7	31.5	2.1	5.9
Bull	6349.6	100.0	13.1	5.0	18.1	–	21.1	10.0	32.7	2.1	5.5
Nokia	1279.9	22.1	–	14.1	37.3	7.1	2.5	4.6	34.3	.4	1.3
ICL-Fujitsu	2862.9	100.0	13.7	12.3	9.1	2.6	9.1	17.2	36.0	.9	N.A.
Cap Gemini Sogeti	1684.9	100.0	–	–	–	–	–	2.0	88.0	.5	1.2

SOURCE: Datamation and IDC.

Finally, in this paper it will be shown that while uniformity prevailed across Europe in the general design of public policy, diversity was also present across European countries in the implementation of these policies due to the specific institutions active in the various countries with their built-in specializations and expertise.

In Section 1 of this paper, the major features of electronics technologies and of their impacts on the organization of the innovative process will be discussed. In Section 2, the different characteristics of the organization of the innovative process and the commercialization of new technologies in the European semiconductor and computer industries will be analyzed. Section 3 focuses on the different positions of European firms in the challenge of globalization, while Section 4 analyzes the role of public policies and institutions.

1 Technological Change in the Electronics Industry and the Organization of Industry

Throughout the 1980s the increase in complexity and the growing importance of system features constituted two of the major characteristics of technological change in semiconductors and computers (Rosenberg 1982 and 1987). The advent of the microprocessor and the very large scale integration in semiconductors (VLSI) has increased the functionality of a unit semiconductor device to 10,000 times what it was 20 years ago. The overall complexity of computers has increased enormously as a consequence of the cumulative changes experienced over the years by their various component parts: central processing units, main memory and mass storage devices, and various peripherals such as terminals, readers, and input and output devices. This complexity has meant that multidisciplinary knowledge has become necessary for the generation and development of new products. In the computer industry, for example, the disciplines involved in the innovation process may range from solid state physics to mathematics, and from language theory to management science. Moreover, systemic features have become more and more relevant not only in computers, but also in semiconductors. A microcontroller, for example, is a single integrated circuit containing a central processing unit, a memory, and an input-output capability. In addition system features have underscored the need for firms to develop new dimensions for their product strategies, such as modularity, compatibility, and integration among various components and subsystems. The problems of creating technical standards and equipment interfaces and the strategic choice to have a proprietary network or to have an open one have become key issues in the industry.

The 1980s also witnessed the growth of new market segments such as application specific integrated circuits (ASIC), general purpose mini-computers, minisupers, workstations, personal computers, application software, and software services. These developments were rooted in three major changes that occurred in the late 1960s–early 1970s: the introduction of the microprocessor, the emergence of software, and the unbundling between hardware and software. In some of these segments such as ASICs, superminicomputers, personal computer clones, print-ers, low and medium speed drives, application software, and software services, the relatively low financial requirements for startup activities have led to the emergence of a variety of firms that were quite different in terms of size, product differentiation, specialization, and diversification (Malerba and Torrisi 1990).

These various market segments now present in the semiconductor and computer industries may be grouped into two major categories: basic components and basic systems, and system applications. The first group includes standard semiconductor devices, computer hardware and system software, while the second group encompasses application specific integrated circuits (ASIC), application software, software services and integrated information technology systems. Within each of these categories the organization of the innovative process is handled differently. For this reason this classification will be used here throughout.

2 Basic Components and Basic Systems

The category of basic components and basic systems is characterized by the increasing cost in the development and introduction of new technologies. The development and production costs for microprocessors, for example, jumped from $20 million for an eight-bit microprocessor, to $110 million for a 32-bit microprocessor. More generally, while semiconductor firms spent on average eight percent of their sales on R&D during the 1970s, they spent close to fourteen to fifteen percent in the second half of the 1980s. The escalating cost of R&D was also associated with an increase in capital spending: from an average of ten percent of sales in 1976–77 to 20–30 percent in 1984. The minimum investment required to set up a new production line has also increased from $11 million in 1978, to $100 million in 1984, to $230 million in the late-1980s and early 1990s. Finally, rising costs have been increased by the short-ening of product life cycles to three to five years from introduction to maturity (EEC 1989).

Increasing costs and the complexity of technology and products have acted as barriers to entry in R&D and production for basic components and basic systems during the 1980s. In spite of the rise of Japanese manufacturers and the relative decline of several major American producers, the same firms (IBM, DEC, Hewlett Packard, etc.) have dominated the industry during the 1980s (allowing for shifts in relative rankings).

3 System Applications

In contrast, the category of system applications is software intensive. In terms of shares of total computer system costs, software moved from less than 30 percent in 1975 to more than 70 percent in 1985 (*Business Week*, May 9, 1988; *Datamation*, 1983). The development of new software products caused major changes in the R&D game and the innovative process when new product development became less of a team effort and more of a kind of "craft" or intellectual work, highly person embodied (Brooks 1972). Software engineering (Computer Aided Software Engineering—CASE and formal methods) is not widely used yet in spite of major developments occurring in the last years. The more one moves from system software to application software to services, the more the cost of product development[1] diminishes, appropriability decreases (because of the lack of standards and the limits to effective protection), and the role of users in the development of application programs increases.

Software also increases the flexibility of information technologies to various applications. End-user applications may range from general (horizontal) applications regarding accounting, payroll, office automation, CAD-CAM, or process control, to specific (vertical) applications regarding, for example, banking, retailing, and insurance. It is important to note that these remarks concern not only computer systems, but also ASICs (which are customized for specific uses or customers). The users of ASICs, moreover, may even specify and eventually design their own chips (Hobday 1988).

The effect of these changes on the organization of the industry was an increase in the heterogeneity of producers in terms of capabilities, specialization, innovativeness, and competitive behavior. In the computer industry, for example, software houses were able to choose different types of specialization: cross-industry programs, industry-specific programs, semi-custom programs, one-shot software programs, professional services, or processing software services. In addition, software houses could become value added retailers and system integrators. Similarly, new independent design houses and user design centers in the

---

[1] The cost of a software development program, including the computer, may range between $1,000 to $15,000. *The Economist*, January 30, 1988.

ASIC industry operated side-by-side with the major semiconductor producers: design houses concentrated their activities on a limited number of users and did not produce on a large scale,[2] while user design centers focused on the design and photomask of circuits, which are then produced externally and finally used in-house (Hobday 1988).

4 The Organization of the Innovative Process and the Commercialization of New Technologies: The European Experience

The distinction between basic components and basic systems and system applications proves quite useful in interpreting the evolution of the organization of the innovative process and the commercialization of new technologies in Europe. It must be noted that most large European firms such as Philips, Siemens, Olivetti, and SGS-Thomson, are present both in basic components or basic systems, and in system applications. These firms, however, have followed a different logic of product development and commercialization in these two product ranges.

5 The Scale Factor in Basic Components and Basic Systems

As far as basic components and basic systems are concerned, in Europe the organization of the innovative process and the commercialization of new technologies was driven by scale considerations with respect to both R&D and production, given the cost of new technologies and the need for firms to reach significant dimensions. In semiconductors, industrial concentration took place through the merger of the Italian company SGS and the French Thomson Semiconducteur in 1987. The SGS-Thomson merger had the goal of up-scaling R&D and production by unifying two firms with complementary assets, technologies, products, and geographical markets. Through the merger, in fact, SGS and Thomson Semiconducteur grew in terms of R&D and production which was better suited to international competition. In 1990, SGS-Thomson spent $263 million (18 percent of sales) on R&D, and reached sales of $1463 million (2.5 percent share of the world market), becoming the second largest European semiconductor producer. Following the merger, moreover, a subsequent rationalization of production facilities (three plants in France and two in the Far East were closed) and of the product portfolio (from 45,000 to 30,000 products) took place. SGS-Thomson became a broad product range supplier, including memories (21 percent of sales in 1990),

---

[2]They may be able to begin their operations with an investment of $50 million, including workstations, EDA software tools, and distribution.

microprocessors (13 percent), ASICs and dedicated semiconductors (36 percent), discrete devices (38 percent), and commodities (7 percent). In particular, it occupied a leading position in power linear integrated circuits, EPROM memories, and CMOS integrated circuits.

In 1989, the new company SGS-Thomson took a further step by acquiring the English firm INMOS. In this case as well the specialization of INMOS in transputer microprocessors complemented that of SGS-Thomson.[3] In computers, industrial concentration increased in various ways. Siemens acquired Nixdorf in 1989, Fujitsu took control of ICL and Nokia, Philips de facto abandoned computers production in 1991, and Bull acquired Zenith in 1989.

Increasing firm size and industrial concentration in Europe was associated with an increase in cooperative agreements, which allowed firms either to share the cost and reduce the risk of new developments, or to obtain complementary expertise (see Table 10.4). SGS-Thomson, for example, established various types of cooperative agreements and alliances. Strategic agreements were of three types: one devoted to fill the gap in certain product ranges (in microprocessors with Zilog and Motorola, in CMOS with Toshiba, in programmable logic devices with GEC-Plessey), one aimed at developing capabilities in a new and growing area, ASICs (with LSI Logic, Lattice, CDI, and Cadence), and one (the most important) with the goal of developing and strengthening vertical links with customers (Bosch, Marelli, IBM, Seagate, Nokia, Alcatel, and Matra-Harris). Other agreements were made with regard to the sale or exchange of technology in areas where SGS-Thomson was at the technological frontier (i.e., power semiconductors).

A major cooperative agreement took place between Siemens and Philips in 1986: the Megaproject. This venture had the goal of bringing these two firms to the technological and productive frontiers in memories. The agreement involved sharing risk and costs and avoiding duplicative R&D. The two firms invested $430 million in a four year R&D joint venture, and received support from the German and Dutch governments (Hobday 1988). The know-how and expertise gained from the project were supposed to be used by the two firms in different product ranges: Philips aimed to develop manufacturing capabilities for its low power consumer applications with the one-megabit static RAM device, while Siemens aimed to develop manufacturing capabilities for its data processing applications with the four-megabit dynamic RAM. In 1989

---

[3]It must be remembered that Thomson had already acquired control of Mostek in 1985.

TABLE 10.4

Agreements per Geographical Area and Type in Semiconductors, Telecommunications, and Computers

Area	N. Agr.	%	SC %	TEL %	INF %	PROD %	TECH %	COMM %	EQUITY %
				1980–1983					
Europe	43	10.5	5.1	14.0	12.0	10.4	14.5	8.4	17.8
Japan	46	11.2	17.9	6.4	11.3	7.6	15.9	6.5	5.1
USA	150	36.8	35.0	38.6	34.7	34.0	35.9	38.5	25.4
Europe-Japan	39	9.6	11.1	8.8	9.7	6.9	4.1	11.6	5.9
Europe-USA	123	30.0	29.0	30.4	29.8	36.8	28.3	32.7	40.7
Europe-ROW	8	1.9	1.9	1.8	2.5	4.3	1.3	2.3	5.1
		100%	100%	100%	100%	100%	100%	100%	
Total N.	409		117	171	124	144	145	275	118
				1983–1986					
Europe	140	1.0	13.7	12.0	15.0	15.9	19.5	10.2	19.8
Japan	84	8.4	9.8	11.2	4.8	5.8	9.9	8.8	16.6
USA	396	39.6	27.4	36.8	45.2	33.6	37.5	38.3	26.1
Europe-Japan	90	9.0	20.3	5.7	8.1	11.2	7.0	10.5	9.2
Europe-USA	252	25.2	26.1	27.5	22.7	24.4	22.0	26.9	23.0
Europe-ROW	38	3.8	2.7	6.8	4.2	9.1	4.1	5.3	5.3
		100%	100%	100%	100%	100%	100%	100%	
Total N.	1000	100.0	153	418	458	295	354	647	283

SOURCE: Cainarca, Colombo, and Mariotti, "Accordi fra imprese nel sistema industriale dell'informazione e della comunicazione," Milan, 1989.

LEGEND: SC: Semiconductors, TEL: Telecommunications, INF: Information Systems, PROD: Production Agreements, TECH: Technological Agreements, COMM: Commercial Agreements, EQUITY: Equity Agreements.

The percentages for equity agreements are calculated with respect to the total number of agreements per area.

the Megaproject was discontinued, because Siemens entered in JESSI and Philips abandoned the large scale production of semiconductors.

It should be noted that the process of learning and capability formation by Siemens took a long time. Siemens had to form a technological alliance with Toshiba for the production of one-megabit memories in order to be able to continue the project as planned and in 1990 it had 2.5 percent of the world market in one-megabit DRAM. In addition, by 1989, Siemens was able to produce 4-megabit memories. Finally, in 1991, Siemens established an agreement with IBM for joint production of sixteen-megabit DRAM (*Business Week*, 1990 and *The Wall Street Journal*, Europe, July 5, 1991).

6 System Integration and Networks in System Applications

In the European industry the organization of the innovative process and the commercialization of new technologies in system applications differed significantly from the case of basic components and basic systems: size did not act as a barrier to entry and did not always represent a critical factor. Rather, other factors were at work. First, the success of new products often resulted from the integration of existing pieces of hardware and software. Commercial success, therefore, did not come from an ability to develop internally all parts of a system, but from an ability to identify, target, and meet the requirements of either specific sectoral applications, or groups of users, or even of a single large customer.

Olivetti is a case in point. Olivetti grew quite rapidly despite the fact that it entered the electronics industry relatively late (late 1970s). In 1981, it began the development of its first personal computer, and in 1984 it moved to the production of IBM-compatible personal computers. Since the early 1980s, Olivetti has been active in introducing new electronics products based on an internal technical capability, an external network of alliances, agreements and acquisitions, and a solid distribution network. Olivetti's strategy to acquire external software expertise and distribution channels gave the firm a high degree of flexibility and speed of movement in an industry in which technology and markets were changing rapidly. Olivetti was able to supply personal computers with software packages to small and local firms through its personal computer dealer networks and retailers; information systems to medium size and large firms through value added retailers and agents; information systems to specific market segments and the public administration through the integration of in-house and external hardware and software.[4]

A second key factor for the success in the commercialization of new technologies in system applications was related to user-producer relationships and to learning from interaction with customers. The introduction of new products, in fact, was increasingly driven by an understanding of the specific characteristics of markets and users. Cap Gemini Sogeti Sesa (CGSS), for example, was able to provide a wide variety of software services, ranging from professional services (both software consultancy and software development) to system integration and software

---

[4]It must also be noted that the increasing relevance of system integration by European electronics firms is quite evident when one observes that several European producers purchase and distribute large and intermediate Japanese systems: Siemens has a link with Fujitsu, Olivetti with Hitachi, and Bull with NEC (and Honeywell).

products, through operational teams specialized in specific local markets or business sectors. These branches had detailed knowledge of the market or sector in which they operated, and were able to interact fully and continuously with customers.

Another example is provided by Nixdorf. In order to focus on specific applications, Nixdorf adopted an end-user orientation characterized by tailored software, customer training, and careful maintenance, with operating units that worked closely with customers. This aim to maintain a close and proprietary interaction with users encouraged Nixdorf to develop a sales and distribution network based mainly on the company's workforce. Proprietary technology and tailored software were developed for specific market segments such as financial institutions, retailing, and small businesses. Only a limited use was made of agreements and cooperation with other firms.

Relatedly, the understanding of the specificities of local and domestic markets in terms of type of demand and characteristics of users became a very important factor for success in Europe, given the major economic, structural, and institutional differences across European countries. A consequence of the importance of local markets and of user-producer interaction for commercial success was the recentralization of firm activities. In computers, during the 1980s, Nixdorf created small and decentralized local operating units. This decentralized operational structure was maintained by continuously splitting up growing units into smaller ones. Similarly, in software services in the late 1980s, CGSS established 237 small decentralized branches, each of which consisted of an average of 50 professionals. This use of the branch as a basic operating level permitted greater flexibility and speed in opening or closing specific branches according to changes in market conditions.

It should be noted that in the CGSS case extensive decentralization was compensated by a very efficient internal transfer of information. All 237 branches of CGSS shared a common knowledge base consisting of a reference database on CGSS' projects, specialized centers with advanced capabilities in specific methods, programs, and standard proposals, as well as an expert club composed of leading specialists in various areas of information technology. In the area of ASICs, SGS-Thomson rapidly developed a number of decentralized design centers located near major markets which were able to offer design capabilities for user requirements. The pan-European venture, European Silicon Structure (ES2), aimed at establishing a truly European firm operating throughout Europe in the growing ASIC market, has also provided advanced customer services and has facilitated the maintenance of close links with semiconductor, users.[5]

Another direct consequence of the major role played by users in the innovation process was the growing importance of well-developed commercialization and distribution networks. Olivetti again represents an excellent example. Olivetti's capability of producing and commercializing personal computers and information technology systems was enhanced by the existence of a well-developed and geographically dispersed sales and distribution network. This network was extended and enriched by the acquisition of, or the participation in, additional retailer and distribution channels, such as authorized personal computer and office automation dealers and complete system distributors. In 1986, for example, direct sales accounted for only 55 percent of Olivetti's revenue, while other retail chains, value-added retailers, and other original equipment manufacturers such as ATT accounted for the remaining 45 percent (International Data Corporation Italia (IDC), *The Olivetti Group*, Milan, 1987). This articulated sales and distribution network was able to offer products with standard and custom-oriented software, maintenance programs, and user training.

A closer link between development, marketing, and commercialization also became a key competitive asset for system application firms. In this case, the link does not go in a single direction, from development, to marketing, to distribution. Rather, feedbacks and interactions in the opposite direction, from marketing and distribution to product development, are also significant: people involved in development have to be as open as possible to useful information on market and sector characteristics and on product performance coming internally from marketing and distribution and externally from users.

In this respect, within European firms there has been an increasing integration of the development, production, marketing, and distribution functions. Philips, in fact, has recently reorganized its R&D activities by increasing its control of the various product divisions at the development stage (at the expense of the central laboratories), by integrating

---

[5] ES2 was established in 1986 by various ASIC users such as Philips, Olivetti, British Aerospace, and Saab Scania, and was supported by the EUREKA program. Its aim was to enter the prototype and low volume ASIC market. The initial investment was $60 million, with headquarters in West Germany, production facilities in France, and various design centers in Europe. ES2 was successful in offering low-cost products by using direct-write electronic beam lithography and fast prototype delivery. In addition, ES2 offered the possibility to customers to design their own chips (Integrated Circuit Engineering Corporation (ICE), *Status: A Report on the Integrated Circuit Industry*, Scottsdale, 1989; *Financial Times*, September 12, 1985).

marketing people into research groups, and by increasing the effectiveness of technology transfer from research to production through specific programs and through the direct transfer of researchers into the production stage. Approximately 21,500 people (80 percent of Philips' researchers) are engaged in development (Sigurdson 1989). Similarly, Siemens has recently moved its 38,000 researchers engaged in development (92 percent of Siemens' researchers) under the control of its operating product divisions. Siemens' management has also engaged the company in a major reorganization in which fifteen units have been created, all of which have a sharp focus in terms of technology and customers. Finally, as of 1989, the Olivetti Group has been divided into three companies, each of which is concerned with a specific product-market area and is in charge of all stages of the innovation process, from design, development, and production to marketing and sales.[6]

The development of articulated networks of long-term alliances, cooperative agreements and acquisitions has also played a key role in allowing firms to keep pace with multidisciplinary and multisource technical change, to manage complexity and systemic features, and to enhance technological, productive and marketing capability and flexibility in system integration (Vaccá 1989, Teece 1989). The extent of these networks and the specific type of links, however, have depended on company strategy and on the characteristics of the line of business. Olivetti represents one of the most successful cases of an effectively articulated network. In addition to its commercialization strategy (described above), Olivetti extended its network through acquisitions (by taking control of Triumph Adler, Acorn, Scanvest Ring, and Bunker Ramo), cooperative agreements, and strategic long-term alliances, all of which provided Olivetti with complementary technologies and productive and commercialization assets. Venture capital was used mainly as a means to acquire technology windows.

The use of networks and the policy of acquisitions or control of foreign companies played a role in the competition in software services. Here, in order to gain knowledge and expertise about local markets, European firms reorganized their marketing and distribution channels by acquiring, taking control, or participating in, firms with already developed competences regarding consumers or demand. CGSS, for example, acquired five software firms in 1986, three in 1987, four in 1988, and two

---

[6]Olivetti Office (typewriters, text processors, printers, etc.), Olivetti Systems and Networks (personal and minicomputers, local area networks, terminals, workstations, etc.), Olivetti Information Services (software, value-added network services, etc.).

in 1989 (up to May). These firms were located in both Europe and the United States.

It must be noted that the policy of developing articulated networks of alliances was launched by Olivetti as a catch-up strategy in the early 1980s. Later on, given the developments of technology, demand, and competition, it became a dominant strategy among all leading computer firms, including IBM.

System integration, user-producer interaction, and networks of cooperative agreements have not displaced internal R&D to a secondary position. R&D, in fact, still plays a major role in system applications. Suffice it to note that several European computer firms operating in information technologies continue to be highly R&D intensive. In the late 1980s, the R&D intensity (measured by the ratio between R&D expenditures and sales) of Nixdorf, for example, was between nine percent and ten percent, Bull between ten percent and twelve percent, STC-ICL between eight and nine percent, and CGSS between eight and nine percent.

System integration, user-producer interaction, and cooperative networks, however, have changed the role of firms' internal R&D. Internal R&D has become increasingly important for the identification, selection, and absorption of external knowledge and external technology, as well as for the search for partners in the development or exchange of technology.

European firms have been quite successful in system applications. Small and medium size software firms have targeted national or local markets for the development of specific applications. Among large firms, Olivetti and CGSS are cases in point. Between 1984 and 1988 Olivetti's sales growth, market shares, and profitability were quite high. In this period the Olivetti Group increased its information systems sales 170 percent (*Datamation*, 1989), reaching a world market share of 1.8 percent and a European market share of 4.3 percent. Olivetti also became the leading European firm in personal computers. Between 1983 and 1985 the Olivetti Group's net income over sales remained above 7.5 percent, declining only in 1987 and 1988 to 5.5 percent, still a high value for European standards. Similarly, between 1984 and 1990 CGSS increased its sales by 139 percent. In 1990, it reached a share of the European information system of 1.2 percent, becoming the largest European independent software producer.

Nixdorf, on the other hand, moved from a position of growth and profitability to one of losses in a very short time span, so that it was finally acquired by cash-rich Siemens in 1989. Nixdorf's strategy of remaining a specialist in particular information technology applications and of offering internally developed proprietary hardware and software tailored to specific market segments proved successful up to 1987. Between 1983 and 1987, for example, Nixdorf's sales grew 170 percent and net profits grew from 3.7 percent of sales in 1983 to 5.2 percent in 1987. These assets became liabilities in 1988 and 1989, however. Proprietary technology, an extreme fragmentation and customization of products, and a geographical and a sectoral concentration on a single market (banking in West Germany) were no longer suited to a situation in which open standards and compatibility became increasingly dominant and in which applications required the concentration of increasing resources and witnessed the entry of major multinational firms. As a result, Nixdorf experienced a crisis in 1988 and 1989. Sales growth rate declined to 11 percent, and net profits turned to losses in 1988 and 1989. Before its acquisition by Siemens in 1989, Nixdorf had already changed strategy by concentrating its efforts on markets such as insurance, increasing its cooperative agreements, opening its systems to UNIX, and offering personal computers with standard software.

7 Globalization in Basic Components and Basic Systems and National Markets in System Applications

The 1980s were also characterized by an increase in the globalization of competition in basic components and basic systems. This trend pushed European firms to try to move away from the supply of a wide range of products at the domestic level in order to be able to supply specific products at a world level.

The timing and extent of the internationalization strategies of different European firms varied significantly throughout the 1980s. A first group of European firms was already internationalized at the beginning of the 1980s. Philips, for example, was historically one of the most internationalized electronics firms in Europe, with production and development activities geographically dispersed around the world near the various product divisions. Only its headquarters, central R&D laboratory, and a few production facilities were located in The Netherlands. As Table 10.5 shows, in 1988 less than seven percent of Philips' total sales were directed to the home market, while 73 percent were absorbed by the European market. Similarly, Olivetti has long been an internationalized company in terms of activities and sales. In 1988, for example, the domestic market absorbed only 38 percent of Olivetti's total sales, and only 50 percent of employment was concentrated in Italy.

A second group of firms began to augment the pace of internationalization during the 1980s. The internationalization of SGS-Thomson

TABLE 10.5

Sales Distributions of Main European
Semiconductor and Computer Firms

(total sales)

| | 1984 | | | |
Firms	Total Sales mill. $	Home %	Rest of Europe %	Rest of the World %
SGS-Thomson	636.0	63.0[a]	18.0[a]	19.0[a]
Siemens	16076.8	49.0	27.0	24.0
Philips	5221.2	7.0[a]	49.0[a]	44.0[a]
Olivetti	2891.9			
Nixdorf	1147.4	50.0	39.5	10.5
STC (ICL)	1222.7	70.5[b]	14.0[b]	15.5[b]
Cap Gemini Sogeti	206.0	44.0	29.0	27.0
Nokia	1569.2	60.0	40.0	–
Bull	1555.6	63.4	–	36.6

| | 1988 | | | |
Firms	Total Sales mill.$	Home %	Rest of Europe %	Rest of the World %
SGS-Thompson	1222.0	28.0	30.0	42.0
Siemens	59681.0	52.0	25.0	23.0
Philips	28365.7	6.5	72.5	21.0
Olivetti	6445.5	37.5	43.5	19.0
Nixdorf	3044.9	54.0	40.0	6.0
STC (ICL)	4193.6	69.0	12.0	19.0
Cap Gemini Sogeti	976.5	46.0	37.0	17.0
Nokia	5212.4	64.0	36.0	–
Bull	5296.7	39.9	36.1	24.0

SOURCES: *Datamation*, 1984 and 1988; Balance sheets, Electronic Business.
NOTES: [a] 1982, [b] 1986

during the 1980s, for example, was impressive: in 1982, 63 percent of the company's sales were concentrated in Italy and France, while non-European outlets absorbed only nineteen percent. In 1990, these shares were respectively 20 percent and 45 percent. A third group of firms, traditionally strong in the domestic market, has only recently begun to move towards a more internationalized structure of sales and activities. Even a large firm like Siemens, with 63 percent of total employment located in West Germany in 1988 and 52 percent of its sales absorbed by the German market in 1988, has only recently begun to move towards a higher degree of internationalization.

The reliance on the domestic market proved a liability for firms such as Nixdorf, STC-ICL, and Nokia during the 1980s. In 1988, (before its acquisition by Siemens) Nixdorf had 54 percent of its sales concentrated in West Germany; in 1988 (before its acquisition by Fujitsu) STC-ICL had 69 percent of its sales concentrated in the United Kingdom; in the same year the Scandinavian Nokia had 64 percent of its sales concentrated in Scandinavia (and had its computer division purchased by Fujitsu). Bull represents an interesting case in point. In 1988, 63 percent of Bull's sales were concentrated in the French market. The creation in 1987 of the international company Bull-Honeywell-NEC (and the subsequent control by Bull in 1988), however, resulted in a decline of the share of the French market for the Groupe Bull to 39 percent (*Datamation*, 1989). The decline is even more evident if the acquisition by Bull of the American Zenith Data System is taken into consideration.

On the other hand, in system applications, given the significance for successful innovation of user-producer interactions and of a full understanding of the characteristics of markets, differences across countries and regions have played a major role in determining the international location of R&D and production activities and the internationalization of networks of alliances, agreements, and acquisitions. These differences concerned socio-cultural aspects, institutions, and regulations, the sectoral composition of demand, and the technological capability and size of users. For example CGSS transformed itself from a French-based company into an international actor through its acquisition policies: in 1988 only 46 percent of CGSS's sales was directed to France (versus 37 percent to the rest of Europe) and only 42 percent of total employment was located in France (versus 37 percent in the rest of Europe).

It must be noted that the globalization of competition in basic components and basic systems, and the need for a local presence in system applications has led American multinationals to augment their degree of "Europeanization." Firms such as IBM increased their presence in Europe and their interaction with local firms and institutions during the late 1980s.

8 The Differentiated Role of Public Policy and Institutions

As mentioned in the introduction, public policy and institutions have played a major role in affecting the organization of innovative activities and the commercialization of new technologies. This role has been quite different for basic components and basic systems, and for system applications. These two quite different cases are examined in the following pages.

9 Basic Components and Basic Systems: From Top-Down to Bottom-Up Mission-Oriented Policies

During the 1980s, most major European countries launched mission-oriented programs in basic components and basic systems. These programs maintained several of the characteristics of traditional European R&D support policies: R&D incentives have been granted to leading domestic firms for the development of specific projects in semiconductors and computers. In semiconductors, for example, approximately one-third of the total development cost of the Megaproject involving Siemens and Philips was supported by the Dutch and German governments. In information technologies, major programs such as the French "Plan d'Action pour la Filiere Electronique" (1982–87), the German "Informationstechnik programme" (1983–88), the Schemes and Programs of the English DTI, and the Swedish "National Microelectronic Program" (1983–88), directed financial support to large established firms for catching-up or for innovation at the technological frontier.

These programs are difficult to evaluate. They have granted unconditional support to often non-competitive national champions. Yet, they have also contributed to the continuation of research and production in some key basic technologies by domestic firms. It should also be noted that as far as defense is concerned, both France and the United Kingdom have moved towards more competitive defense procurement policies. Similarly, countries such as West Germany and France have explicitly shifted toward more innovation-oriented public procurement (Guy and Arnold 1987).

Compared to the mission-oriented programs of the 1970s, however, the more recent programs have shown new relevant characteristics. Public policies in various European countries have paid increasing attention to linkages and interdependencies among technologies and industries. The recognition of links among sectors and technologies at the policy level (quite pronounced in the French case, in which the concept of "filiere" guided the "Plan d'Action pour la Filiere Electronique") reflects the increase in the system features of electronics technologies. It should be noted, however, that even those policies that expressly mentioned interdependencies and sectoral links encountered serious problems at the implementation stage because of the difficulties of coping with complex technological links, feedbacks, and cumulative effects. Moreover, in the French case general support for a better integration among industries within the Plan was weakened at the implementation level by the existence of a large number of governmental agencies that often intervened in an uncoordinated way in implementing specific segments of the plan.

In addition, the policies of various European countries have strongly supported R&D cooperation among domestic firms. West Germany is the most advanced country in this field, with programs and institutions that have been successful at supporting and implementing cooperation: its "Informationstechnik programme," for example, has strongly encouraged cooperation in R&D. Recently, France has also moved in this direction, with the reorientation of the CNRS, the relaunching of ANVAR, and the support of cooperative R&D. Finally, in Italy, the Ministry of Scientific and Technological Research has launched several National Research Plans (1987) to support joint projects between research organizations, universities, and industry.

One European program designed to support cooperative research, the Alvey Program,[7] provides a useful example of the organizational structure of public programs for cooperative R&D. Alvey lacked a strong set of administrative procedures and deadline systems. Rather, coordination and integration was provided by the Alvey Directorate, composed of the Department of Trade and Industry, the Department of Defense, the Science and Engineering Research Council (SERC), industrialists, and academics. This structure worked satisfactorily because it granted a greater degree of flexibility to the program structure than would otherwise have been the case (Georghiou and Guy 1987, Guy and Arnold 1988). Results in terms of new product developments have been meager, however, because the Alvey Program targeted mainly research and not development and did not try to involve users in the research activities.

Policies of support of international cooperation in R&D at the European level have been particularly important. These policies supported

---

[7]The English Alvey program (1983–1990) aimed to support pre-competitive cooperative R&D in areas such as intelligent knowledge base systems, software engineering, and men-machine interface, and to support development as well as precompetitive research in very large scale integration semiconductor devices, in particular process technologies, CAD, and architecture for ASICs. Alvey has been able to create a research environment consisting of firms, universities and research organizations in information technologies, and to reorient part of academic research to industrial needs and applications. The recent reorientation of English government policy of R&D support away from direct funding of industry R&D, toward the funding of research-industry links, has drastically cut funding for the possible continuation of the Alvey program, except within the framework of ESPRIT and EUREKA, or collaborative R&D. The same cut occurred to the Microelectronics Support Scheme and the Software Product Scheme (Sharp 1988).

R&D projects between firms, research organizations, and universities of different European countries, both through programs managed at the EEC level and through European-wide programs. These programs include ESPRIT (launched in 1984) and EUREKA (launched in 1985).

ESPRIT and EUREKA have affected the organization of the innovative process in Europe in several ways. First, in line with their original intention, these programs have increased cooperative relationships and opened new links between European institutions. In addition, EUREKA has contributed significantly to the creation of new European firms, such as ES2 in ASICs. The SGS-Thomson merger was initially favored by a EUREKA project between SGC and Thomson to foster joint R&D in EPROM memories. Finally, vertical links have been created between semiconductor equipment producers, semiconductor firms, and semiconductor users through the JESSI program. JESSI (3804 million ECU program over eight years) has involved 32 European firms and institutions and has been divided into four parts: basic technology, equipment and materials, applications, and basic and long-term research.

It must be emphasized that European mission-oriented programs have switched to a bottom-up (rather than a top-down) approach, which allows for programs to be more sensitive to the needs of firms and more responsive to market conditions and technology dynamics. Again, typical is the case of ESPRIT and EUREKA. ESPRIT had a bottom-up origin, because it was established by the EEC at the request of the twelve major European electronics firms with the aim to support pre-competitive R&D in information technology between firms and institutions from different EEC countries. The program, however, was then organized in a top-down way, with funds divided according to national shares. EUREKA, on the other hand (with nineteen participating countries), was organized in a bottom-up way. It aimed to support cooperation in the development, application and commercialization of new technologies between firms and institutions from different European countries.

10 System Applications: The Emergence of Policies of Technical Standards, Human Capital Formation, Information Provision, and Technology Diffusion

In system applications a set of policies, different from the ones discussed above for basic components and basic systems, was launched in Europe. Instead of directly supporting firms through the financial support given to specific firms for specific innovation projects, European governments provided indirect general support to firms involved in system applications. These policies were more in tune with the knowledge base underpinning innovative activities, the condition of demand and the type of competition of system applications.

Indirect general policies of support were focused on the technological environment, the service infrastructure, and the skill level of system application firms. They pushed for the establishment of open technical standards, for the wide dissemination of technological information, for the establishment of an advanced communication and data exchange infrastructure, and for the supply of skilled human capital. The basic rationale for these policies was that open technical standards would provide a reference framework for would-be innovators, the dissemination of information would supply useful inputs regarding the use and integration of new technologies, the diffusion of new system applications to downstream firms would create major incentives to innovation (by enlarging the size of potential demand), and relevant feedbacks on product performance to upstream producers, and the supply and training of advanced human capital would play a key role in the building-up of advanced competence of system application firms.

In Europe, various policies of setting technical standards in terms of both quality and compatibility have been launched. Quality standards have been linked to the development of metrics in order to assess quality, reliability, and process productivity in software. Great Britain's Ministry of Defense has been particularly active in this respect. Compatibility standards have been the object of EEC programs, such as the European Common Tool Environment program in software (Brady and Quintas 1991). A major choice of UNIX as the standard environment for European software development has been done in ESPRIT. It must be noted that UNIX was also previously adopted by the British Alvey Program. Interestingly enough, 28 projects in ESPRIT regarded standards, either adopted as international standards or elaborated by an international standards working party (ESPRIT, 1988). In this respect, the most important result was a portable common tools environment (PCTE), developed by Bull, GEC, Nixdorf, Olivetti, Siemens, and ICL, which is becoming the basis for an international standard. This environment consisted of basic mechanisms for interprocess communication and interface linking software development tools (*Datamation*, January 1, 1989).

During the 1980s there was substantial uniformity throughout Europe in the launching of public policies in support of technical and scientific education and human capital training; in the development of infrastructure in terms of technical information provision, consultancy, and the transfer of technology; and in the adoption and diffusion of information technologies, particularly with respect to small- and medium-size firms. The rationale for this last intervention has been that small- and medium-size firms may encounter major difficulties in the innovation and adoption process as a consequence of the lack of trained and experienced employees, detailed technical and market information, and financial resources (Braunling 1985, Meyer Krahmer 1985). Among the various programs launched by European countries, the French "Plan d'Action pour la Filiere Electronique" included support for technical and scientific education and training, the encouragement of technology transfer through national projects in areas such as software engineering, image processing, workstations and CAD for VLSI. In addition, France launched various diffusion-oriented policies managed by ANVAR and technical information provision projects supplied by the ARIST network. In the Federal Republic of Germany the "Informationstechnik Programme" and the "Special Program on the Application of Microelectronics" (1982–84) included incentives for information technology awareness and for the use of information technology in education. It also provided support for the development of a technical infrastructure and for information technology diffusion policies. In the United Kingdom, the DTI launched a series of programs such as the Microelectronics Industry Support Scheme and the Microelectronics Application Project. In Sweden, the "National Microelectronic Program" supported education in computer aided circuit design in technical universities and colleges, while the Renewal Funds (1985) provided incentives for workers' training programs within firms. In The Netherlands, the "Information Stimulation Plan" (1984–88) focused on technical education, information infrastructure, and information technology diffusion. Finally, in Denmark the "Technological Development Program" supported the diffusion of information technology.

Many of these programs have been relatively successful in terms of information provision, human capital formation, and technology diffusion. The German "Special Program on the Application of Microelectronics" and the British "Microelectronics Application Project," for example, were quite successful in fostering microelectronics applications in products by small and medium size firms. They were characterized by relatively simple administrative procedures, which made applications and assessment procedures easy to build into the program (OECD 1989).

Some of the factors that led to the emergence of these new indirect types of policies also produced some changes in the features of the major mission-oriented programs for basic components and basic technologies. In the JESSI program, for example, close cooperation at the equipment level between semiconductor manufacturers (providing specifications and technical guidelines) and equipment manufacturers (developing the technical concept and the system design) was aimed at prototype development. At the system application level, close cooperation between producers and users has targeted the development of CAD tools and environments which emphasize too, interaction, standardized reference systems, user-friendliness, configurability, and an openness to new tools for various levels of implementation. The interaction between suppliers, producers, and users has also been made evident by the composition of the JESSI board: three semiconductor producers (Philips, Siemens, SGS-Thomson), three major semiconductor users in different market segments (Olivetti, Alcatel, Bosch), one equipment manufacturer (Electrotech Equipment), and one public research institution (the Dutch STW). Also the new ESPRIT II (launched in 1988) has moved towards a stronger emphasis on industrial applications, focusing on ASICs and information processing systems. Particular emphasis has been placed on technology integration projects, regarding large database applications and high performance computers with engineering and scientific applications. EUREKA, being more expressly focused on the development, application, and commercialization of new technologies, has also generated several results of industrial significance.

11 Diversities in Public Policies and Institutions

Despite these broad similarities across European countries, the main type and the general goals of the policies supporting basic components and basic systems, and system applications, major differences existed in the institutional architecture and in government competence and expertise as far as the implementation of these policies was concerned.

In mission-oriented programs major differences persisted among the European countries as a consequence of the historical, socio-political, and economic differences discussed by Nelson (1984) and Ergas (1986). French and British programs have been heavily influenced by the inclusion of defense projects. As far as civilian programs were concerned, French projects have tended to be more interventionist in spirit than German ones, while the English projects have been characterized by discontinuities and uncertainties (Guy and Arnold 1987, Sharp and Holmes 1989).

With respect to public programs in support of human capital formation, information provision, and technology diffusion, major differences in the role and competence of bridging institutions existed between countries such as West Germany, Netherlands, and Denmark on

TABLE 10.6
Indicators of Microelectronics Diffusion

Extent of Use of Microelectronics: 1983
(Weighted percentage of all manufacturing establishments)

	West Germany	France	United Kingdom
Product Applications	13	6	10
Process Applications	47	35	43
All Applications	51	38	47
No Applications	49	62	53

*Extent of Use of Microelectronics by Employment
Size of Establishment: 1983*
(Percentage of all manufacturing establishments)

	West Germany	France	United Kingdom
Product Applications			
20–99	8	5	7
100–199	18	9	12
200–99	25	11	15
1000 and more	45	33	35
Process Applications			
20–99	39	27	32
100–199	59	47	54
200–999	72	70	71
1000 and more	93	90	94

SOURCE: OECD (1989).

the one hand, and the other European countries on the other. This is because large multisectoral agencies such as the Fraunhofer Gesellschaft in West Germany, the TNO in the Netherlands, and the Technological Service Network in Denmark, with their advanced technical and managerial capabilities accumulated over the years, have played a major role in creating a link between basic research and industry requirements through contract and cooperative research with industry.

The German-Dutch-Danish model has been successful in the development, application, and diffusion of new electronic technologies, especially among small and medium size firms (see for example the aggregate data of Table 10.6). German, Dutch, and Danish agencies focused a large part of their budget on the commercialization of electronics technologies by transforming public knowledge and basic technologies (partly developed internally and partly acquired externally) into specific applications. These agencies diffused technical knowledge about microelectronics and information technologies to industry, particularly to small- and medium-size firms. They maintained a market orientation: particular attention

TABLE 10.7

Taxonomy of Technology Development and Commercialization Arrangements

RESEARCH ORGANIZATIONS

Big Science and University

Bridging Institutions

BASIC

knowledge generation

{
Universities

Technical Schools

Big Science National Networks
Max Planck Gesellschaft (D)
CNRS (F)
CNR (I)
CSIC (E)
Research Council (S)
}

APPLIED

a) *Specialized Governmental Agencies*
(Military, nuclear, health, agriculture environment, geology-mines, telecommunications, etc.)

13 National Research Centres (D)
CEA, CNET, CNES, INRIA,
    INRA, Pasteur (F)
4 DTI labs, UKAEA   (GB)
ENEA, INFN, ISS   (I)
RISO   (DK)
FOA   (S)
JEN, INIA, INSALUD   (E)
ECN, NLR, WL   (NL)

knowledge application/ diffusion

b) *Diffusion Agencies/Centres/Associations*
b1 – National multi-sectoral: e.g., ANVAR (F), TNO (NL), FhG (D), ATV and Technical Institutes (DK), STO (S), CDTI (E)
b2 – National mono-sectoral: e.g., AIF (D), SERC Directorates and Res. Assoc. (GB), 4th Group of Institutes (DK), Microelectronic Centres (NL)
b3 – Local: e.g., Scottish Dev. Ag. (GB), CRITT (F), Grunder und Technologiezentren (D), CSATA (I)

DEVELOPMENT

c) *Transfer units at Universities and Technical Schools*

d) *International Agencies/Centres*
CERN, ESA

e) *Private Research Institutions* :
(multinationals, national)

SOURCE: Onida and Malerba 1989.

was devoted to the specific requirements of firms. A large share of their budget was financed by firms: in the late 1980s an average of 30 percent of the funds of the Fraunhofer Gesellschaft came from industrial contractors (versus 15 percent ten years earlier); whereas 80 percent of the funds of the Danish ATV of the Technological Service Network came from private funding and research contracts; and 60 percent of the total research funds of the TNO came from contract research in industrial technologies (Onida and Malerba 1989).

These agencies were also able to interact with users on a long-term basis by having them participate in research planning. Moreover, their technologically diversified structure and their departmental organization facilitated the successful performance of multidisciplinary research. Finally, the maintenance of flexibility in the organization and management of research allowed them to cope with a continuously changing technological environment. A case in point is provided by the Fraunhofer Gesellschaft, which closed one of its existing 35 institutes and opened a new one every year of the last decade. Similarly, the Danish Technological Service Network frequently opened new institutes and frequently closed existing ones, if they were not necessary anymore.

Other European countries are now moving in this direction, but their newly created agencies do not yet have the accumulated skills and capabilities of the German-Dutch-Danish agencies. France reformed the structure of its CNRS in 1982 to allow for more applied and cooperative research with industry: a new agency (DVAR) was created with the goal of finding industrial applications for the results of basic research, while the CNRS-industry committees within the Comité de Relations Industrielles (CRIN) aimed to develop links with industry. In 1982, ANVAR was also relaunched with the aim of fostering the diffusion of technology as well as performing cooperative research with industry (especially small- and medium-size firms). Similarly, during the 1980s the Italian ENEA greatly increased cooperative R&D and contract research with firms, and supported the diffusion of microelectronics as well as other technologies.

It should be noted that within each European country these major bridging institutions have been inserted into complex networks of national, regional, and local agencies in charge of technology diffusion. They include university-industry liaison offices, science parks, regional development agencies, and innovation centers linked to specific universities (see Table 10.7). In this case as well, it is possible to identify a German-Dutch-Danish model, with the possible inclusion of Switzerland and Sweden, in which the link between Universities, Polytechnics, and

Technical Schools, and industry is extensive in terms of liaison offices, joint research, and the mobility of personnel (Onida and Malerba 1989).

12 Summary and Conclusions

During the 1980s a well-articulated and differentiated organization of innovative activities and public policies emerged in the European semiconductor and computer industries. In basic components and basic systems the increasing costs of R&D and production resulted in an increase in firm size through mergers and acquisitions. A wave of agreements among firms took place driven by cost saving, knowledge sharing, and technology exchange considerations, and centered on those technological areas in which Europe lagged behind. In system application, by contrast, the relevance of software and of flexibility and systemic features in technology, and the presence of variety in demand, generated heterogeneity among firms and led to agreements among firms in which complementarity of knowledge about applications and technology was the driving force.

In addition, the globalization of competition in basic components and basic systems acted as a powerful engine of change for those European firms that were traditionally concentrated in the domestic market, pushing them to increase their degree of internationalization. For European firms active in system applications, internationalization meant also that firms had to pay attention to the specificities and differences in the various national and local markets.

Public policies in favor of basic components and basic systems were different from the ones in favor of system applications. In basic components and basic systems, mission-oriented programs were launched. They were more bottom-up than top-down and increasingly focused on interdependencies among technologies. Public policies for system applications, on the other hand, were centered on standard setting, human capital formation, and technology diffusion, and paid special attention to small- and medium-size firms both as suppliers and as users of system applications.

As a consequence, one of the salient and distinctive features of European public policy (compared to American and Japanese policies) almost completely disappeared: the unconditional public support of at least one national champion per country in strategic industries through top-down mission-oriented programs. On the other hand, European public policies have apparently moved closer to the Japanese model by paying increasing attention to the support of R&D cooperation. Many of the key elements of Japanese research associations have been reproduced in the ESPRIT, EUREKA, and the other European programs. In addition, European governments have increased their role in the coordination of firms' technology efforts and in information provision, a role played very effectively by MITI in Japan.

The process of adaptation to change by European firms has been somewhat successful in basic components and basic systems, products in which European firms had previously experienced major losses. Through restructuring, concentration, and renewed R&D efforts, some European firms have been able to reduce their losses, and even return to profitability. In this respect, the major public programs launched by the EEC and by various European countries have contributed to a greater internationalization of European firms by creating informal European research networks. They have also fostered vertical links between semiconductor equipment suppliers, producers and users, as well as between basic research and applications.

It is in system applications, however, that European firms and public policies seem to have fared relatively better. Several European firms have forcefully implemented a wide range of strategies that proved extremely successful during the 1980s in various local markets or specific vertical applications: a major focus on interacting with users, understanding user requirements and market characteristics, and integrating both internal and external competences (in terms of components and subsystems) into specific applications. They have also fostered the introduction of decentralized firm structures, the implementation of close links between different stages of the innovation process (such as development, marketing, and distribution) and the development of an extensive network of alliances, agreements, and acquisitions. Interestingly enough, some of these strategies were launched by lagging European firms (such as Olivetti) as a catch-up strategy during the early 1980s. But later on, because of developments in technology, demand, and competition, they became the dominant strategies followed by the leaders in the industry as well (such as IBM). European public policies did not intervene directly in firms' R&D policies. Rather, they supported the establishment of technical standards, formation of human capital, the provision of technical information, and the diffusion of new technologies. These were policy measures aimed to affect firms indirectly, by acting on skills and information, leaving the heterogeneous European firms free to introduce autonomously new products and processes.

This does not mean that the well-known differences across European countries disappeared completely. This analysis has shown that diversity is still present in terms of institutions active in public policy, and levels and ranges of capabilities and expertise within each national policy system. It is therefore still possible to speak of a French-British model of institutions linked to mission-oriented policies and of a German-Dutch-Danish model of institutions linked to diffusion policies. Undoubtedly, however, the process of globalization, the launching of major European-wide programs of support, and the development of a single European market will attenuate these differences in the near future.

References

Arnold, E., and K. Guy. 1986. *Parallel Convergences: National Strategies in Information Technologies*. London: Frances Printer.

Brady, T., and P. Quintas. 1991. Computer Software: The IT Constraint, in *Technology and the Future of Europe*, C. Freeman, M. Sharp, and W. Walker (eds.). London: Pinter.

Braunling, G. 1985. *Policy Instruments for Technology Transfer and Their Effects*. Paper presented at the German-Japanese Symposium Kassel, October 4–11, 1985.

Brooks, F. 1972. *The Mythical Man-Month*. Reading, MA: Addison Wesley.

Business Week, Innovation: The Global Race, 1990 Special Issue.

Cainarca, G., M. Colombo, S. Mariotti. 1989. *Accordi tra imprese nel sistema industriale dell'informa zione e della comunicazione*. Milan: Comunità.

Dosi, G. 1984. *Technical Change, Industrial Transformation and Public Policies*. London: Macmillan.

EEC. 1988. *L'espace scientifique et technologique european dans le contexte international. Resources et conditions de la competitivité de la communauté*, Bruxelles.

EEC. 1988. ESPRIT *1987 Annual Report*, Luxembourg.

EEC. 1989. *European Community Policies for Semiconductors*, Brussels.

Ergas, E. 1986. *Does Technology Policy Matter?* Center for European Policy Studies (CEPS) papers, no. 29.

EUREKA. 1988 and 1989. *Progress Report*.

Foray, D., M. Gibbons, and G. Ferné. 1989. *Major R&D Programmes for Information Technologies*, OECD, Paris.

Georghiou, L. and K. Guy. 1987. *Evaluation of the Alvey Program: Interim Report*. London: H.M.S.O.

Guy, K. and E. Arnold. 1987. *Government IT Policies in Competing Countries*, SPRU, University of Sussex.

Hobday, M. 1988. *Trends in the Diffusion of Application Specific Integrated Circuits*, Science Policy Research Unit (SPRU), University of Sussex.

Hobday, M. 1989. Corporate Strategies in the International Semiconductor Industry, *Research Policy*, vol. 18, no. 4 (August).

Lesourne, J., R. Leban, K. Oshima, and T. Takushiti. 1988. *Europe and Japan Facing High Technology. From Conflict to Cooperation?* Paris: Euro-Japan Project on High Technology.

Malerba, F. 1985. *The Semiconductor Business*. Madison: University of Wisconsin Press.

Malerba, F. and S. Torrisi. 1990. Complessità tecnologica, organizzazione industriale e diversità tra imprese: il caso delle grandi imprese informatiche, *Economia e Politica Industriale*, 65.

Malerba, F., S. Torrisi, and N. von Tunzelmann. 1991. Electronic computers, in C. Freeman, et al., *Technology and the Future of Europe*. London: Pinter.

Massow, V. von. 1986. *Organization and Promotion of Science in the Federal Republic of Germany*. Bonn: Inter Nationes.

Meyer Krahmer, F. 1985. Government Promotion of Linkages between Research Institutions and Industry in the Federal Republic of Germany, in *Institutional Linkages in Technological Development*. Proceedings of an International Seminar, San Paolo.

Mowery, D. 1988. *International Collaborative Ventures in U.S. Manufacturing*. Cambridge: Ballinger.

Nelson, R. 1984. *High Technology Policies*. Washington: American Enterprise Institute (AEI).

Nelson, R. 1989. Capitalism as an Engine of Progress, *Research Policy*, vol. 19, no. 3 (June).

OECD. 1989. *The Changing Role of Government Research Laboratories*. Paris.

OECD. 1989. *Government Policies and the Diffusion of Microelectronics*. Paris.

Okimoto, D., T. Sugano, F. Weinstein, eds. 1984. *Competitive Edge: The Semiconductor Industry in the U.S. and Japan*. Stanford, CA: Stanford University Press.

Onida, F. and F. Malerba. 1989. R&D Cooperation between Industry, Universities and Research Organizations in Europe, *Technovation*, vol. 9, nos. 2–3 (June).

Patel, P. and K. Pavitt. 1988. Measuring Europe's Technological Performance: Results and Prospects, CEPS papers, no. 36.

Pistella, F. 1989. Dai progetti al mercato, in *Technology Review*, 11 (July).

Rosenberg, N. 1982. *Inside the Black Box*. Cambridge: Cambridge University Press.

Rosenberg, N. 1987. Science, Technology and Economic Growth, Center for Economic Policy Research (CEPR) papers, no. 101. Stanford University.

Sharp, M. 1988. *R&D Cooperation Among Firms, Universities and Research Organizations in the United Kingdom*, SPRU.

Sharp, M. and P. Holmes, eds. 1989. *Strategies for New Technologies*. New York: Phillip Allan.

Shimada, H. 1989. *Humanware Technology and Industrial Relation*, presented at the OECD International Conference on Science, Technology and Economic Growth, June 1989. Paris.

Sigurdson, J. 1989. Industrial R&D in Europe. Paper presented at the

Conference on Changing Global Patterns of Research and Development, Stockholm.

Steinmueller, W. E. 1986. Industry Structure and Government Policies in the U.S. and Japanese Integrated Circuit Industries, CEPR papers, no. 105.

Steinmueller, W. E. 1988. International Joint Ventures in the Integrated Circuit Industry, CEPR paper, published in D. Mowery, ed., *International Joint-Ventures in U.S. Manufacturing*.

Teece, D. 1989. Technological Development and the Organization of Industry. Paper presented at the International OECD Seminar on Science, Technology, and Economic Growth, Paris.

Vaccá, S. 1989. *Scienza e tecnologia nell'economia dell'impresa*. Milano: F. Angeli.

Part IV

The International Environment and the Commercialization Process

11

Dollar Devaluation, Interest Rate Volatility, and the Duration of Investment in the United States

RONALD MCKINNON AND DAVID ROBINSON

The general report of the M.I.T. Commission on Industrial Productivity (Dertouzos et al. 1989) was based on a detailed study of the strengths and weaknesses of eight major American manufacturing industries against their foreign counterparts in Europe and Japan. Being concerned with the lack of productivity growth in the American economy since 1973, and with the consequential stagnation of real wages (by most measures), the report uncovered a wide variety of potential contributing factors. American formal education and on-the-job training were less rigorous. The use of engineers and other technically competent people in American managerial ranks has thinned out, leading to inadequate product design procedures, greater estrangement of the work force, and so on. The problem is indeed complex and multifaceted. However, the Commission did identify one recurrent theme:

American industry has also been handicapped by shrinking time horizons and a growing preoccupation with short-term profits. There have been many recent instances in which U.S. firms have lost market share to overseas competitors despite an early lead in technology or sales, or both. Often these firms cede a potential market by not "sticking to their knitting;" instead they diversify into activities that are more profitable in the short run. (Berger et al. 1989, p. 43)

In attempting to explain this greater American "impatience," the Commission did note some aspects of comparative financial structures. The cost of capital, in the sense of some pretax real rate of return that companies require on new investment projects, could be substantially higher in the United States than in Japan. The Commission also worried about the concentration of American shareholding in mutual and pension funds, whose managers themselves have a very short-run orientation toward growing profitability in each current accounting period. To avoid undervaluation of their firms, making them vulnerable to a takeover, corporate executives then respond by trying to keep profits high in the short run at the expense of profitability in the more distant future.

But the impatience problem has another dimension. After the breakdown of the Bretton Woods system of fixed exchange rates in 1973, the American capital market was transformed from one where interest rates were generally low and stable in the 1950s and 1960s, compared to Japan and Germany, to one where they became higher and more volatile in the 1980s. From 1973 to 1988, American price inflation was about 50 percent greater than Japan's or Germany's, with a comparable devaluation of the foreign exchange value of the dollar in terms of yen and marks. Because of higher anticipated inflation, whether or not American real interest rates were higher is uncertain.

But the higher level of nominal interest rates in the United States could shorten the time horizon of American corporate decision making. For a bond of given maturity, its duration, a weighted average of the times in the future when interest and principal payments are to be received, shrinks with the level of nominal interest rates. For example, the duration of a twenty-year corporate bond with a high coupon rate of fifteen percent is a little over seven years, whereas that with a coupon rate of five percent is over thirteen years. For the high coupon bond, the cost of debt service comes relatively quickly. Will the shorter effective duration of American versus Japanese bonds (of the same maturity) also shorten the required payback time on investment projects undertaken by American in comparison to Japanese companies?

The answer depends on what financial risks accompany the higher nominal interest rates and how they interact with the corporate income tax. We identify two distinct financial effects that could well lead to greater corporate impatience as nominal interest rates increase and become more volatile. The first is the inducement to greater leverage arising from the asymmetric treatment of debt and equity financing by the tax system. The effective tax rate on equity finance rises rather sharply with inflation while the effective tax rate on debt may, if anything, decline as inflation increases. This encourages increased reliance on debt financing raising the overall level of bankruptcy risk in the economy. The second is the overborrowing syndrome, where firms find that, to finance a given investment project, they must initially borrow more than the initial outlays for purchasing the physical plant and equipment. Both effects shorten the optimal time horizon, that is, the optimal duration, of investment projects; these key microeconomic issues are taken up in Section 2 of the paper.

In Section 1, however, we explore macroeconomic data and theories on how exchange rate regimes and interest rate behavior are related. For the major industrial economies, we compare how interest rates behaved when exchange rates were fixed (the classical gold standard, 1879–1913), to when they were virtually fixed (Bretton Woods, 1950–70), to when they floated with no official parities after 1973. Then we explore why, relative to those in Japan and Germany, American price inflation and nominal interest rates became higher in the 1970s and 1980s in comparison to the 1950s and 1960s. The answer is important in understanding what might be done to reduce the current level and volatility in U.S. interest rates, and in assessing whether or not continually devaluing the dollar really does increase the international competitiveness of American industry.

1 Interest Rates Under Alternative Exchange Rate Regimes

After the Bretton Woods System of pegged exchange rates broke down during 1971–73, the surprising violence and essential unpredictability (randomness) of exchange-rate movements is now well established. To better understand the implications for the efficiency of the international capital market, the resulting exchange rate risk has two important aspects.

1.1   Two Components of Exchange Rate Risk

The first is real exchange rate risk: the extent to which the nominal exchange rate differs from purchasing power parity. PPP is defined as that exchange rate which would align national price levels of a very broad basket of internationally tradeable goods. During past periods when a common monetary standard with fixed exchange rates was operative—under the classical gold standard or under the fixed rates dollar standard of the 1950s and 1960s—broad wholesale (producer) price indices remained remarkably well aligned internationally (McKinnon and Ohno 1989).

Under no-par floating, in contrast, short-term volatility in real exchange rates is high on a month-to-month or quarter-to-quarter basis: nominal exchange rates fluctuate while national price levels change relatively little (Frenkel and Mussa 1981). But the more striking characteristic of the modern experience is the extent to which national price levels

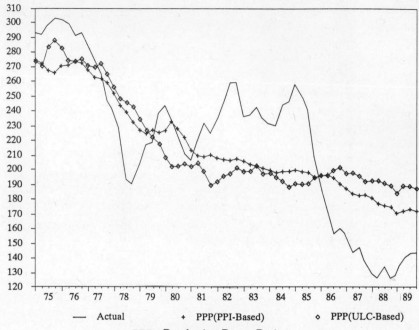

— Actual   + PPP(PPI-Based)   ◇ PPP(ULC-Based)

PPP: Purchasing Power Parity
PPI: Producer Price Index
ULC: Unit Labor Cost

SOURCE: Kinichi Ohno, "Estimating Yen/Dollar and Mark/Dollar
Purchasing Power Parities," *IMF Staff Papers* (forthcoming), 1990.

FIGURE 11.1   Actual and PPP Yen/Dollar Rates

have been misaligned for prolonged periods. For 1975 to 1989, Figures
11.1 and 11.2 (taken from Ohno 1990) compare movements in the actual yen/dollar and mark/dollar exchange rates to the path(s) of their
estimated purchasing power parities (PPPs)—those rates which would
have aligned either producer price indices or unit labor costs between
the United States and Japan or Germany.[1]

Deviations from PPP of twenty percent in either direction have been
sustained, sometimes for years, with individual spikes ranging up to 40
percent. In the now worldwide markets for goods and many services,

---

[1] How best to estimate (absolute) PPP exchange rates remains contentious, but alternative statistical techniques lead to estimates similar to those appearing in Figures
11.1 and 11.2 (McKinnon and Ohno 1989, Ohno 1990). More basically, substituting
long-term moving averages of the yen/dollar and mark/dollar rates for the PPP rates
in Figures 11.1 and 11.2 would show similarly large deviations of current exchange
rates from these long-term trends.

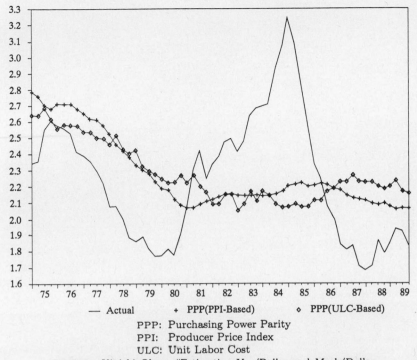

SOURCE: Kinichi Ohno, "Estimating Yen/Dollar and Mark/Dollar
Purchasing Power Parities," *IMF Staff Papers* (forthcoming), 1990.

FIGURE 11.2 Actual and PPP Yen/Dollar Rates

these are tremendous swings in the international competitiveness of any
country's merchants and manufacturers. Manufacturers whose country's
real exchange rate suddenly appreciates can experience a sudden decline
in profitability, as did those in Japan from the sudden appreciation of
the yen in 1986–87, and did those in the United States in 1981–85 with
the appreciation of the dollar.

But can't such real exchange rate risk be effectively hedged if forward markets for foreign exchange are fairly complete? Unfortunately,
no. Even if international financial markets permit industrialists to freely
adjust their net foreign exchange positions to virtually any term to maturity, forward commodity markets for brand-name manufactured goods
remain seriously incomplete. Thus, industrialists planning long-term investments in plant and equipment in various parts of the world cannot
fully hedge real-exchange-rate risk (McKinnon 1979 and 1988). Without
forward commodity contracts in hand, industrialists find they must live

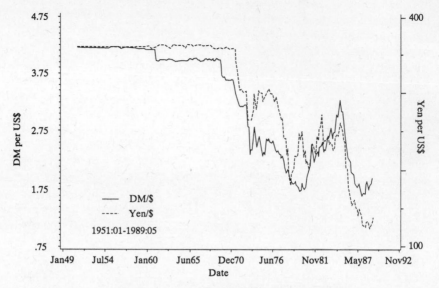

FIGURE 11.3 Nominal Exchange Rates

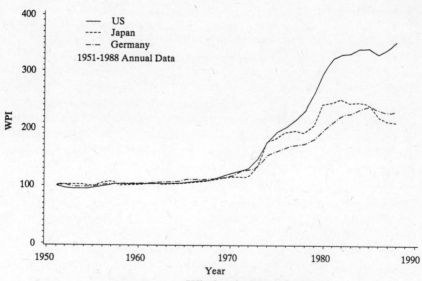

FIGURE 11.4 Wholesale Price Indices

with substantial residual exchange risk even when their forward positions in foreign exchange are optimized ex ante.

The second, and less well recognized, component is what I shall call relative price level risk. Even if we suppose that exchange rates equaled their PPP values at all points in time so that real exchange rate risk was absent, the choice of a currency with which to finance a long-term investment is still non-trivial if there is no common international monetary standard. Country A's price level could still evolve quite differently, and unpredictably, from that in country B. From 1973 to 1988, Figures 11.3 and 11.4 show that price inflation in the United States was more than 50 percent higher than that in Germany and Japan while the dollar was depreciating—albeit very erratically—by a similar order of magnitude. Whereas, in the earlier period of fixed exchange rates from 1951 to 1969, Figures 11.3 and 11.4 also show that inflation in all three countries' wholesale price indices was virtually the same—less than one percent per year.

Relative price level risk hampers business firms trying to decide whether to issue long-term bonds denominated in dollars, marks, yen, or in some other currency. A company that issued dollar bonds in the 1980s, which then bore a substantially higher nominal interest rate than yen bonds, could be saddled with a higher real debt burden if American price inflation relative to that in Japan now turns out to be substantially lower—with correspondingly lower exchange depreciation of the dollar—than predicted by the long-term interest differentials prevailing in the 1980s. Conversely, a company fortunate enough to issue yen bonds with low nominal yields could now find itself with a substantially lower real debt burden. Nor can this uncertainty about the relative evolution of national price levels be hedged in forward markets for foreign exchange.

In summary, because neither real exchange rate risk nor relative price level risk can be fully hedged, floating exchange rates introduced new net risk into world trading relationships and into the international capital market after 1973. Thus investment was inhibited while becoming shorter term and more defensive. Increased protectionism was one unfortunate consequence. Governments often respond with protectionist measures to reduce some of the risks seen by domestic entrepreneurs. Quotas, voluntary export restraints (VERs), and other cartel-like market-sharing arrangements—such as the semiconductor agreement between the United States and Japan—seal off domestic markets from exchange rate fluctuations. (Even more vulnerable is each country's agricultural sector where pass through effects of exchange rate changes are more immediate, and thus lead to more intense protectionist pressure.) Hence,

TABLE 11.1
Bilateral Fluctuation Bands for Exchange Rates

Currency		Parity	Lower Limit	Upper Limit
		Gold Standard (1879–1913)		
Sterling	($/pound)	4.866	4.827	4.890
Franc	(FF/$)	5.183	5.148	5.215
Mark	(DM/$)	4.198	4.168	4.218
		Bretton Woods Dollar Standard (1950–1970)		
Sterling	($/pound)	2.8	2.772	2.828
Franc	(FF/$)	4.937	4.887	4.986
Mark	(DM/$)	4.2	4.158	4.242
Yen	(Yen/$)	360.0	356.4	363.6
		European Monetary System (1979–1990)		
Franc	(FF/DM)	2.310	2.258	2.362

SOURCE: A. Giovannini, "How Do Fixed Exchange Rate Systems Work: The Evidence from the Gold Standard, Bretton Woods and the EMS" in M. Miller, B. Eichengreen, and Richard Portes, *Blueprints for International Monetary Reform*, Academic Press, 1989.

increased exchange risk per se could help explain the well-documented slowdown in productivity growth in all the major industrial economies of Europe, North America, and Japan after 1973.[2]

1.2 Interest-Rate Volatility in Historical Perspective

But this exchange risk also affects investment indirectly through its impact on long-term interest rates. Relative to eras when a common worldwide monetary standard was operative, long-term interest rates in the current regime of no-par floating have been remarkably volatile. During the classical gold standard from 1879 to 1913,[3] nominal exchange rates were confined within their gold points—a band width of less than one percent (Table 11.1), which was expected to last into

---

[2]Whether the worldwide slowdown after 1973 could be attributed to the change in monetary regimes, i.e., to the breakdown of the fixed rate dollar standard, is an interesting question that is seldom asked by people who study the sources of productivity growth. For example, in their careful assessment of the slowdown in "Potential Output in Major Industrial Countries," Adams, Fendon, and Larson (1987) do not mention increased exchange risk in their catalog of possible explanations. However intriguing, analyzing the global slowdown in productivity growth is beyond the scope of this paper.

[3]I define the period 1879–1913 to be that of the "classical" gold standard because (1) it was a truly international standard encompassing all the major industrializing countries, and (2) the "rules of the game" for maintaining the fixed exchange rate system more or less indefinitely were well established. (See McKinnon 1989a for a further discussion of what these rules were.) After being interrupted by World War I, the gold standard was never re-established in this classical format.

the distant future. Figure 11.5 shows that long-term interest rates remained low, close together, and did not change much through time. With capital even freer to move internationally in the late nineteenth century than now, yields on prime grade U.S. railway bonds remained in the neighborhood of four percent, with those on British consuls about three percent or slightly less; French rentes (representative of yields on other European governments' securities) were sandwiched in between.

In sharp visual contrast to Figure 11.5 on the classical gold standard, Figure 11.6 portrays the modern experience with long-term interest rates on the government bonds of Japan, the United States, Germany, and the United Kingdom. From 1973 to 1988, interest rates levels were much higher, often over twelve percent, volatility was much greater (best seen by the monthly plots in the lower panel), and the dispersion of interest rates across countries was high: a differential of five to six percentage points between the United States and Britain on the one hand, and Germany and Japan on the other, was not unusual. However, during the fixed-rate dollar standard of the 1960s (which could be projected back into the 1950s), Figure 11.6 also shows that long-term interest rates were somewhat closer together and less volatile—at least until 1969–71, when the pegged rate system came under pressure and finally broke down.

Using historical U.S. data on average month-to-month and year-to-year changes, Table 11.2 more precisely compares interest-rate volatility under the classical gold standard, the Bretton Woods dollar standard, and the present period of no-par floating. Measured on either a monthly or annual basis, the mean percentage point change in long-term interest rates on U.S. bonds was about ten times as high under floating as under the classical gold standard, and about four times as high under floating as under Bretton Woods.[4] (For short-term interest rates, Table 11.2 also shows that differences in volatility across exchange-rate regimes were considerably less pronounced: lowest under Bretton Woods, considerably higher under the gold standard, and highest during the recent float.)

---

[4]Using more formal statistical testing procedures, Robert Teh (1990) shows that U.S. interest rate volatility increased sharply after 1973 in comparison to the period from 1956 to 1973. The hypothesis that interest rate volatility remained the same is conclusively rejected. Teh goes on to show that volatility in long rates after 1973 increased by even more than was "warranted" by the increased volatility in short rates—remembering that long rates are strongly influenced by the average of short rates expected in the future.

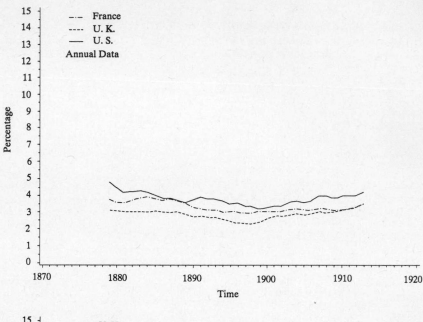

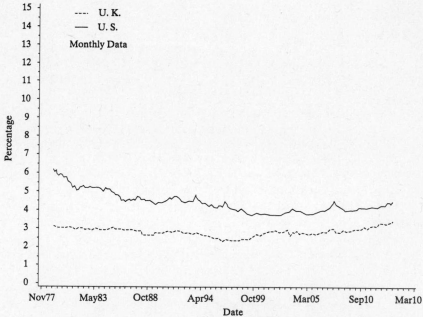

FIGURE 11.5 Long-Term Interest Rates: Gold Standard

TABLE 11.2
U.S. Interest Rate Volatility
(summary statistics of historical comparisons)

Interest	Period	Annual Data Mean Absolute Percentage Change	Standard Deviation
Short-term	Gold Standard	0.90	0.91
	Bretton Woods	0.69	0.77
	Floating	1.72	2.03
Long-term	Gold Standard	0.12	0.11
	Bretton Woods	0.29	0.30
	Floating	1.25	1.51

NOTE: The volatility measures are based on annual change in percentage points. Gold standard 1879–1913. Bretton Woods 1950–1970. Floating 1973–1988.

Interest	Period	Monthly Data Mean Absolute Percentage Change	Standard Deviation
Short-term	Gold Standard	0.40	0.57
TB	Bretton Woods	0.16	0.24
	Floating	0.47	0.77
FFR	Bretton Woods	0.18	0.26
	Floating	0.53	0.95
Long-term	Gold Standard	0.03	0.04
	Bretton Woods	0.07	0.11
	Floating	0.27	0.37

NOTE: The volatility measures are based on monthly changes in percentage. TB: Treasury Bill. FFR: Federal Funds Rate.

SOURCE: See Data Appendix.

The unusually high variance in nominal (and real) exchange rates since 1973 is now well recognized. In historical perspective, however, the modern period's extraordinary volatility in long-term interest rates— and possible implications for stock prices—is sometimes forgotten.

Could the success of the late nineteenth-century system in keeping long-term interest rates low and stable be because the common price level was well anchored? In fact, the common price level, as measured by national wholesale price indices (WPIs), was highly cyclical—as a glance at Figure 11.7 quickly confirms. Indeed, Richard Cooper (1982) calculated that the standard deviation of annual changes in the American WPI from 1879 to 1913 was 5.4, which Table 11.3 shows is high in comparison to a standard deviation of just 1.7 from 1950 to 1969 and 3.8 from 1973 to 1988.

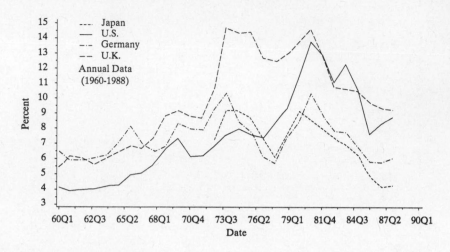

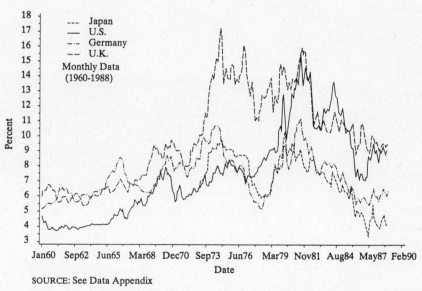

SOURCE: See Data Appendix

FIGURE 11.6 International Long-Term Government Bond Yields

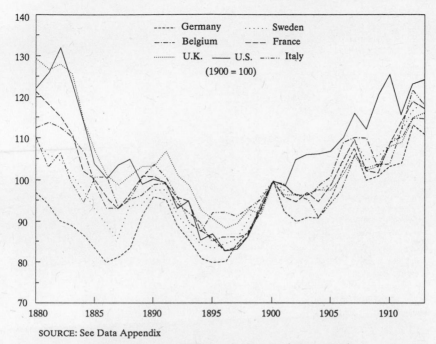

SOURCE: See Data Appendix

FIGURE 11.7 Wholesale Prices (1900=100)

However, Table 11.3 also shows that U.S. price level exhibited virtually no trend over the whole 1879–1913 period: increasing at 0.1 percent
per year, which is indistinguishable from measurement error. Superim-
posed on the classical gold standard's sharp cyclical fluctuations, however, wholesale prices drifted down from the 1870s to 1896 by about
two percent a year; and upward thereafter to 1913 by about the same
amount, as shown in Figure 11.6. Whether or not any of these price-
level movements should have been expected and thus registered in interest rates has been a matter of long-standing curiosity: the so-called
"Gibson paradox" (Fisher 1930, Barsky 1987). But they were not so
registered, as comparing Figure 11.5 to Figure 11.7 makes clear. Instead, the nature of the exchange rate regime, rather than the classical
gold standard's success in anchoring the price level(s), could be more important in minimizing volatility in long-term interest rates (McKinnon
1989). First, the classical gold standard was fully international, encompassing all the financially important countries. No alternative liquid
financial asset—with significantly different monetary characteristics—
existed into which international investors could conveniently and continually shift. Bonds denominated in sterling, dollars, francs, and so on,

TABLE 11.3

Price Level Volatility: United States, West Germany, and Japan

Period		Annual Change	Standard Deviation
United States			
Bretton Woods	1951–1969	0.9	1.7
Floating	1973–1988	5.6	3.8
West Germany			
Bretton Woods	1951–1969	0.7	1.5
Floating	1973–1988	3.6	3.0
Japan			
Bretton Woods	1951–1969	0.5	2.4
Floating	1973–1988	2.3	6.2

SOURCE: IMF, *International Financial Statistics*.

NOTE: Annual change is calculated as the time coefficient of the log of the variable on a time trend. Standard deviation is the standard error of the estimate of the fitted equation $lnP_t = a + blnP_{t-1}$ where P_t is the Wholesale Price Index in year t.

Cooper (1982) performs similar calculations for the United States during the Gold Standard period of 1879–1913. He derives an annual average change of 0.1 and a standard deviation of 5.4.

bore the same risk of general price-level inflation or deflation. Unless some obviously persistent general price inflation seemed in the offing, arbitrage between bonds (with worldwide markets) and relatively illiquid nonmonetary assets, such as real estate or inventories of commodities, was too inconvenient to appeal to international portfolio managers.

Second, nations made firm long-term commitments to return to their traditional gold parities even if gold convertibility had to be temporarily suspended because of some panic or run. Thus investors had long-term confidence in, say, what the dollar/sterling exchange rate was expected to be twenty years hence, although occasionally they might face some near-term uncertainty as to whether or not the exchange rate would be contained within the gold points next week or next month. Thus did the classical gold standard avoid the continual churning of asset portfolios that we now associate with free international capital mobility. Because of the absence of long-term exchange risk across otherwise economically and politically diverse countries, long-term interest rates remained low and uniform.

In contrast, under floating exchange rates with no official parity commitments, news of how the exchange rate might evolve in the future is immediately telescoped back into portfolio choices in the present; thus both the spot exchange rate (Frenkel and Mussa 1980) and long-term interest rates (McKinnon 1989) become more volatile. Insofar as the market judges that some currencies are likely to depreciate against others, and that these weak currencies also suffer from more general uncertainty about what their governments' long-run monetary intentions are, rates of interest on bonds denominated in weak currencies will be higher, as will volatility in the interest rates of all nations with open financial markets.

1.3   The U.S. Dollar in the Postwar: From Strength to Weakness

The postwar financial evolution of the United States vis-à-vis Japan nicely illustrates how this weak-currency, high-interest-rate syndrome can develop. In the 1950s and 1960s, the United States was the anchor of the world dollar standard, with relatively low interest rates and a stable price level. From 1950 to 1969, the yen was fixed at 360 to the dollar, and the mark at 3.92 to the dollar, save for a five percent mark appreciation in 1961. Thus the three countries' price levels for tradeable goods, as approximated by their WPIs, remained remarkably well aligned (Figure 11.4), and increased less than one percent per year from 1951–69 with relatively little variability around these trends (Table 11.3). Indeed, the Bretton Woods period of fixed exchange rates had a much better overall record of price-level stability than did the classical gold standard.

Then, over 1971–73, the commitment to the common monetary standard based on fixed exchange rates broke down. Although the subsequent course of the yen/dollar and mark/dollar exchange rates has been very erratic (Figure 11.3), the dollar generally declined against important "hard-money" trading partners to as low as 123 yen and 1.7 marks in late 1988, before increasing to about 145 yen and 1.97 marks by the third quarter of 1989. (However, the dollar still remained slightly below reasonable estimates of purchasing power parity—based on aligning wholesale or producer price indices—of about 170 yen and 2.05 marks for 1989 (McKinnon and Ohno 1990).)

Since 1973, this exchange depreciation was accompanied by substantially higher price inflation in American tradeable goods. From 1973 through 1988, the U.S. WPI increased 5.6 percent per year, the German WPI by 3.6 percent, and the Japanese WPI by only 2.3 percent (Table 11.3). Because of this higher price inflation in the United States, continual dollar devaluations have not sustained equivalent increases in America's international competitiveness. Offsetting domestic price inflation tends to restore PPP in the long run. (See Table 11.4.)

However, an important analytical question remains: did some natural tendency toward higher inflation in the United States cause the dollar to decline; or, did some policy-induced push to devalue the dollar increase

TABLE 11.4
Inflation Rates: United States, Germany, Japan 1951–1988

Year	United States	Germany	Japan
Fixed Par Values for Dollar Exchange Rates			
1951	11.3	18.7	21.1
1952	−2.7	2.4	2.0
1953	−4.4	−2.6	0.7
1954	0.2	−1.6	−0.7
1955	0.2	1.8	−1.8
1956	3.4	4.4	4.4
1957	2.7	4.7	3.0
1958	1.5	−0.3	−6.5
1959	0.2	4.4	1.1
1960	0.2	1.1	1.1
1961	−0.4	1.5	−1.6
1962	0.2	1.2	0.4
1963	−0.4	0.5	1.6
1964	0.2	1.2	0.4
1965	2.0	2.4	0.7
1966	3.4	1.9	2.4
1967	0.2	−1.0	1.7
1968	2.4	0.7	1.0
1969	3.9	1.7	2.0
1970	3.6	5.1	3.6
Breakdown of Par Value Obligations			
1971	3.3	4.0	−0.8
1972	4.4	2.7	0.8
1973	13.1	6.8	15.8
No-Par Floating			
1974	18.8	13.2	31.1
1975	9.3	4.7	3.0
1976	4.6	3.7	5.1
1977	6.1	2.9	1.9
1978	7.8	1.1	−2.6
1979	12.5	7.5	17.8
1980	14.0	7.5	17.8
1981	9.1	7.8	1.4
1982	2.0	5.9	1.8
1983	1.3	1.5	−2.2
1984	2.1	2.8	−0.2
1985	−0.5	2.5	−1.2
1986	−2.9	−2.5	−9.2
1987	2.6	−2.5	−3.6
1988	1.0	1.3	−1.0

SOURCE: IMF, *International Financial Statistics.*

the inflationary pressure in the United States above that of Germany and Japan? Clearly, the positive correlation between dollar devaluation and higher American price inflation so evident in Figure 11.7 doesn't, itself, indicate the direction of causation.

How the exchange rate, national monetary policies, and price levels were (are) simultaneously determined is a complicated issue. Even with the help of a large-scale econometric model encompassing the major industrial countries, the direction of causation may prove virtually impossible to establish. Instead, by taking the more stable-price policies in hard-currency Europe and Japan as given,[5] let us focus more narrowly on American monetary and foreign exchange policies since the breakdown of fixed par values for exchange rates after 1970. Consider two competing hypotheses.

Hypothesis I: Within the American economy, indigenous increases in aggregate demand generated inflationary pressure with rising domestic prices and exchange depreciation.

Treating monetary policy, M^{US}, as the principal control variable, Hypothesis I can be represented schematically as

(I)

$$M^{US} \quad \begin{matrix} P^{US} \\ \\ \\ Y^{US} \end{matrix} \quad E^{US}$$

The solid lines trace out the dominant path of causation as if the American financial markets were somewhat "insular" (McKinnon 1981). Expansionary monetary policy raises the demand for domestic output (Y^{US}) that induces higher price inflation (P^{US}) and eventually forces the foreign exchange value of the dollar (E^{US}) to decline as net imports increase—at least incipiently.

Still under Hypothesis I, the dashed lines represent another route by which higher price inflation and dollar devaluation might develop. Instead of being insular, suppose that American financial markets are fully open to those in Europe and Japan. Then, by lowering short-term interest rates, domestic monetary expansion induces (incipient) capital outflows that quickly depreciate the exchange rate, which then stimulates foreign demand for American goods that feeds back into American

---

[5]This simplifying assumption is satisfactory for considering long-run trends in national monetary policies. Throughout the 1970s and 1980s, however, the Bundesbank and the Bank of Japan have responded quite strongly to cyclical fluctuations in the mark/dollar and yen/dollar exchange rates (McKinnon 1984).

prices and output. For either route, however, the initiating force is domestic monetary expansion—perhaps in the (false) pursuit of lower domestic unemployment and higher real output.

But there is an alternative causal hypothesis based on exchange-rate-led inflation. Dollar devaluation—or talking down the dollar—has often been a policy objective in its own right. Beginning in the 1960s, several American economists began to argue that the dollar might have to be devalued because of accounting deficits in international payments (Solomon 1976). By August 1971, the overwhelming majority of economists supported President Nixon's successful efforts[6] to force a devaluation as a means of reducing America's emerging trade deficit. The push continued through the early Carter administration, when Secretary of the Treasury Michael Blumenthal stated in June 1977 that the dollar was overvalued—although it was not misaligned by the purchasing power parity criterion (McKinnon 1984). But this official pronouncement helped provoke another run against the dollar in 1977–78 (Figure 11.3).

Even during a period of dollar strength from the largely accidental (and unanticipated) dollar overvaluation of 1981–85, private market participants expected, albeit incorrectly before March 1985, the dollar to fall in order to restore America's international competitiveness (Frankel and Froot 1987). Worried about the growing American trade deficit, in 1987 when the dollar was already well below purchasing power parity, Treasury Secretary Baker warned foreign central banks not to prop the dollar up, thus provoking an exchange crisis where inflows of private foreign capital dried up (McKinnon 1989b). In 1988, central banks intervened quite heavily to keep the dollar within a range of about 125 to 135 yen, and the foreign exchanges were relatively tranquil. Nevertheless, with the persistent American trade deficit in mind, influential economists such as Martin Feldstein (1989) continue to advocate that the dollar should be devalued. (In 1989 and early 1990, a significant respite in this devaluation syndrome seems to have occurred. Worried about American inflation, the Fed tightened up money growth sufficiently to drive the dollar up in the foreign exchanges closer to its purchasing power parity.)

Suppose we summarize this alternative explanation of higher U.S. price inflation as

Hypothesis II: By continually targeting a lower foreign exchange

---

[6]President Nixon ended the Bretton Woods system of "virtually" fixed exchange parities by closing the gold window and imposing an import surcharge until the other industrial countries orchestrated a joint appreciation of their currencies against the dollar.

value for the dollar, the American government and its professional economic advisors (inadvertently) induced the U.S. Federal Reserve System to follow a policy of easier money and higher price inflation.

The key idea underlying Hypothesis II is that current or prospective domestic monetary policy dominates how a floating exchange rate is determined against stable-valued foreign monies. Starting from a position of asset equilibrium in the foreign exchanges, suppose the American government becomes dissatisfied and wishes to induce devaluation because it is, say, worried about the trade deficit. Either the central bank can increase the current money supply, reducing short-term interest rates and provoking a capital outflow; or the government can try to "talk down" the exchange rate without changing the current money supply. For the latter to be successful, foreign exchange traders must be convinced that money will eventually become easier and that the future exchange rate will be lower. Then, within the asset market approach to the exchange rate, the dollar's lower value expected in the future is immediately telescoped back to the present: the dollar falls right away. A more formal model of how current exchange-rate devaluations lead or induce (expected) future monetary expansions can be found in McKinnon 1990, but the exchange-rate led process underlying Hypothesis II can be summarized in the following schematic diagram:

(II)

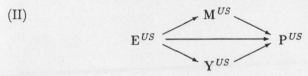

For our financially open economy, the solid arrows emanating from E^{US} now show how targeting the exchange rate influences the money supply and prices. The effect of dollar devaluation on increasing American prices in the future is strong because of (1) direct arbitrage in international commodity markets (the middle arrow), and (2) inducing the Fed to be more expansionary in order to sustain the devaluation (the upper arrow), and (3) through buoying real output (the lower arrow).

Putting the matter a bit more weakly, one could say that having a low target for the dollar exchange rate (below PPP) inhibits the Fed from tightening monetary policy once the inevitable price inflation begins to appear. For example, in 1989-90, the Fed responded to inflationary pressure—the lagged effect of the earlier period of dollar undervaluation cum easy money—by tightening monetary policy. Then, as the tighter monetary policy induced a higher foreign exchange value for the dollar (but one still below purchasing power parity), the Fed was criticized for aggravating the trade deficit.

Parenthetically, the traditional argument that devaluation will correct a trade deficit is not generally valid (McKinnon 1981 and 1990). Suppose that the economy is open so that arbitrage in goods and financial markets with the outside world is unrestricted. Then, as we have seen, a devaluation can only be effected by easing monetary policy, either current or prospective, thereby increasing inflationary expectations. Because it signals that the domestic price level will be higher in the future, a devaluation may well induce current domestic absorption to rise relative to current output: the "perverse absorption effect" (McKinnon 1990). Although domestic goods become cheaper on world markets, total domestic spending for both investment and consumption also increases. Thus, the monetary value of the trade balance is as likely to worsen as to improve from a policy of devaluation cum easy money.

In a highly open economy, only policy that operates directly on the savings-investment balance can reduce a current account deficit. In the 1980s, the American fiscal deficit created a saving deficiency in the American economy, resulting in the need to borrow abroad as reflected in the current account deficit. Exchange-rate changes can neither substitute for nor complement the obviously needed fiscal correction for improving the monetary value of the trade balance, which dominates the current account.

However, participants in the financial markets themselves might still believe in the traditional elasticities model of the balance of trade (Robinson 1937): a devaluation is likely to improve the trade balance.[7] Alternatively, financiers may anticipate that government officials will be influenced by economists, the majority of whom still adhere to the elasticities model. Either way, as long as the American fiscal and trade deficits remain substantial, the threat of further dollar devaluation is aggravated.

Because of this continual risk of further devaluation, international investors in the 1980s demanded much higher yields on dollar assets than on those denominated in yen or marks. This dramatic transformation of

---

[7]In an insular economy with capital controls and limited arbitrage in international goods markets, exchange-rate determination can be separated from present and prospective monetary policy. While keeping the domestic money supply constant, the government could undertake a controlled devaluation through (sterilized) intervention in the foreign exchanges. As output (and exports) are stimulated, domestic absorption would then be constrained by increased interest rates—unlike in the case of a financially open economy (McKinnon 1990).

the United States from a strong-currency country in the 1960s to a weak-currency country in the 1980s is most apparent in Figures 11.8 and 11.9 and Figures 11.10 and 11.11, where nominal interest rates facing American firms are plotted against those prevailing in Japan and Germany respectively. For the Bretton Woods period, Figures 11.8 and 11.9 show that the official prime rate for short-term lending by Japanese banks, and the (controlled) yield on long-term industrial bonds, were between six and eight percent.[8]

In contrast, throughout the 1950s (not shown) into the mid-1960s, American prime loan counterparts were lower and more stable: in the range of four to five percent until the commitment to fixed parities began to fray with the threatened dollar devaluation towards the end of the 1960s.

By the 1980s, however, a major reversal had occurred. Figures 11.8 and 11.9 show that U.S. nominal interest rates had become much higher than Japanese—although the differential was variable. (Figures 11.10 and 11.11 show the same reversal between U.S. and German interest rates.) At the beginning of 1989 with the dollar having fallen to about 130 yen, the Japanese prime lending rate was 3.5 percent in comparison to 11.5 percent in the United States; high-grade industrial bonds were over 9.5 percent in the United States and less than five percent in Japan. By the third quarter of 1989, however, U.S. interest rates came down a point or more, while those abroad rose as the dollar strengthened in the foreign exchanges. Apparently, the Fed's very recent policy of tighter money and a stronger dollar has successfully reduced expectations of future American price inflation and devaluation.

Of course, higher American interest rates in the 1980s were consistent with either Hypothesis I or Hypothesis II. Once higher price inflation in the United States became fully anticipated, international investors boosted the nominal interest rates they required to hold dollar assets vis-à-vis those on assets denominated in yen or marks. However, which hypothesis one accepts is important in determining how best to stabilize the American price level while bringing interest rates down.

If one accepts Hypothesis II, that higher American inflation and higher nominal interest rates were the inadvertent result of devaluing the dollar, then an explicit policy change toward putting a floor

---

[8]Because both the prime loan rate and the rate on industrial bonds were more or less controlled by the Japanese government, there is good reason to believe that the actual cost of capital to Japanese industry was substantially higher (Horiuchi 1984). For example, a bank often required its loan customers to maintain a noninterest-bearing compensating balance with it.

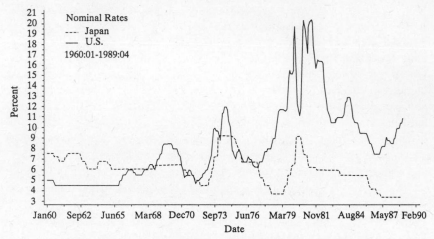

FIGURE 11.8 Prime Rates: United States and Japan

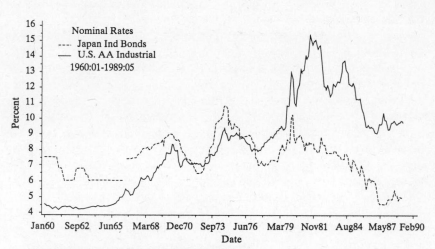

Prior to October 1966 the prime rate series is used
as a proxy for the Japanese industrial bond rate.

FIGURE 11.9 Corporate Bond Rates: United States and Japan

under the dollar in terms of yen and marks is the natural linchpin for stabilizing the American price level. If the American government were to disavow any further resorting to dollar devaluation (and disavow protectionism[9]) as a presumed corrective for America's current account deficit, such a policy could be highly credible in keeping American nominal interest rates close to those prevailing in hard-currency Europe or Japan. True, such a pact would require a major conceptual change in the professed views of government officials and economists who see dollar devaluation as an instrument for reducing the current account deficit. But once over that, there would be no deep-rooted political or economic opposition. Indeed, most people don't like having their currency devalue. The Fed would have a clear mandate to adjust American monetary policy, in consultation with the Bank of Japan and Bundesbank, to put a floor under the foreign exchange value of the dollar.

However, if one accepts Hypothesis I of domestic demand-led inflation, the stabilization problem is somewhat muddier. The domestic political or economic forces pushing for inflationary excess demand—possibly to help balance the budget (within Gramm-Rudman) or seeking overfull employment—would need to be identified and reined in. If so, what domestic monetary indicators—M_1, M_2, nominal GNP, short-term interest rates, and so on—should the authorities use to target the American price level that would serve them better in the future than in the past? Although still useful, the mark or yen value of the dollar would no longer serve as such a clear external benchmark of how well or badly the Fed is doing.

Whatever the macroeconomic remedy might be, nominal interest rates in the United States in the 1980s were higher and more volatile than those prevailing in Japan or Germany. Without claiming that real interest rates were higher in the United States, can one demonstrate that the syndrome of higher price inflation (comparing WPIs) and higher nominal interest rates will tend to shorten the time frame of corporate decision making in the United States relative to that abroad?

2 High Interest Rates, Taxation, and the Cost of Capital

A natural place to start is to review the extensive literature on how inflation and higher nominal interest rates affect the overall burden of the corporate income tax across countries; and, second, to consider the

---

[9]Higher American price inflation could continue behind quota restraints on imports—such as those now prevailing in automobiles, steel, and textiles.

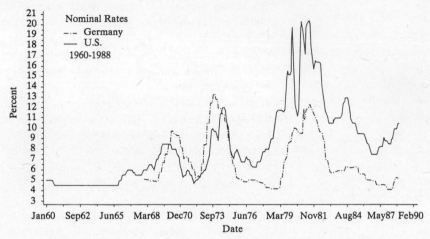

FIGURE 11.10 Prime Rates: United States and Germany

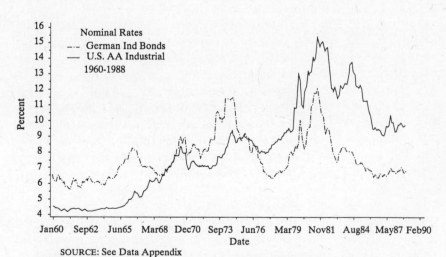

SOURCE: See Data Appendix

FIGURE 11.11 Corporate Bond Rates: United States and Germany

incidence of this tax on physical investments of different maturities depending on the degree of leverage in corporate financial structures.

2.1   The Effective Rate of Taxation

The appropriate overall rate of taxation is that pertaining to the marginal investment rather than an average rate of taxation. Hence simply comparing how much revenue the corporate tax garners as a share of GNP (which incidentally suggests a higher effective rate of taxation in Japan (Ando and Auerbach 1988)) is not satisfactory. However, computing a single marginal effective tax rate requires that the separate provisions applying to type of asset, source of finance, and type of investor be all accounted for.

Shoven and Tachibanaki (1988) provide a five-country comparison of effective marginal tax rates evaluated at actual inflation rates for 1980. They suggest a substantially lower effective tax rate in Japan than in the United States. Their estimate of 4.4 percent for Japan is, however, somewhat misleading. The estimate is calculated assuming an inflation rate of nine percent which was the average (CPI) inflation rate during the 1970s. Actual inflation in Japan has been much lower than this and has now virtually disappeared. Shoven 1989 contains similar calculations for Japan in 1985 over a range of inflation rates. Assuming an expected inflation rate of two percent for Japan in 1985, we derive an effective marginal tax rate in the order of 26 percent. This is much closer to Shoven's estimate of 37.2 percent for the United States, and even closer to Fullerton's (1987) estimate of 29.4 percent for the United States prior to the 1986 tax reform. Any difference in "the" effective rate of taxation of American and Japanese corporations seems not, itself, to have been sufficient to explain the difference in impatience between the two countries.

Before the tax reform of 1988, the sensitivity in Japan of estimated effective tax rates to the imputed value for inflation was very pronounced. The dominant effect of inflation was to decrease the overall tax burden on new corporate investment. Nominal interest payments were deductible from the corporate income tax base, while interest on personal savings in Japan had been effectively exempt from taxation. Thus the value of interest-rate deductions against the corporate income tax increased as inflation and nominal interest rates rose—without an offsetting increase in revenue from the personal income tax. The effects of inflation on the effective rate of corporate taxation were (are) not, however, unidirectional. The value of historic cost depreciation allowances will be decreased and the pure increase in the level of nominal profits will tend to raise effective corporate tax rates.

Both the United States and Japan have recently undertaken important tax reforms. The U.S. tax reform of 1986 basically eliminated the investment tax credit on equipment and stretched out depreciation allowances more consistently across the useful lives of corporate assets. Fullerton (1987) estimates that this results in a jump in effective tax rates from 29.4 percent to 41.1 percent assuming a four percent inflation rate in both cases. Japanese tax reforms of 1988 have been more sweeping: lowering the statutory rate for corporate income tax, lowering individual income tax rates but including interest on personal saving in the tax base, increasing tax thresholds, but introducing taxes on capital gains on securities as well as a type of value added tax. From Shoven 1989, the net effect of these changes is to decrease the sensitivity of the tax rate to inflation and to increase the zero-inflation effective tax rate (probably the best forecast of current inflation) to 31.8 percent.

So, for our puzzle of why American firms have become more impatient, what can we conclude from this literature on corporate taxation? Reduced to a single number, the effective or "real" rate of corporate taxation in the United States seems not substantially different from the average of Japanese and European competitors. Indeed, as of 1990, corporate tax rates, depreciation provisions, deductibility of interest, and so on, seem quite similar across countries.

2.2 The Inducement to Greater Leverage

While higher inflation and higher nominal interest rates need not lead to an unambiguously higher effective rate of corporate taxation, there is an unambiguous effect on the financing decisions of firms. For debt-financed projects, the tax deductibility of interest payments combined with the reduction in the real value of the outstanding debt principal largely offsets the falling real value of depreciation allowances so that the net effect of inflation will be quite small. However, the same project financed by either equity issues or internal funds will see a sharp decrease in value as inflation rises and merely erodes the real value of depreciation allowances. Thus the high inflation high nominal interest rate economy will display a clear preference for debt rather than equity financing.

To help fix these ideas consider the following simple numerical example. A firm has a four-year investment project with a known series of real returns. It must decide whether to finance the project by issuing a four-year bond or by using internal funds (this case is equivalent to the use of equity with no dividend payouts until the termination of the project). The choice between these options will be based on a Net Present Value calculation.

All interest payments are tax deductible, depreciation allowances are calculated on a straight-line basis, and corporation tax is levied at 50 percent. For simplicity, the examples are constructed in such a way that the projects yield positive pre-tax returns in all cases so that we do not have to deal with the complications arising from potential carryover of corporate losses. Post-tax profits are assumed to be placed in a sinking fund which yields the market rate of interest.

(A) No Inflation

(i) Debt Finance

Year	1	2	3	4
Gross Operating Income	50	50	50	50
+ Interest on Sinking Fund	0	1.75	3.54	5.38
− Interest Payment	5	5	5	5
− Depreciation Shelter	25	25	25	25
Taxable Income	20	21.75	23.54	25.38
Post-tax Income	10	10.88	11.77	12.69
+ Depreciation	25	25	25	25
Net Post-tax Cash Flow	35	35.88	36.77	37.69
Sinking Fund	35	70.88	107.65	145.34
− Principal Repayment				100
Value of Project				45.34
NPV				37.30

(ii) Internal Funds/Equity Financing

Year	1	2	3	4
Gross Operating Income	50	50	50	50
+ Interest on Sinking Fund	0	1.88	3.80	5.77
− Depreciation Shelter	25	25	25	25
Taxable Income	25	26.88	28.80	30.77
1ex] Post-tax Income	12.50	13.44	14.40	15.39
+ Depreciation	25	25	25	25
Net Post-tax Cash Flow	37.50	38.44	39.40	40.39
Sinking Fund	37.50	75.94	115.34	155.73
− NPV of initial investment	−100			
NPV of Project				28.12

Now consider the implications of introducing fully anticipated inflation into the model. We assume that pre-tax nominal interest rates adjust to fully reflect the anticipated inflation so that nominal interest rates are determined from the Fisher relation: $1+i = (1+r)(1+\pi)$ where r, the real interest rate is held constant. In the example below the pre-tax real interest rate is held at five percent so when we introduce inflation at ten percent nominal interest rates increase to 15.5 percent. Note that this assumed constancy of the pre-tax real interest rate implies that the real post-tax interest rate will vary with inflation. The alternative assumption of full Feldstein-Darby-Tanzi effects (so that the post-tax real interest rate is held constant) is also feasible. However, our results are robust to the choice of which real interest rate to hold constant.

(B) 10 percent Inflation

(i) Debt Finance

Year	1	2	3	4
Gross Operating Income	55	60.50	66.55	73.21
+ Interest on Sinking Fund	0	5	10.81	17.54
− Interest Payment	15.50	15.50	15.50	15.50
− Depreciation Shelter	25	25	25	25
Taxable Income	14.50	25	36.86	50.25
Post-tax Income	7.25	12.50	18.43	25.13
+ Depreciation	25	25	25	25
Net post-tax Cash Flow	32.25	37.50	43.43	50.13
Sinking Fund	32.25	69.75	113.18	163.31
− Principal Repayment				100
Value of Project				63.31
NPV of Project				35.58

(ii) Retained Income/Equity Finance

Year	1	2	3	4
Gross Operating Income	55	60.50	66.55	73.21
+ Interest on Sinking Fund	0	6.20	13.31	21.43
− Depreciation Shelter	25	25	25	25
Taxable Income	30	41.70	54.86	69.64
Post-tax Income	15	20.85	27.43	34.82
+ Depreciation	25	25	25	25
Net post-tax Cash Flow	40	45.85	52.43	59.82
Sinking Fund	40	85.85	138.28	198.10
− Initial Investment				−100
NPV of Project				11.32

To a certain extent the above comparison is unfair. Even in the no inflation economy debt finance will be preferred. To illustrate that our results do not depend on this initial ranking we consider an additional project yielding higher real returns but which can only be financed by the use of internal funds. For example this could be a project whose success required a great deal of effort by the corporate managerial team and this effort level would not be forthcoming if external funding was used—an extreme form of the principal-agent problem. We construct our example in such a way that the internally financed project will be chosen in the zero inflation regime, whereas the debt financed project is selected when inflation is at ten percent.

(1) No Inflation, Internal Funding

Year	1	2	3	4
Gross Operating Income	56	56	56	56
+ Interest on Sinking Fund	0	2.03	4.10	6.23
− Depreciation Shelter	25	25	25	25
Taxable Income	31	33.03	35.10	37.23
Post-tax Income	15.50	16.52	17.55	18.62
+ Depreciation	25	25	25	25
Net Post-Tax Cash Flow	40.50	41.52	42.55	43.62
Sinking Fund	40.50	82.02	124.57	168.19
− Initial Investment				−100
NPV of Project				38.37

(2) 10 percent Inflation, Internal Funding

Year	1	2	3	4
Gross Operating Income	61.60	67.76	74.54	81.99
+ Interest on Sinking Fund	0	6.71	14.42	23.25
− Depreciation Shelter	25	25	25	25
Taxable Income	36.60	49.47	63.94	80.24
Post-tax Income	18.30	24.74	31.97	40.12
+ Depreciation	25	25	25	25
Net Post-tax Cash Flow	43.30	49.74	56.97	65.12
Sinking Fund	43.30	93.04	150.01	215.13
− Initial Investment				−100
NPV of Project				20.89

The central message of these examples is that the introduction of inflation imparts a significant advantage to the use of debt as a means of financing a project. Thus there is a clear incentive to increase leverage which will result in a higher level of bankruptcy risk in the economy and correspondingly higher risk premia in interest rates. (In the above numerical examples, however, firms could borrow at unchanging real interest rates even when leverage or inflation increased.)

2.3   The Cost of Capital for Investment Projects

To what extent though do the documented higher U.S. nominal interest rates and higher risk premia translate into a higher real cost of capital— inclusive of the overall burden of taxation—for U.S. corporations? The cost of capital is, broadly speaking, the cost to the corporation of raising external finance. Less succinctly, the cost of capital can be defined as "the minimum pre-tax real rate of return that an investment must earn in order to give an investor a competitive or equilibrium post-tax return" (Shoven and Tachibanaki 1988). Thus the objective is to produce a single number characterizing the cost to the firm of raising finance for investment projects. This still omits a very important dimension of corporate decision-making—the time horizon of the firm. The cost of capital to a firm that considers very short-term investments could differ from one that focuses on long run growth; the natural analogy (and a partial explanation) is the existence of a term structure in interest rates where rates of return on short term bonds differ from those on long term bonds (unless the term structure is flat).

International differences in the cost of capital can arise not just from real interest rate differentials but also from alternative tax structures and differing legal and financial institutions. Note however that the existence of international differences in the cost of capital does not pre-suppose some kind of inefficiency in international capital markets. In the Appendix, we demonstrate that even under perfect foresight and no impediments to trade in either goods or asset markets, the cost of capital is not in general equated across nations. In practice, of course, the presence of risk or currency premia in international capital markets will tend to exacerbate this separation of international costs of capital. For example, although nominal interest rates were, in the 1980s, very low in Japan relative to the United States, American firms could not finance their North American operations in yen without incurring unac-ceptable exchange risk. In the absence of a common monetary standard, quite different costs of capital could persist across countries even when financial flows are unrestricted—and when gross international flows of finance capital are as massive as they are today.

The existing literature concludes almost unanimously that the cost of capital (reduced to a single number) is higher in the United States than Japan. Hatsopoulos and Brooks (1987) argued that the U.S. figure in the 1980s was twice that of Japan. By emphasizing the much lower earnings-price ratios on Japanese equities compared to American, Bernheim and Shoven (1989) suggest that "the cost of capital for an equity financed plant is approximately 12.6 percent in the United States, but only 5.1 percent in Japan." But other authors suggest narrower differentials, as do Bernheim and Shoven in earlier work. Of course, this higher (real) cost of capital will itself reduce American firms' time horizons for investing, as well as reducing the number of investments they are willing to undertake.

One interesting feature of these studies is that they often suggest that the cost of capital (a real rate) is in fact a decreasing function of the rate of inflation. When nominal interest rates can be expensed, marginal effective tax rates could decline as inflation increases. It is at this point that the omitted time dimension becomes vital. High inflation rates will be accompanied by high nominal interest rates, which will result in a shrinking of corporate time horizons. Firms are forced to focus on projects producing high income streams in the near future to meet the initial heavy burden of high nominal interest rate payments (to be shown below), and to prevent erosion of the present value of income tax depreciation allowances (as described above). Inflating faster is hardly likely to stimulate productive investment.

Although divining the relative cost of capital between Japan and the United States from market data alone remains an uncertain art, one can appeal directly to the informal rules of thumb used by major corporations in deciding whether or not to undertake new investments. For screening new projects, nominal "hurdle" rates of return were, in the 1980s, substantially higher in the United States for companies that, to avoid exchange risk, must finance mainly in dollars. American companies demanded before-tax hurdle rates of return—fifteen to twenty percent; whereas Japanese companies financing mainly in yen were satisfied with less than ten percent.[10] But are "real" hurdle rates higher in the United States? Over the typical five- to twenty-year life spans of fixed plant and equipment, the (largely implicit) inflationary expectations of American corporations vis-à-vis their Japanese counterparts can only be roughly estimated.

Unlike the 1950s and 1960s when there was a common monetary standard and virtually the same rates of inflation (less than one percent per year in the Japanese and American WPIs as shown in Table 11.3), inflation has since been much higher in the United States. Indeed, since the par value system for exchange rate broke down altogether in 1973, Table 11.3 shows U.S. WPI inflation averaging 5.6 percent per year compared to 2.3 percent in Japan to the end of 1988. But these trends

---

[10]I am indebted to conversations with Ralph Landau for these numbers.

TABLE 11.5
NPV of Straight-Line Depreciation Allowances

		Nominal Interest Rate						
		0	5	10	15	20	25	50
	1	100	95.2	90.9	87.0	83.3	80.0	66.7
	2	100	93.0	86.8	81.3	76.4	72.0	55.5
Asset	5	100	86.6	75.8	67.0	59.8	53.8	34.7
Life	10	100	77.2	61.4	50.2	41.9	35.7	19.7
	20	100	62.3	42.6	31.3	24.3	19.8	10.0
	50	100	36.5	19.8	13.3	10.0	8.0	4.0

NOTE: The entries in the table are computed as $(NPV/I_0) \times 100$.

have been neither smooth nor persistent. Since the traumatic, albeit fairly successful, worldwide disinflation of 1982, the Japanese WPI has fallen absolutely, albeit gently, by two percent per year through the end of 1988; whereas the American WPI has continued to increase at about two percent per year with more rapid increases in the last year or two (Table 11.5). Consequently, the difference in real hurdle rates is considerably less than the spread in nominal hurdle rates. However, if the anticipated inflation differential—expected dollar devaluation— remains of the order of five percent per year, the real hurdle rate in the United States still seems to be significantly higher. The Bernheim-Shoven (1989) risk-adjusted estimates of the difference in the cost of capital between Japan and the United States of the order of seven to eight percent would seem to be consistent with this informal hurdle-rate approach.

In contrast, if one looked only at nominal interest-rate differentials on government or high grade corporate bonds—9.5 percent in the United States and five percent in Japan at the end of 1988 (Figures 11.8 and 11.9)—this spread is about the same order of magnitude as any plausible projected differential in inflation rates and long-term rate of dollar devaluation. Apparently, on safe long-term assets, one can't establish a significant difference in real interest rates between the two countries.

However, real hurdle rates are point estimates of the overall real cost of capital—including equity, bank financing, and lower grade debt issues—that in turn reflect perceived risk in the American financial environment relative to the Japanese. Indeed, Table 11.6 shows the very high yields on unrated American "junk" bonds for which there is no Japanese equivalent. From the above analysis, the American junk-bond binge of the 1980s could simply reflect the incentive to raise the ratio of

TABLE 11.6

Long-Term Interest Rates: United States and Japan

Year	Government Bonds		High Grade Corporates		Junk
	Japan	U.S.	Japan	U.S.	U.S.
1978	6.09	8.41	7.23	8.74	10.92
1979	7.69	9.41	8.03	9.65	12.07
1980	9.22	11.46	9.02	11.99	13.46
1981	8.66	13.91	8.37	14.19	15.97
1982	8.06	13.00	8.16	14.03	17.84
1983	7.42	11.11	7.82	12.00	15.71
1984	6.93	12.44	7.18	12.95	14.97
1985	6.34	10.62	6.90	11.57	15.40
1986	4.94	7.68	6.09	9.63	14.45
1987	4.21	8.38	4.85	9.59	12.66
1988	4.27	8.85	4.96	9.62	13.95

SOURCES:

Japanese long-term Government bond series (7 years) from IMF *International Financial Statistics*.

U.S. ten-year Government Bond from *Federal Reserve Bulletin*.

Japanese industrial bond yield (12 years) from Bank of Japan *Economic Statistics Monthly*.

U.S. AA Industrial bond series from *Moody's Bond Survey*.

U.S. junk bond series is taken from Altman and Minowa 1989.

debt to equity in a high-inflation, high-interest-rate environment. And the price-earnings ratios on American common stocks are less than one fourth of their Japanese counterparts.

Thus, by this broader criterion, American investors seem to be demanding a higher risk premium on new investments than do Japanese. Besides the greater corporate leverage as reflected in the mania for leverage buyouts, however, the higher risk premium in the American capital markets could also reflect such diverse issues as product liability suits, environmental suits, and the general increase in uncertainty over what corporate property rights actually are. Most of these are well beyond the scope of this paper. But we can't resist speculating that the greater macroeconomic uncertainty—as manifest in less predictable inflation, more volatility in interest rates, and more exchange risk—arising out of continual threats to devalue the dollar is an important contributing factor.

In summary, some American impatience, the shortening of time horizons for new physical investments, in the 1980s can be explained by an effectively higher risk premium. American corporations seem to be using a higher internal real rate of interest—if one could really reduce the decision problem to a single number—for discounting the future.

2.4 High Nominal Interest Rates and the "Overborrowing" Syndrome

Whatever "the" effective real cost of capital in the United States might be, let us pursue further the consequences of having nominal interest rates that are higher and more volatile than those of major industrial competitors.[11] We have already established an important tax effect: the increased discounting of the present value of depreciation allowances (Table 11.5); this creates a bias against longer lived corporate investments financed by equity or internal funds, which induces greater corporate leverage and thus increases the risks seen by outside lenders.

But even in a tax-free world, high and volatile nominal interest rates could create an overborrowing syndrome. To reduce risk, firms find that they need to borrow more than the actual cash outlays for the physical investment being undertaken. And this overborrowing then militates against longer-lived investments so as to reduce the average duration of projects undertaken in the higher inflation, higher nominal interest rate country.

The standard textbook definition of duration for any asset, say a coupon-bearing bond with n years to maturity, is

$$(11.1) \qquad D = \frac{(C_t \circ t)/(1+i)^t}{C_t/(1+i)^t} \qquad \text{where } t = 1, 2, \ldots, n.$$

C_t is the coupon rate including the final repayment of principal, and i is the effective nominal yield to maturity. In order to compare the value of coupon payments through time, equation (11.1) weights each payment by its present value. Only for a zero-coupon or deep-discount bond, with but one payment (equal to its face value) after n years, would duration and time to maturity be equal—as shown in Table 11.7. Otherwise, with positive coupon payments throughout the life of the bond, the effective duration will be much less than the time to maturity. For example, Table 11.7 shows, on the basis of (11.1), that the duration of a ten-year bond with a five percent coupon rate is 8.11 years; whereas if the coupon rate was fifteen percent, the duration shrinks to 5.77 years. In effect, the higher coupon rate (in the higher inflation country) implies that the real costs of debt service come sooner rather than later: the burden of the final repayment of principal falls relative to the burden of the initial interest payments.

---

[11]John Hussman of Stanford University greatly helped with the formulation of arguments in this section.

TABLE 11.7
The Level of Interest Rates, Basis Risk,
and Duration of Bonds with Differing Maturities

Time to Maturity	Coupon Rate	Yield to Maturity	Bond Price ($100 face value)	Duration[1] (years)	Basis Risk[2]
5 years	5%	5%	100.0	4.55	4.33%
	0%	5%	78.4	5.00	4.76%
	10%	10%	100.0	4.17	3.79%
	0%	10%	62.1	5.00	4.55%
	15%	15%	100.0	3.85	3.35%
	0%	15%	49.7	5.00	4.35%
10 years	5%	5%	100.0	8.11	7.72%
	0%	5%	61.4	10.00	9.52%
	10%	10%	100.0	6.76	6.14%
	0%	10%	38.6	10.00	9.09%
	15%	15%	100.0	5.77	5.02%
	0%	15%	24.7	10.00	8.70%
20 years	5%	5%	100.0	13.09	12.46%
	0%	5%	37.7	20.00	19.05%
	10%	10%	100.0	9.36	8.51%
	0%	10%	14.9	20.00	18.18%
	15%	15%	100.0	7.20	6.26%
	0%	15%	6.1	20.00	17.39%

SOURCE: John Hussman's computer program.
NOTES:
[1] "Duration" of the bond follows F. R. Macauly's definition:

$$D = \frac{\sum_{t=1}^{n} \frac{C \cdot t}{(1+i)^t}}{\sum_{t=1}^{n} \frac{C_t}{(1+i)^t}}$$

where C_t is coupon and/or principal payment, t is time elapsed, i is yield to maturity, and n is length of time to final maturity.

[2] Basis risk is the sensitivity of the percentage change in the bond price to a one percentage point change in the yield to maturity.

$$BR = \frac{\partial B/\partial i}{B}$$

where B is the market value.

But need this foreshortening of the duration of bonds of a given term to maturity in the high interest rate environment similarly shorten the time span of physical investments made by firms? After all, financing arrangements can be flexible as long as the underlying investment project is profitable in real terms.

Consider the ultrasimple case of a single physical investment which is to be debt financed by the issue of some kind of bond. To further simplify, suppose that this physical investment is of the classic textbook sort where the time to fruition (duration) itself is variable: how long to age wine or whiskey, or how long to age a stand of timber before harvesting, and so on. (Otherwise distinct shorter-lived projects would have to be compared to longer-lived ones.) Take the particular textbook example (Copeland and Weston 1988: 51) of a stand of trees growing according to the square root principle[12] such that the revenue from harvesting at any point in time, Rev_t, is

$$(11.2)   Rev_t = e^{\pi t} \circ S \circ \cdot (1+t)$$

where S is a scale factor connoting the value of the trees in today's dollars, and π is the (expected) rate of price inflation. Assuming that the rate of price inflation is known, then—in the standard textbook model—the optimal time to cut and the net present value of the trees depends only on the *real* rate of interest, r, prevailing in today's financial markets. Moreover, this real rate of interest is independent of the rate of inflation if the latter is always fully reflected in nominal interest rates through a "Fisher effect" according to

$$(11.3)   1 + i = (1+r)(1+\pi)$$

Then from knowledge of r, and for any π, one can determine the time to harvest by simply maximizing the project's net present value (NPV) with respect to t.

$$(11.4)   \text{NPV} = e^{\pi i t} \circ e^{-\pi t} \circ S \cdot (1+t) = e^{rt} \circ S(1+t)$$

For example, if $r=.05$, then NPV is maximized if cutting takes place in nine years irrespective of how high today's nominal interest rates might be—see Copeland and Weston 1988.

But what risk factors have been left out of this standard analytical approach? Even if the rate of inflation can be projected into the future, (real) interest rates can still fluctuate through time. How then has the firm—implicitly in the standard NPV maximization—immunized itself against future interest rate fluctuations? The implicit financial instrument behind the NPV calculation in (11.4) is a zero coupon bond that pays its face value to the holder nine years hence. In this way, the duration of the bond is exactly matched to the duration of the investment project: when the trees are cut nine years hence, the bond holders are paid off without receiving prior coupon payments.

But this immunization strategy assumes no inhibitions on the issuance of zero-coupon bonds, which are seldom observed in private financings. One reason is that basis risk, the sensitivity of the capital value of the bond to changes in market interest rates, is highest for zero coupon bonds. Suppose we define basis risk narrowly as

---

[12]The trees themselves need not have square roots.

TABLE 11.8
Interest Rate Volatility in the 1980s

Nominal Interest Rate Volatility	Mean	Standard Deviation
United States		
Commercial Paper Rate 90-day	0.63	0.85
3-Month Treasury Bill	0.55	0.76
AAA Corporate Bonds	0.32	0.27
Government Bonds (10-year)	0.38	0.34
Japan		
3-Month Gensaki	0.22	0.37
Industrial Bonds (12-year AA)	0.12	0.16
Government Bonds (7-year)	0.25	0.25

NOTE: Data refer to the mean and standard deviation of absolute changes.

(11.5) $BR = \frac{\partial B/\partial i}{B}$ where B is the bond's market value.

Table 11.7 shows how basis risk varies inversely with coupon rates—the more so for longer term bonds. At the extreme, for a twenty-year bond with a fifteen percent coupon rate, the basis risk is 6.25 percent; but this basis risk rises to 17.39 percent if the coupon rate is zero. (In general, Table 11.7 and the BR measure understate the market-value sensitivity of bonds with high nominal yields to interest-rate fluctuations. Countries with higher nominal interest rates tend to experience greater percentage point fluctuations: in the 1980s, interest rates have fluctuated twice as much in the United States as in Japan—as per Table 11.8.)

Because of this basis risk to the holders of zero coupon bonds which could induce them to demand a risk premium, firms more commonly opt to issue long dated bonds with fixed coupon rates spread evenly through time more or less at the prevailing nominal interest rate. If the firm in our tree-growing example followed this strategy, then the duration of coupon bonds that mature in nine years would be less than the duration of the stand of trees—possibly much less if nominal interest rates were high (Table 11.8). Thus, to finance the purchase of trees and the early interest payments to bondholders, the firm has to overborrow—the more so when nominal interest rates are high. If the firm faces some capital constraint on the amount it can borrow, this would then force it to either forego the project or cut the trees earlier. In short, it would begin to exhibit "impatience."

It is straightforward to demonstrate how a firm operating in an inflationary environment might be forced to overborrow compared to a firm in a no inflation regimen. Consider a five-year investment project yielding the known pattern of real returns:

Year	1	2	3	4	5
Real Return	10	10	10	10	10+100

The project is to be financed by means of a five-year bond with a face value of 100. Assume a real interest rate of five percent, and that the coupon rate on the bond is equal to the nominal interest rate. Nominal interest rates are determined via the Fisher relationship:

$$(1 + i) = (1 + \pi)(1 + r)$$

where i is the nominal interest rate, π is the rate of inflation, and r is the real interest rate.

(I) No Inflation : $r = .05$, $\pi = 0$, $i = .05$

Year	1	2	3	4	5
Revenue	10	10	10	10	10+100
Debt Service	5	5	5	5	5+100
Net Return	5	5	5	5	5

(II) Inflation : $r = .05$, $\pi = .1$, $i = .155$

Year	1	2	3	4	5
Revenue	11	12.1	13.31	14.64	16.11+161.05
Debt Service	15.5	15.5	15.5	15.5	15.5 +100
Net Return	−4.5	−3.1	−1.19	−0.86	61.66

The effect of introducing inflation is quite clear. A project which yielded positive net returns in an o inflation regime is transformed into one generating net losses during the first few periods of its operation. Note though that we have not changed the real revenue stream or the real value of debt service so both the duration of the project and its net present value remain unchanged.

Suppose now that the firm cannot simply absorb these early losses and is forced to maintain a non-negative net income stream. Thus the firm has to arrange external finance to cover the losses occurring during the first four periods. Assume that the firm simply increases the amount initially borrowed using the additional funds in excess of the initial investment as a cash reserve to be used to meet debt service obligations. The unused portion of the cash reserve is assumed to earn the nominal interest rate. Then we can show that the amount of this overborrowing is given by the following

$$X = (1 + i)^T \Sigma_1^T (iF - N_t)/(1 + i)^t$$

where X is the amount of overborrowing, F the face value of the bond, i the nominal interest rate, T the number of initial periods of deficits, and N_t is the nominal return at time t.

For our case (II) the firm must borrow an additional 14.86 so initial borrowing becomes 114.86.

(III) Inflation and non-negative value for project

Year	1	2	3	4	5
Revenue	11	12.1	13.31	14.64	16.11+161.05
Debt Service	17.8	17.8	17.8	17.8	17.8 +114.86
Cash Balance	17.16	11.96	7.23	3.16	0.0
minus transfer	10.36	6.26	2.74	0.0	

Note that the assumption that borrowing and lending rates are equal, so that the interest rate earned on unused cash balances equals the rate charged on the bond, implies that there is no change in the net present value of the project.

Another way out of this financial dilemma, which might seem particularly attractive if the firm itself cannot predict the general rate of inflation with which the value of its trees is correlated, would be to finance the long-term project by simply rolling over short-term debt. Then, if the rate of inflation varied through time, and this was more or less accurately reflected in short-term nominal interest rates on commercial bills or bank loans, the firm would have something of a hedge against changes in inflation. But again this forces the firm to pay debt service before it has yet cut any trees and thus to overborrow. Although in year one it need just borrow enough at short term to purchase the young stand of trees, at the time the loan is rolled over in year two, the firm will have to borrow even more in real terms in order to meet the debt service (principal plus interest) on the loan made in year one. And overborrowing cumulates faster when nominal interest rates cum price inflation are high.

In addition, by rolling over short-term loans, the firm would now be exposed to real interest rate risk insofar as short-term nominal interest rates were not well correlated with ongoing inflation. Both would impinge on the firm's capital constraint, a constraint that could be relaxed somewhat by cutting the trees earlier.

In summary, as long as financial risk discourages firms from issuing deep-discount bonds to finance long-term projects of the same duration (as seems to be the case in practice), then high nominal interest rates, even those reflecting high inflation, will aggravate the need to overborrow and inhibit longer term investments.

2.5   Implications for Macroeconomic Policy: A Concluding Note

In response to some apparent decline in America's international competitiveness and increasing trade deficits over the past twenty years, economists have often advocated devaluing the dollar in order to make American goods cheaper on world markets. But continual dollar devaluation had, in the 1980s, the unlooked-for consequence of making nominal interest rates, and inflationary expectations, higher and more volatile in the United States than in Japan or hard-currency Europe, with the further consequence of shortening the time horizon of corporate decision making in the United States. Indeed, the increased leveraging of American corporations—through leveraged buyouts, mergers and acquisitions, and so on—is naturally accentuated when nominal interest rates are high, and this in turn raises the riskiness of external finance. The resulting shortening of time horizons of American firms is hardly likely to improve America's international competitiveness in any long-run sense.

What are the implications for U.S. macroeconomic policy? To bring American nominal interest rates down toward German and Japanese levels, the most efficient monetary strategy is to give international investors confidence that the dollar will not depreciate against hard foreign monies over the next twenty years as it did in the past. The U.S. Federal Reserve System, in cooperation with the Bank of Japan and the Bundesbank, could be instructed to put a convincing floor under the dollar in terms of yen and marks. To maximize credibility, a return to a par-value system for setting exchange rates (within narrow bands) among the European bloc, Japan, and the United States could be a rather dramatic signal to international investors into the indefinite future that the dollar was most unlikely to depreciate. And Table 11.9 suggests that purchasing power parity exchange rates in early 1990 would value the dollar between 155 and 175 yen, and about 1.9 to 2.1 marks. These rates would roughly align the producer price levels of the three blocs, without imposing relative inflation or deflation on any one of them.

Such a pact would require full-fledged monetary cooperation among the three major blocs of the major industrial countries, as described by McKinnon (1984, 1988, and 1990b). The Fed would tighten up when the dollar tended to fall below these target ranges for the yen/dollar and mark/dollar exchange rates, and loosen up when the dollar was unduly strong by this PPP criterion. By stabilizing the dollar exchange rate against hard foreign monies, which have had a better recent record of price level stability than the United States itself, international (and domestic) investors would get much of the reassurance that they need to

TABLE 11.9
PPP Estimates for 1989: Q4

Method and Base Used	Index:	Yen/Dollar				Mark/Dollar			
		PPI	ULC	CPI	GNP defl.	PPI	ULC	CPI	GNP defl.
Long-run averaging with base 1975:Q1–1989:Q4		177	193	183	181	2.05	2.21	1.91	2.01
Long-run averaging with base 1980:Q1–1989:Q4		174	186	177	177	2.17	2.27	2.04	2.12
Morrison-Hale method with base 1985, updated using PPI		203	–	–	–	2.36	–	–	–
Price pressure method with the sample period of 1975:Q1–1988:Q2, updated using PPI or ULC		171	187	–	–	2.06	2.15	–	–

SOURCE: Kenichi Ohno "Estimating Yen/Dollar and Mark/Dollar Purchasing Power Parities," IMF Staff Papers (forthcoming) 1990.

NOTE: Indices used are producer prices (PPI), unit labor costs (ULC), consumer prices (CPI), and GNP deflator, all taken from OECD Main Economic Indicators. Exchange rates are from IMF International Financial Statistics.

	Yen/Dollar	Mark/Dollar
Suggested Target Ranges for 1990 (PPI-based estimates)	155 to 175	1.9 to 2.1

reduce nominal interest rates on dollar assets. In addition, the triumvirate could then choose to stabilize their common producer price level as the natural "nominal anchor" for the system overall.

However, as long as the U.S. fiscal deficit remains so large, such a monetary pact is unlikely to succeed in completely equalizing the cost of capital across countries. The saving deficiency in the American economy created by the fiscal deficit makes a corresponding deficit in the current account of the balance of payments (net foreign borrowing) a virtual necessity, with the unfortunate side effect that devaluationists become more difficult to silence. In the more distant future, the fiscal deficit (and the rapid build up of government debt) increases the likelihood, in the minds of international investors, that the American government will eventually resort to the use of the inflation tax. On both counts, American interest rates would be inhibited from falling all the way to Japanese levels. Even if an international monetary pact was in place, some risk premium in the American cost of capital relative to Japan and hard-currency Europe would remain until the U.S. public finances were brought under better control.

Nevertheless, even in the presence of U.S. fiscal and trade deficits over the next several years, a formal monetary pact, or even an informal understanding, putting a floor under the dollar could still facilitate a substantial reduction in the level and volatility of nominal interest rates in U.S. capital markets that would be valuable in lengthening the time horizons of American companies.

Appendix: International Parity Conditions with Taxation

Consider a two-country economy producing a single good under conditions of perfect foresight with no impediments to trade (including transport costs) in either goods or assets. Under these conditions arbitrage in goods markets will ensure that Purchasing Power Parity holds while asset market equilibrium will be characterized by the equalization of post-tax real returns. These assumptions enable us to analyze the structure of international capital markets in the presence of taxation but abstracting from both risk and currency premia.

Portfolio equilibrium of the domestic resident investor requires that the post-tax real return on domestic securities is equal to the post tax real return in domestic currency units of foreign currency denominated securities.

$$(1) \quad (1-\theta)i - \pi = R = (1-\theta)i^* + (1-\lambda)\hat{s} - \pi$$

where i is the yield on domestic corporate debt, θ is the rate of taxation on income, λ is the rate of taxation on capital gains, $\hat{s}$ is the anticipated change in the exchange rate, and a $*$ signifies the equivalent foreign variable. The left hand side of (1) represents the post-tax real return to domestic investors on domestic corporate debt and the right hand side that on foreign debt.

The change in the exchange rate is determined via the Purchasing Power Parity condition:

$$(2) \quad \hat{s} = \pi - \pi^*$$

Substituting (2) into (1) and rearranging yields the parity conditions:

$$(3a) \quad i = r^* + \frac{1-\lambda}{1-\theta}\pi - \frac{\theta-\lambda}{1-\theta}\pi^*$$

$$(3b) \quad r = r^* + \frac{\theta-\lambda}{1-\theta}(\pi - \pi^*)$$

$$(3c) \quad R = \frac{1-\theta}{1-\theta^*}R^* + \lambda^* \frac{1-\theta}{1-\theta^*}\pi^* + \frac{(1-\lambda^*)(1-\theta)-(1-\theta^*)}{1-\theta^*}\pi$$

Equation (3b) serves as a reminder that two economies can have differing pre tax real rates of interest even if their post-tax rates are equal. In general capital gains are taxed favorably compared to interest income so that $\theta > \lambda$. Thus the high inflation economy will have higher pre tax real interest rates than the low inflation economy.

The relationship between post-tax real interest rates is captured in (3c) which is also derived in McClure 1988. In general there is no reason to suspect that post-tax real interest rates are equalized despite the highly simplified model under consideration. (It could, however, be argued that in the long run the place of residency of investors for tax purposes is endogenous so that if $R > R^*$ the domestic economy would experience a net inflow of investors until post tax real interest parity is restored.) Suppose that personal tax rates are the same in both countries so that $\theta = \theta^*$. Then:

$$R = R^* + \lambda^*(\pi - \pi^*)$$

Once more, the high inflation economy will have the higher real rate of interest.

Now, assuming that marginal investments are debt financed, no depreciation allowances and no investment tax credits, the real cost of capital to the domestic corporation is given by:

(4) $C = (1 - \tau)i - \pi$

where τ is the domestic rate of corporate taxation. Manipulating (4) and using (1) we obtain:

(5) $C = \frac{1-\tau}{1-\tau^*}C^* + \frac{(1-\tau)(1-\lambda)-(1-\theta)}{1-\theta}\pi - \frac{(1-\tau^*)(1-\lambda)-(1-\theta)}{1-\theta}\frac{1-\tau}{1-\tau^*}\pi^*$

Thus, even in a world of perfect foresight and no arbitrage costs there is no tendency for the real cost of capital to be equalized.

References

Adams, C., P. R. Fenton, and F. Larsen. 1987. Potential Output in Major Industrial Countries, *Staff Studies of the World Economic Outlook*. Washington, D.C.: International Monetary Fund.

Altman, E. Z. and Y. Minowa. 1989. Analyzing Risks, Returns, and Potential Interest in the U.S.: High Yield Corporate Debt Market for Japanese Investors. *Japan and the World Economy*, 1, pp. 163–86.

Ando, A., A. J. Auerbach. 1988. The Cost of Capital in the United States and Japan: A Comparison, *Journal of the Japanese and International Economies*, 2, pp. 134–58.

Barsky, Robert B. 1987. The Fisher Hypothesis and the Forecastability and Persistence of Inflation. *Journal of Monetary Economics*, vol. 19, no. 1 (January), pp. 3–34.

Berger, B., M. Dertouzos, R. Lester, R. Solow, L. Thurow. 1987. Toward a New Industrial America, *Scientific American*, vol. 260, no. 6 (June), pp. 39–47.

Bernheim, D. and J. Shoven. 1989. Comparison of the Cost of Capital in the U.S. and Japan: The Roles of Risk and Taxes, Stanford University, September.

Cooper, R. 1982. The Gold Standard: Historical Facts and Future Prospects, in *Brookings Papers on Economic Activity*, I, pp. 1–45; reprinted in R. N. Cooper, *The International Monetary System*, The M.I.T. Press, 1987.

Copeland, T. and J. Fred Weston. 1988. *Financial Theory and Corporate Policy*, 3rd ed., Addison Wesley.

Dertouzos, M., R. Lester, and R. Solow, *Made in America: Regaining the Productive Edge*. Cambridge, MA: The M.I.T. Press.

Feldstein, M. 1989. The Case Against Trying to Stabilize the Dollar, *American Economic Review*, vol. 79, no. 2 (May), pp. 36–40.

Fisher, I. 1930. *The Theory of Interest*. New York: MacMillan.

Frankel, J. and K. Froot. 1987. Using Survey Data to Test Standard Propositions Regarding Exchange Rate Expectations, *American Economic Review*, March, pp. 93–106.

Frenkel, J. and M. Mussa. 1980. The Efficiency of Foreign Exchange Markets and Measures of Turbulence, *American Economic Review*, 70 (May), pp. 374–81.

Fullerton, D. 1987. The Indexation of Interest, Depreciation, and Capital Gains and Tax Reform in the United States, *Journal of Public Economics*, 32, pp. 25–51.

Giovannini, A. 1989. How Do Fixed Exchange Rate Systems Work: The Evidence from the Gold Standard, Bretton Woods, and the EMS, in Miller, Eichengreen, and Portes, eds., *Blueprints for Exchanqe Rate Management*, New York: Academic Press, pp. 13–43.

Hatsopoulos, G. N., and S. H. Brooks. 1987. The Cost of Capital in the United States and Japan. Paper presented to the International Conference on the Cost of Capital, Kennedy School, Harvard, November, 19–21, 1987.

Horiuchi, A. 1984. The Low Interest Rate Policy and Economic Growth in Post-War Japan, *The Developing Economies*, vol. 22, no. 4 (December).

King, M., and D. Fullerton. 1984. *The Taxation of Income from Capital: A Comparative Study of the U.S., the United Kingdom, Sweden, and West Germany*. Chicago: University of Chicago Press.

McCauley, R. N. and S. A. Zimmer. 1989. Explaining International Differences in the Cost of Capital, *Federal Reserve Bank of New York Quarterly Review*, Summer, pp. 7–28.

McClure, J. H. 1988. PPP, Interest Rate Parities and the Modified Fisher Effect in the Presence of Tax Agreements: A Comment, *Journal of International Money and Finance*, 7, pp. 347–50.

McKinnon, R. 1981. The Exchange Rate and Macroeconomic Policy: Changing Post-War Perceptions, *Journal of Economic Literature*, 19 (June), pp. 531–57.

McKinnon, R. 1984. *An International Standard for Monetary Stabilization.* Washington, D.C.: Institute for International Economics.

McKinnon, R. 1988. Monetary and Exchange Rate Policies for International Financial Stability: A Proposal, *Journal of Economic Perspectives*, Winter.

McKinnon, R. 1989a. *Money in International Exchange: The Convertible Currency System.* Oxford: Oxford University Press.

McKinnon, R. 1989b. Alternative International Monetary Systems. Unpublished manuscript, Stanford University, May.

McKinnon, R. 1990a. The Exchange Rate and the Trade Balance: Insular Versus Open Economies, *Open Economies Review*, vol. 1, no. 1, pp. 17–37.

McKinnon, R. 1990b. Interest Rate Volatility and Exchange Risk: New Rules for a Common Monetary Standard, *Contemporary Economic Issues* (forthcoming).

McKinnon, R. and K. Ohno. 1989. Purchasing Power Parity as a Monetary Standard, in Hamouda, Rowley, and Wolf, eds., *The Future of the International Monetary System.* Hampshire, England: Edward Elgar, pp. 42–64.

Ohno, K. 1990. Estimating Yen/Dollar and Mark/Dollar Purchasing Power Parities, IMF Staff Papers (forthcoming).

Robinson, J. 1950. The Foreign Exchanges, in H. Ellis and L. A. Metzler, eds., *Readings in the Theory of International Trade.* Philadelphia: The Blakiston Company, pp. 83–103.

Shoven, J. B. 1989. The Japanese Tax Reform and the Effective Rate of Tax on Japanese Corporate Investments, in L. H. Summers, ed., *Tax Policy and the Economy*, Vol. 3, Cambridge: M.I.T. Press.

Shoven, J. B. and T. Tachibanaki. 1988. The Taxation of Income from Capital in Japan, in J. Shoven, ed., *Government Policy Towards Industry in the U.S. and Japan.* Cambridge: Cambridge University Press.

Solomon, R. 1977. The International Monetary System 1945–1976: An Insider's View. New York: Harper and Row.

Teh, R. 1990. Have Long-Term Interest Rates Been More Volatile Since Bretton Woods? Investigations and Explanations. Unpublished Manuscript, Department of Economics, Stanford University, March.

12

Japan's Management of Global Innovation: Technology Management Crossing Borders

KIYONORI SAKAKIBARA AND D. ELEANOR WESTNEY

In the industries where Japan has become a major global competitor (such as computers, semiconductors, automobiles, consumer electronics), Japanese companies have developed the reputation for rapid commercialization of new product ideas and for effective and efficient incremental innovations in existing products (Rosenberg and Steinmueller 1988; Dertouzos, Lester, and Solow 1989, pp. 48–49; Stalk and Hout 1990). They have done so with relatively low investments in basic research, and with a very high degree of geographic concentration of their research and development organizations in Japan. Indeed, the success of these companies in designing in their home country laboratories products that meet the needs of customers in many different national markets has been greatly envied by U.S. firms.

These same Japanese firms, however, are now under increasing pressures to internationalize their research and development organizations and to increase their basic research activities. Japanese research managers are beginning to formulate new technology strategies to deal with these pressures and to assess the implications for institutionalized patterns of technology management.

1 New Dimensions of Technology Strategy

In the latter half of the 1980s, Japanese firms have increasingly confronted demands that they put more of their technology development activities overseas. Some of these pressures come from the governments and business communities within the countries which are Japan's major markets, the United States and Europe in particular, where policy makers are increasingly critical of Japanese firms' low level of local value-added, not only in manufacturing but also in product development and research (Ishikawa 1990). Western policy makers and businessmen also contend that Japan is not "pulling its weight" in investment in basic research to increase the global stock of scientific and technical knowledge (*The Economist*, 1989).

However, much of the pressure for dispersing R&D geographically is self-generated: Japanese firms want to become "true" international companies, on the model of leading Western multinationals like IBM. In addition, Japanese managers anticipate a growing shortage of scientists and engineers within Japan itself, as the aging of the Japanese population lowers the numbers of university graduates and as they must increasingly compete for those graduates with the financial services sector (which is hiring more and more scientific and technical graduates) and with foreign firms establishing R&D facilities in Japan. A survey of 177 leading Japanese firms in 1988 (*Nihon Keizai Shimbun*, September 13, 1988) found that over 80 percent of the respondents were either actively working to establish R&D bases abroad or interested in doing so.

Japanese companies are by no means alone in seeing the internationalization of their technology development capacity as an important strategic challenge. A 1985 Booz Allen study of technology management found widespread agreement within the sixteen U.S, European, and Japanese multinationals they surveyed on the perception that

"New technologies and the specialized talent that produces them will continue to develop locally in 'pockets of innovation' around the world. Nurturing those technologies, uprooting them, and cross-fertilizing them for commercialization and global distribution will continue to be major challenges in technology management." (Perrino and Tipping 1989, p. 13)

Tapping geographically dispersed "pockets of excellence"—which include government laboratories, cooperative R&D projects, and universities—can be undertaken in a variety of ways: technology scanning, cross-licensing, strategic alliances, and joint ventures. However, managers are increasingly realizing that even these can be most effectively supported by a local technology development capacity: that is, by a wholly owned critical mass of credible researchers who are part of the multinational corporation but who can function as insiders within the national technology system. More than five years ago, the Booz Allen study found a consensus that

"What we have called the 'global network' model of technology management is clearly the 'wave of the future' when it comes to competing globally. This model consists of a network of technology core groups in each major market— the United States, Japan, and Europe–managed in a coordinated way for maximum impact." (Perrino and Tipping 1989, p. 13)

In a growing number of firms, this perception is leading to increasing efforts to internationalize R&D organization, and, in those U.S. and European multinationals that already have local product development centers, to enhance and integrate these dispersed facilities (Perrino and Tipping 1989, Herbert 1989, Hakanson and Zander 1986, De Meyer and Mizushima 1989).

Within Japan, the drive to expand basic research is rooted in internal pressures that are as strong as those behind the push for internationalization, if not stronger: Japanese managers increasingly emphasize the need to generate new technology within the company, as external sources become scarcer and harder to tap (MITI 1989). In contrast to the dominant patterns in the United States and Europe, industrial companies, not government or the universities, have assumed the primary role in expanding Japan's basic research activities (Sakakibara 1988). Between 1980 and 1985, over forty of the companies listed on the Tokyo Stock Exchange have established new research facilities, many of which are oriented to basic research rather than the more traditional product-oriented R&D. In a dramatic break from tradition, these have been built away from existing corporate research and manufacturing sites, to emphasize the autonomy of the new labs.

For some companies, the drive to expand basic research is integrally linked to their internationalization strategies. Otsuka Pharmaceuticals Co. Ltd., for example, has set up research facilities in Maryland and Seattle in the United States and in Frankfurt, West Germany; their mandates cover basic research as well as clinical development. NEC has established a basic research facility in Princeton, New Jersey. Ricoh Co. Ltd. recently established a center for research in artificial intelligence in California's "Silicon Valley."

Other firms are pursuing the two agendas simultaneously but independently, expanding basic research activities at home and setting up R&D facilities overseas with less ambitious mandates. Hitachi, for example, has set its top priority on establishing the basic research facility in Japan that it set up in 1985, but it is also working on plans to build research facilities in the United States and the United Kingdom.

2 International Technology Strategy of Japanese Firms

Theories of the multinational corporation (MNC) developed in the 1960s and early 1970s, based primarily on the experience of U.S. firms, made the internationalization of production the defining criterion of the MNC. Japanese business scholars, reflecting their own national experience, have tended to focus more broadly on the internationalization of the firm, using a three-stage model (see, for example, Saito and Itami 1986). In the first stage, marketing and distribution organizations are set up offshore but other functions, including manufacturing, remain concentrated at home. The second stage centers on the establishment of production facilities in the firm's major markets abroad. And the third—the globalization of corporate management—involves the internationalization of core corporate functions such as finance and R&D. In terms of this model, Japan's leading industrial companies have moved beyond the first stage of internationalization. They are well advanced into the second stage ("classic" foreign direct investment in production facilities abroad), and they are now moving toward the third stage (Saito and Itami 1986).

The internationalization of the technology development function of large Japanese firms has been somewhat more complex, involving five stages. The first stage, technology scanning, is associated with the first stage of internationalization, in which Japanese companies manufacture products in Japan for sale abroad. In this stage the company focuses on developing organizational systems to collect scientific and technical information and product information for use in the product development organization back in Japan. Some companies have relied heavily on sending individual "scouts" on specific technology-gathering missions; others established separate offices in the United States and Europe that were explicitly charged with technology scanning. The companies relied primarily on their own nationals in staffing these offices.

The second stage involves the creation of an organizational system to support the transfer of technology to production facilities overseas. Most production transplants have set up a technology department, following a standard Japanese pattern, in which each major factory is supported by a technology department or laboratory capable of process technology development and some incremental product improvement. In some companies these technology departments have been capable of minor modifications of product technologies to suit local markets, although new product development remained concentrated in Japan. These offshore technology departments were usually heavily staffed by Japanese, although of course local engineers were also recruited (Trevor 1988, pp. 143–44).

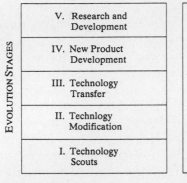 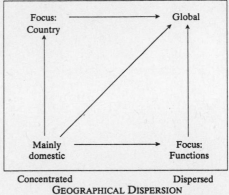

FIGURE 12.1 The Dimensions of International Technology Strategy

The formal establishment of R&D laboratories marks the beginning of the third stage. However, for many Japanese companies, the overseas laboratory, despite being called an R&D center, has done very little actual research or product development. Instead it has been a base for performing a range of other activities: technical cooperation with suppliers, support for technology transfer into production facilities, cross-licensing support, and the supervision of contract research. The overseas research laboratories of several of the leading Japanese pharmaceutical companies exemplify this stage. Instead of carrying out research directly, they contract out research to independent laboratories and specialized drug testing companies that supervise the clinical trials necessary to satisfy local regulatory requirements. In addition, they monitor technological trends and evaluate emerging technologies and new products.

In the fourth stage, overseas research laboratories embark on new product development, which becomes their central mission. These laboratories epitomize what is generally defined as the internationalization of R&D. The fifth stage extends the strategic mandate to encompass basic research, where the laboratory participates in an advanced, global division of technology development within the company.

Figure 12.1 summarizes the previous discussion and provides a way for mapping various companies' technology strategies, using two variables: the five stages of internationalization, and the geographic dispersion of the technology function (measured by the number of regions in which the company has developed a technology capability). The Y axis represents the stages of development of the strategic motivation behind the internationalization of technology strategy, and the X axis indicates the degree of geographical dispersion of technology-generating activities. The upper left corner maps a case in which all R&D for the company, including basic research, is carried out within one country; the lower right corner represents a case in which the activities associated with technology scanning are distributed among many countries. The upper right corner, where research activities are geographically dispersed and varied, represents the "global network" of technology development described in the Booz Allen study cited above (Perrino and Tipping 1989, p. 13).

Roughly speaking, most Japanese firms are in the process of shifting from Stage II to Stage III on the Y axis and are moving gradually out on the X axis. In the computer industry, for example, NEC has advanced farthest on the Y axis, with the establishment of its basic research facility in New Jersey in the spring of 1988. Hitachi is advancing into Stage IV and is moving farther than NEC along the X axis, with its plans to set up full-scale research facilities in both the United States and the United Kingdom. Fujitsu has relied primarily on its equity partnership with Amdahl to penetrate the North American technology system, and on its non-equity strategic alliance with ICL in the United Kingdom to penetrate the European system. To date, it has not made public any plans to support these activities through a wholly owned basic R&D presence in either region. Toshiba and Mitsubishi, in comparison, have yet to propose publicly any clear internationalization strategy for technology development in their computer businesses.

Otsuka Pharmaceutical Company is another of the handful of Japanese companies who have approached Stage V on the Y axis. Otsuka has built an international system for pharmaceutical product development, including basic science, that incorporates research institutes in Maryland (near the National Institutes of Health) and Seattle in the United States and in Frankfurt, West Germany.

The model of internationalization portrayed in Figure 12.1 differs somewhat from the model proposed by Robert Ronstadt on the basis of his research into the internationalization of technology development in U.S. multinationals, which has to date been the dominant typology in studies of the internationalization of R&D. Ronstadt identified four types of overseas technology facilities in the U.S. multinational corporation: the TTU (Technology Transfer Unit), which supported the transfer of product and production technologies from the parent to a manufacturing subsidiary; the ITU (Indigenous Technology Unit), with a capacity for new product development for the local market; the GTU (Global Technology Unit), which had the capacity to develop new products for worldwide markets; and the CTU (Corporate Technology Unit), which carries out basic research to develop advanced technology that will be applied elsewhere in the multinational (Ronstadt 1978). Ronstadt was careful to say that his typology was not a model of evolutionary stages: some of the companies he studied showed a gradual progression over time from the TTU to the GTU or CTU, while others did not.

The typology of the internationalization of technology development in Japanese firms has some similarities with Ronstadt's model. Stage II corresponds to Ronstadt's TTU, and Stage V to his GTU. However, the Japanese internationalization trajectory has been strongly influenced by two factors: the much longer reliance on export-based strategies for penetrating international markets, and Japan's lengthy experience as a technology follower. In consequence, the earliest stage of technology internationalization for Japanese firms was the development of technology-gathering outposts in the highly industrialized nations. The long period of success in gathering technology for use in their home countries laboratories has not only given Japanese firms the capacity to develop products at home for world markets; it has become something of an impediment in their ability to develop local product development capacity. Even when top management develops a strategic commitment to develop a local product development capacity, the patterns institutionalized in Stages I and II lead the home country R&D organization to treat the developing research center as a "listening post" whose function is to report on local developments in technology and to host visiting technology scanners. In consequence, the facility finds it difficult to hire and keep good local technical people. This difficulty reinforces the belief of the home country organization that the local technical organization can never match the product development capacity in Japan.

Moving on to Stage IV and V therefore requires a very strong commitment on the part of top-level R&D managers as well as top-level general management. It often follows the development of a strong local production and marketing organization that is capable of breaking the long-established loop of gathering technical and local market information and transferring it back to Japan for embodiment in new products or product modifications, which are then transferred again to the local production organization. The lag in product development and improvement time inherent in this loop has been a major factor in stimulating Japanese and local managers to a commitment to develop a local product development capacity.

3 Intrafirm Coordination of Dispersed R&D

The establishment of R&D centers abroad expands the range of technological expertise on which the firm can draw effectively. However, establishing the relationships among the dispersed centers to ensure synergies in technology development strategies raises problems of internal coordination.

Figure 12.2 presents a range of configurations for intra-function coordination that develop in response to an international technology strategy. The circles represent country R&D centers (the numbers within the circles indicate different countries). The rectangles are groups of countries. In general, coordination becomes more difficult as the number of countries increases; the figure presents only a simplified set of models.

The first model is a "country-centered" approach, which concentrates all R&D activities in one country. Strictly speaking, it is not part of an international technology strategy, even though R&D is undertaken on a global scale for multiple countries. This model makes for the easiest type of intrafirm coordination, and it preserves economies of scope and scale in R&D. Many Japanese companies still pursue this approach.

The second model is "pooled," when R&D activities are conducted at several overseas bases, and half of the research is initiated by each base, making for simultaneous, parallel R&D within the company. In this mode, some firms clearly mandate a division of labor, so that each research base has a distinctive mandate (either by product or by project segment). Others permit some duplication of R&D among their overseas bases, but usually in the same way that they permit project duplication within their home country R&D organization: to select the most promising outcome for the corporation's technology pool.

This approach is relatively simple, and close horizontal coordination across the R&D bases is not a complex problem. However, it puts a heavy load on management control systems to prevent unnecessary duplication of R&D investment. In the extreme case, it might give rise to a multidomestic strategy in which each R&D base develops a complete set of products for the national market in which it is located.

Otsuka Pharmaceutical is one example of a firm which has adopted this approach. Each overseas R&D base conducts its own basic and advanced research in the pursuit of original new pharmaceutical products. Interdependence with the home research organization is not expected; the products developed at each lab become part of the worldwide product line of the firm.

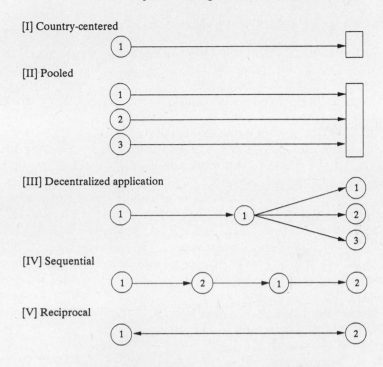

[I] Country-centered

[II] Pooled

[III] Decentralized application

[IV] Sequential

[V] Reciprocal

NOTE: Circles represent countries. Each number represents a different country. Rectangles represent plural countries.

FIGURE 12.2 Configurations of R&D Activity Flow

The third model is "decentralized application," in which the firm concentrates roughly half its R&D activities in Japan (particularly basic and advanced product development), and distributes the remaining half in offshore R&D centers, which focus on applied product development. The centralized part of R&D emphasizes the expansion of the basic technology portfolio of the firm; direct contact with local markets and associated local product development are pursued offshore. The actual ratio of centralized to decentralized R&D is a matter of strategic choice. If the centralized part grows too large, the R&D pattern approaches the·first model described above ("country-centered"); if the offshore R&D centers come to dominate, it approaches the second model ("pooled").

This third model leads to increased complexity in managing the interdependence of the home country and local research centers. Nevertheless, many Japanese companies are taking this route. For instance, many IC makers put their custom IC development facilities abroad, where they can be closer to the customer. Several of the pharmaceutical companies test their new drugs in laboratories in Europe and the United States, close to local markets and local regulatory authorities.

The fourth model is the "sequential" strategy, in which dispersed R&D centers share their results on a continuous basis. A typical example is the joint development of software by Xerox in the United States and Fuji-Xerox in Japan. Since 1986, these companies have built up a satellite telecommunications network. At the end of each day, Fuji-Xerox engineers in Japan electronically send their files to their U.S. counterparts. The work then continues in the United States, and at the end of the U.S. working day, the process is reversed. The goal, obviously, is to minimize development time by mobilizing development expertise in both countries, and the most important advantage of this approach is speed of development. It requires the project organization and the technology to be highly standardized across locations, and perhaps works best with routine development work, such as debugging in software development.

The fifth and final model is the "reciprocal" approach, which also features a two-way exchange in the R&D process, but which is distinguished from the sequential model by a division of labor across sites. This is ideal for mobilizing complementary expertise, but it is the most difficult in terms of coordination. A good example is the joint development of a laptop computer, the DG One, by Data General in the United States and its subsidiary in Japan, Nippon Data General. The Japanese side was in charge of hardware and the U.S. side developed the software. The project was conceived and refined through interaction between the two sites, and there was a frequent two-way flow of information throughout the development process.

There are very few actual examples of the "sequential" and "reciprocal" modes, and even fewer successful cases. For example, while Data General's laptop computer featured many noteworthy technical accomplishments, the product itself was not a market success. Other firms were able to move quickly to match its distinctive features, and were quicker to produce incremental innovations to reduce its cost and improve its features. The geographic separation between the two parts of the product development project in Data General may have inhibited those subsequent incremental innovations. Nevertheless, the rapid improvement of international data communications networks will continue to ease the technical problems of cross-border communication and may well make these modes more possible and profitable in the future.

Of these five strategies, all but the first require some degree of coordination and interaction across borders within the technology function, and even the first requires cross-border interactions between the technology function and offshore production. One of the key challenges facing the Japanese firm in internationalizing technology development is the extent to which it adapts to other societies the organizational systems which have been so successful within Japan for managing product and process innovation and for linking product development and production. In their overseas production facilities that undertake considerable local value adding (beyond simple assembly), Japanese manufacturers have made greater efforts than their U.S. or European counterparts to transfer and adapt their home country production organization. As they move into higher value adding activities in technology development, efforts to do the same with the organization of product development and technology transfer are likely, in order to maintain their competitive advantages in design for manufacturability and quality, incremental innovation, and short design cycles. The challenges in adaptation are very great, especially in North America, because of the differences in technology management systems, to which we now turn.

4 Technology Management Systems in Japan and North America

The most fundamental difference between Japanese and U.S. technology management is the locus of responsibility for the career of the technical employee: in North America, that responsibility rests primarily with the individual; in Japan, it rests with the company. This difference underlies the patterns of technology transfer, incremental innovation, and human resource development in large Japanese corporations.[1]

5 Recruitment

The difference in systems begins with recruitment. Leading Japanese firms still hire their technical employees from among new university graduates; those hired to work on new product development are likely to have M.S. degrees. Even today, hiring researchers with experience in other companies is rare, and limited to cases where the firm is diversifying

---

[1] The following discussion is based heavily on our study of the careers and organization of engineers in product development in three Japanese and three U.S. computer companies. A questionnaire was distributed to 98 Japanese and 109 U.S. engineers and interviews were conducted with individual engineers and research managers in those companies. The full research report is available as a working paper from the M.I.T.-Japan Science and Technology Program under the title of "A Comparative Study of the Training, Careers, and Organization of Engineers in the Computer Industry in Japan and the United States" by D. Eleanor Westney and Kiyonori Sakakibara.

TABLE 12.1
Comparison of Motives for Taking Courses

Question: "How important was each of the following motives for taking the additional scientific, engineering, or math courses since receipt of last degree?"

Motives	Japan (n=51) Mean (S.D.)	U.S. (n=56) Mean (S.D.)	t
To update existing skills	4.25 (1.01)	4.00 (1.21)	1.18
To add new skills	3.71 (0.97)	4.56 (0.68)	−5.09***
To improve chances of promotion	2.02 (1.09)	2.77 (1.26)	−3.23***
To improve chances of assignment to more interesting activities	2.30 (1.29)	3.18 (1.40)	−3.30***
Assigned to course by company	3.32 (1.44)	1.32 (0.89)	8.31***

NOTES: (1) Mean score of 5-point Likert scale from 1=unimportant to 5=very important. (2) ***$p < 0.01$

into new technology fields. In this process Japanese companies depend heavily on close, long-term relationships with universities and with key professors within those universities. The result is an annual intake of researchers of similar age, experience, academic background, and career orientation. This similarity is reinforced in the "freshman training" given to new recruits. Japanese companies in general maintain a well-organized program for new hires; the first one or two months is usually conducted in a special training facility. The next two years are closely supervised on-the-job training, with designated "mentors" within the research organization.

In most U.S. technology-intensive firms, the technical organization is made up of people who vary widely in age, job experience, academic background, and career orientation. Few U.S. firms use formal entry-level training programs to develop more homogeneous capabilities and orientations. Instead, college graduates are hired into specific research teams and undergo whatever training they receive with the help of team members.

On the other hand, both Japanese and U.S. firms offer considerable mid-career training. In our study of technical organization in the computer industry (Westney and Sakakibara 1985), for example, we found no great differences between the Japanese and U.S. companies in the opportunities for mid-career technical training courses. The motives behind enrollment in these courses, however, were quite different, as Table 12.1 shows.

In both countries, updating existing skills was an important motivation. But many more Japanese said that they took the program because they were assigned to the course by their company. Many U.S. researchers, on the other hand, cited such motives as "to add new skills," "to improve chances of promotion," and "to improve chances of assignment to more interesting activities." They hoped to use such courses for personal growth or career advancement. Their responses reflect the strong interest in—and responsibility for—designing their own careers. U.S. companies are expected to provide opportunities for education and to provide financial and managerial support for continuing training. But the responsibility for taking advantage of those opportunities rests with the individual. Supervisors can make suggestions to individuals; they can rarely assign them to courses, as is routinely the case in Japan.

6 Evaluations and Rewards

How technical employees are evaluated and rewarded differs dramatically between Japan and the United States, as indicated by the data in Table 12.2.

In Japan, evaluations are usually based on daily observations by managers. Some companies have elaborate formal systems that specify evaluation criteria and formats, but in practice evaluations are informal and are carried out with minimal direct feedback to the employee. The annual face-to-face evaluation meetings that play so important a role in U.S. research organizations are conspicuous by their absence in Japan. The opportunity that such interviews give employees to emphasize their accomplishments and for both parties to exchange information and receive feedback are replaced in the Japanese system by an annual written self-assessment that employees are required to give to their managers. But the employee receives no formal feedback during the process.

This kind of evaluation system works well if there is a strong, trust-based relationship between researchers and managers. Without such a relationship, frustration levels can rise, because there is no system to allow researchers to vent their dissatisfaction. Despite the prevailing image of dedicated, company-oriented researchers in Japan, the levels of frustration are growing, especially among younger and more ambitious researchers.

Frustration levels are growing over the reward system in Japanese technical organization. Even though most Japanese companies have adopted formally a more merit-based pay system, managers are reluctant to differentiate significantly among researchers in either base pay

TABLE 12.2

Perceptions of Appraisal Process

Question: "The following questions concern your views about the appraisal of engineers in your organization. Please indicate how much you agree or disagree with the following statements."

Statements	Percent indicating "agreement"	
	Japan (n = 98)	U.S. (n = 109)
a. The performance of engineers is evaluated by the end results of their efforts rather than by the amount of effort itself.	71.4%	55.6%
b. Data collection on the performance of each engineer is highly mechanized, i.e., scoring systems exist for rating specific types of behavior	7.1%	18.9%
c. The evaluation criteria for each engineer are individualized according to special circumstances, job and organizational situations.	64.3%	50.5%
d. In this organization judgmental or subjective appraisal by the engineer's superior is emphasized.	90.8%	59.8%
e. A regular formal face-to-face assessment interview is emphasized in the appraisal of engineers.	28.6%	70.4%
f. The performance of an engineer is evaluated over a period of five to ten years so that his potential capabilities can be taken into account	48.0%	10.6%
g. In this organization encouragement and rewards usually outweigh criticism or negative sanctions.	34.7%	57.0%
h. In this organization people get financial rewards in proportion to the excellence of their job performance.	15.3%	55.1%

or bonuses. And they have no freedom to differentiate across functions: base pay in research laboratories is set by the same criteria as in manufacturing or marketing within that company. In consequence, job performance has relatively little impact on salary and bonus for the individual researcher (Table 12.2, item h). On the other hand, the Japanese system does evaluate researchers over a long time period (Table 12.2, item f), and enables the company to assess and utilize the talents of each member of the technical organization.

One of the most significant rewards and incentives for Japanese researchers, therefore, is assignment to interesting and potentially important projects. But such assignments are firmly controlled by research managers in Japan, whereas in the United States researchers can seek out and volunteer for specific projects. Some U.S. companies have gone so far as to adopt an "internal market" whereby project positions are posted within the company and individual researchers encouraged to volunteer.

The greater voluntarism in the U.S. companies also leads to the possibility that researchers will leave team assignments during the project. Asked whether there was a chance of leaving their current project teams during the course of the research, 77.7 percent of the Americans said "of course" or "probably, although I might hesitate a bit." On the other hand, 71.4 percent of the Japanese respondents replied either "of course not" or "yes, but only under special conditions," meaning that in principle such departures were rare.

7 Career Patterns

Most Japanese researchers expect to move eventually into line management positions, even those whose interests and abilities lie primarily in technology and research. In interviews, we have found that Japanese researchers are more oriented than their U.S. counterparts to acquiring general managerial skills (as opposed to research management skills). In large part this is due to a widespread perception that the rewards and opportunities are much greater in the management track, and in part it is attributable to highly homogeneous expectations about the direction of careers.

A typical career path for a Japanese researcher who joins a corporate research facility is to remain there for the first seven to ten years as a researcher, usually on new product development, and then to move to a product division to work on incremental product improvement for a few years. The next step, after a few years in the divisional laboratory, is a promotion into line management in the product division. The move from the corporate to the divisional lab is usually an integral part of the technology transfer process: the researcher moves with a project on which he or she has been working, and personally follows the project through manufacturing to the market. In the divisional lab, the next few years are often spent in incremental product improvements on that same product, or a related product family. In the United States, in contrast, outlining a "typical" career for a researcher is an almost impossible task. But most researchers spend most of their working lives in the technical organization; relatively few move out into line management positions in production or general management, and even fewer aspire to do so.

8 Implications of Changing Technology Strategies

Clearly the Japanese system of technical management has worked extremely well in the past. The standardization of careers and the relatively low input from technical people in shaping those careers has meant that the company can move technical employees from product development into close interaction with production and assign them the responsibility for ongoing incremental improvements, regardless of individuals' personal interests and preferences. Resistance from employees' is rare, because the pattern is so strongly institutionalized and taken for granted. The rewards are long term: they are enhanced prospects for advancing into the ranks of upper management. Only a very small number of the most outstanding senior researchers have a long-term career in the research organization. In consequence the corporate research organization is constantly renewed by the entry of new graduates and the movement outward of more experienced researchers.

On the other hand, the system faces challenges from the changing technology strategies of Japanese firms. Internationalization raises questions about whether the system can be transplanted and adapted to other societies. The move to basic research raises the question of whether a system so well-suited to close linkages with production and to continuous improvement of products can accommodate the different individual and managerial orientations required by a major commitment to basic research. One of the key strengths of the system is the homogeneity of technical careers. Japanese firms are beginning to struggle with a difficult dilemma: can they change their existing systems slightly to accommodate the new demands of internationalization and basic research, or must they add new units whose patterns are in sharp contrast to those in the parent technical organization? And if they choose the latter strategy, can they keep the newer patterns from eroding the patterns that have been so successful in the past?

The answers to these questions will only emerge over the next decade. Managers and scholars alike are becoming increasingly aware of the scope of the problems; the solutions still lie in the future.

References

De Meyer, Arnoud and Atsuo Mizushima. 1989. Global R&D Management. *R&D Management*, vol. 19, no. 2.

Dertouzos, Michael L., Richard K. Lester, and Robert M. Solow. 1989. *Made in America: Regaining the Productive Edge*. Cambridge, MA: M.I.T. Press.

Hakanson, Lars, and Udo Zander. 1986. *Managing International Research and Development*. Stockholm: Sveriges Mekanforbund.

Herbert, Evan. 1989. Japanese R&D in the United States. *Research-Technology Management*, vol. 32, no. 6, pp. 11–20.

Ishikawa, Kenjiro. 1990. *Japan and the Challenge of Europe 1992*. London: Pinter Publishers.

MITI (Tsusho Sangyosho Sangyo Seisaku Kyoku). 1989. *NichiBei no Kigyo Kodo Hikaku* (A Comparison of Japanese and U.S. Enterprise Behavior). Tokyo: Nihon Noritsu Kyokai.

National Research Council. 1989. *Learning the R&D System: University Research in Japan and the United States*. Washington, D.C.: National Academy Press.

Perrino, Albert C., and James W. Tipping. 1989. Global Management of Technology. *Research-Technology Management*, May/June, pp. 12–19.

Ronstadt, Robert. 1977. *Research and Development Abroad by U.S. Multinationals*. New York: Praeger.

Rosenberg, Nathan, and W. Edward Steinmueller. 1988. Why Are Americans Such Poor Imitators? *American Economic Review*, vol. 78, no. 2 (May), pp. 229–34.

Saito, Masaru and Hiroyuki Itami. 1986. *Gijutsu Kaihatsu no Kokusai Senryaku* (International Strategies for Technology Development). Tokyo: Toyo Keizai Shimposha.

Sakakibara, Kiyonori. 1988. Increasing Basic Research in Japan: Corporate Activity Alone Is Not Enough. Working Paper No. 8802, Graduate School of Commerce, Hitotsubashi University.

Stalk, George, Jr., and Thomas M. Hout, 1990. Competing Against Time: How Time-Based Competition Is Reshaping Global Markets. New York: The Free Press.

Trevor, Malcolm. 1988. *Toshiba's New British Company*. London: Policy Studies Institute.

Westney, D. Eleanor, and Kiyonori Sakakibara. 1985. A Comparative Study of the Training, Careers, and Organization of Engineers in the Computer Industry in Japan and the United States. M.I.T. Working Paper.

13

International Collaborative Ventures and the Commercialization of New Technologies

DAVID C. MOWERY

During the past two decades the relative economic and technological strengths of the United States have declined. These strengths are now more evenly distributed among U.S. and foreign firms. Combined with changes in product and process technologies and in the industrial and trade policies of the United States and foreign governments, this new international economic environment has contributed to growth in the number and importance of collaborative ventures between U.S. and foreign firms in the development and commercialization of new technologies.

International collaborative ventures are not completely novel undertakings for U.S. firms, but those into which U.S. firms have entered during the past decade differ in some significant ways from the joint ventures of the 1950s and 1960s. European and U.S. participation in these ventures has grown simultaneously with expansion in domestic

An earlier version of this paper was presented at the Conference on Economic Growth and the Commercialization of New Technologies sponsored by the Technology and Economic Growth Program of the Center for Economic Policy Research at Stanford University, September 11–12, 1989. I am grateful to conference participants for useful comments and suggestions. Preparation of the paper was aided by support from the Technology and Economic Growth program at the Center for Economic Policy Research at Stanford University, the Alfred P. Sloan Foundation, Berkeley, and the American Enterprise Institute. Portions of the paper draw on Mowery 1989 and on the case studies in Mowery 1988.

research collaboration in both the United States and Western Europe. Examples of privately financed domestic cooperative research ventures in the United States include the Microelectronics and Computer Technology Corporation (MCC) and others involving universities, such as the recently announced IBM-AT&T initiative in high-temperature superconductivity at M.I.T. In Europe, many regional cooperative research programs receive partial support from public funds, often under the sponsorship of the European Communities; current EC programs include ESPRIT, BRITE, RACE, and the older Airbus Industrie consortium, which is not directly sponsored by the EC and which focuses on product development and manufacture, rather than research. Interfirm cooperation in precommercial research, often supported in part with public funds, has been a prominent feature of Japanese science and technology policy since the 1960s and 1970s.[1]

Much of this intra-Japan, intra-U.S. and intra-EC collaboration is focused on activities that are somewhat further "upstream" than those included in most of the international collaborative ventures that have developed in the last decade.[2] Most recent technology-centered international collaborative ventures appear to focus on product development, manufacture, and/or marketing.

International collaboration is in part a response to the same factors that have led to increased domestic collaboration in research among U.S. firms. How have these factors contributed to growth in international collaborative ventures? How do they influence the types of collaboration between U.S. and foreign firms that have developed in recent years? Does recent experience provide lessons for the organization and management of these ventures? These issues are addressed in this paper, which deals primarily with international collaborative ventures that focus on technology commercialization. This topic largely excludes detailed discussion

---

[1] See Sigurdson 1986, Okimoto 1989, and Mowery and Rosenberg 1989b for discussions of U.S. and Japanese domestic research collaboration.

[2] In a discussion of collaborative ventures within the EC and those involving U.S. and EC firms, Mytelka and Delapierre (1987) note that "Intra-EC agreements are oriented far more towards knowledge production than are agreements between EC and American firms. Thus agreements embodying a knowledge component account for 41 per cent of total intra-EC agreements but only twenty per cent of the EC-USA agreements, and knowledge production agreements made up 21 per cent of intra-EC compared with 14 per cent of the EC-U.S. agreements. While intra-EC agreements stressed knowledge production, EC-U.S. agreements [emphasized] the need for greater market access. 37 per cent of the EC-U.S. arrangements involved the commercialization of products and a further seventeen per cent the production of goods. The comparable figures for intra-EC agreements were 27 per cent and 9 per cent respectively..." (p. 240).

of interfirm collaboration focused on precommercial or fundamental research, and also devotes little attention to interfirm collaboration that is confined to production or marketing.

1 International Collaborative Ventures: Definition and Recent Trends

Collaboration between U.S. and foreign firms assumes numerous forms. Many collaborative ventures fit a narrow definition of a joint venture, including separate incorporation as an entity in which equity holdings are divided among the partners. Others, such as partnerships between "risk-sharing" subcontractors and prime contractors or the purchase by one firm of an equity share in another, do not. In some industries, joint ventures and other forms of collaboration are extensions of subcontracting relationships that cover product development and manufacture; others focus on the marketing of products manufactured largely by one partner. This paper uses the terms "collaboration" and "joint venture" interchangeably and therefore occasionally denotes as joint ventures organizational structures that do not fit the legal definition of this entity.[3]

An international collaborative venture may be defined as *interfirm collaboration in product development, manufacture, or marketing that spans national boundaries, is not based on arm's-length market transactions and includes substantial and continual contributions by partners of capital, technology, or other assets.* In many cases, management responsibility is shared among the partner firms. This definition excludes other forms of international economic activity, such as export, direct foreign investment (which implies complete intrafirm control of production and product development activities), and the sale of technology through licensing.

One can distinguish among at least four types of technology-focused collaborative ventures.[4] The first involves collaboration among firms in research alone—these ventures and consortia, however, generally involve only domestic firms (defining "domestic" to include firms from

---

[3]The formal structure of collaborative ventures, however, often has little to do with either their management or their success, as Gullander (1976) has noted: "There are indications that the difference between a contractual and an equity relationship is highly exaggerated; sophisticated 'cooperators' seem to downplay the importance of ownership control as compared to management control or control through other means" (p. 86). Porter and Fuller (1986) make a similar point.

[4]Chesnais 1988 provides a taxonomy of collaborative ventures on which I have drawn in defining these categories.

throughout the EC when discussing European domestic research collaboration). A second category of technology-focused collaboration that includes many international ventures is the exchange of "proven" technologies within a single product line or across multiple products. These ventures have been particularly prominent in the global microelectronics industry, perhaps because of the long-established practice of cross-licensing within the industry, and are also widespread in robotics. A third category of international collaboration involves joint development of one or more products—this type of venture typifies international collaboration in commercial aircraft and engines and in some segments of the telecommunications equipment, microelectronics, and biotechnology industries. Finally, a number of international collaborative ventures in biotechnology, pharmaceuticals, steel, and automobiles involve collaboration across different functions, with one firm providing a new product or process for marketing, manufacture or application in a foreign market (or, in the case of biotechnology, marketing within the home-nation economy) by another. An important issue for future research concerns the extent of differences in stability or success among these forms of collaboration.

Joint ventures have long been common in extractive industries such as mining and petroleum production (see Stuckey 1983) and account for a significant share of the foreign investment of U.S. manufacturing firms since World War II.[5] Several features of recent collaborative ventures, however, differentiate them from earlier cases. The number of collaborations, both those involving only U.S. firms and those between U.S. and foreign enterprises, has grown. Joint ventures now also appear in a wider range of industries.[6]

---

[5]Hladik's analysis (1985) of data from the Harvard Multinational Enterprise Project concluded that 39 percent of the number of foreign subsidiaries established by U.S. manufacturing firms during 1951–75 were joint ventures.

[6]Harrigan (1984) concluded that domestic joint ventures involving U.S. firms had grown during the previous decade. In the 1960s, joint ventures were concentrated in the chemicals; primary metals; paper; and stone, clay, and glass industries, but now extend beyond these sectors (see also Harrigan 1985). Hladik (1985) found significant growth from 1975 to 1982 in the number of international joint ventures involving U.S. firms, a growth trend that has almost certainly continued through the present. Hladik's conclusions disagree with those of Ghemawat, Porter, and Rawlinson (1986), who compiled a time series of "international coalition announcements" (joint ventures, license agreements, supply agreements, and "other long-term interfirm accords" [p. 346]) covering 1970–82 that displays no upward trend. The differences are likely to be more apparent than real; the Ghemawat et al. database includes a number of interfirm mechanisms of collaboration that are excluded by Hladik. The possibility thus exists that a shift in the mix of the different forms considered by

U.S. firms have entered into international alliances in industries in which direct foreign investment or multinational corporate organization were the exception rather than the rule. Commercial aircraft, for example, is an industry in which direct foreign investment historically has been minimal (Hirsch 1976). Yet this industry now is a leader in international collaboration (Mowery 1987). Other U.S. industries in which these ventures have assumed considerable importance in recent years include both "mature" (e.g., steel and automobiles) and "young" (biotechnology) industries.

In U.S. industries characterized by a small number of large "system integrators" and a large group of suppliers, such as commercial aircraft or automobiles, the increasingly global scale of the activities of the domestic and foreign firms in the top tier (global scale that may have been achieved through collaborative ventures or through direct foreign investment in the United States or abroad) has "cascaded" downward, forcing supplier firms to contemplate or undertake international collaborative ventures. Such ventures have increased rapidly in the U.S. automotive supply industry in the past five years and have also grown in the aerospace supplier sector.[7]

The central activities of many recent collaborative ventures, including research, product development, and production for world markets, were absent from most of the ventures of the pre-1975 era, which focused primarily on production and marketing for the domestic market of the non-U.S. partner firm. The central activities of many of these recent ventures, R&D and product development, may result in significant technology transfer. Several assessments have expressed concern about the long-term implications of international collaboration for the competitiveness of U.S. industry.[8] Other accounts assert that collaborative ventures will produce "hollow corporations," not engaged in producing goods, employing a small number of highly paid financial and marketing executives and few if any production workers (*Business Week*, March 3, 1986). In some cases (e.g., commercial aircraft), most of the technology transfer within international alliances does consist of exports of techno-

---

Ghemawat et al. occurred during 1970–82, as joint ventures and other collaborative agreements increased in importance relative to such alternatives as licensing.

[7]See Donne 1987 or Phillips 1989.

[8]Reich and Mankin (1986) argue that "...the U.S. strategy [in joint ventures with Japan] seems shortsighted. In exchange for a few lower skilled, lower paying jobs and easy access to our competitors' high-quality, low-cost products, we are apparently willing to sacrifice our competitiveness in a host of industries—autos, machine tools, consumer electronics, and semiconductors today, and others in the future" (pp. 78–79).

logy from the United States. In other industries, however, such as steel or autos, these ventures are associated with significant U.S. imports of technology.

International collaborative ventures that deal with technology development and commercialization are based on an exchange relationship, but the relationship covers different commodities from those included in licensing or export transactions. Some collaborations allow U.S. and foreign firms to pool their technological capabilities in a single product without merging all of their activities into a single corporate entity. As was noted earlier, other forms of collaboration support the exchange of technologies across products. In most of these cases, the participants in the collaborative venture are competitors in one or more product markets. Still other collaborative ventures combine one firm's technological capabilities with the marketing or distribution assets of another for a single product. These ventures more frequently involve firms that are not direct competitors. Why are collaborative ventures, rather than alternative mechanisms such as licensing, employed for these exchanges?

2   Comparing Joint Ventures and Other Channels for the Exploitation of Technological and Other Firm-Specific Assets

In recent years, economists have emphasized the importance of firm-specific assets such as technological capabilities in explaining the growth of direct foreign investment and multinational firms.[9] Direct foreign investment, however, obviously is not the only mechanism for reaping the economic returns to these assets. Alternatives to direct foreign investment for the exploitation of firm-specific assets include licensing and collaborative ventures. These different channels for reaping the returns to an asset may be substitutes or complements, and firms frequently utilize multiple channels. The advantages and disadvantages of each of these methods depend on the characteristics of the asset in question. These factors also affect the activities undertaken within collaborative ventures, an issue that is discussed shortly.

Markets for the licensing of advanced technologies that exploit firm-

---

[9]Caves (1982) notes that "...as indicators of these [firm-specific] assets, economists have seized on the outlays for advertising and research and development (R&D) undertaken by firms classified to an industry. That the share of foreign subsidiary assets in the total assets of U.S. corporations increases significantly with the importance of advertising and R&D outlays in the industry has been confirmed in many studies..." (p. 9).

specific knowledge often are very thin, with few buyers or sellers. The dearth of alternative outlets for sellers or alternative sources for buyers means that opportunistic behavior may hamper the operation of markets for technology licenses.[10] Other impediments to licensing technological assets include the tacit nature of much of the knowledge necessary to exploit the technology, the need to reveal a great deal about a technological advance in order to convince a prospective licensee of the value of the license (Arrow 1962, Teece 1986), and the problems of regulating licensor and licensee behavior in a dynamic and uncertain world.

Licensing is likely to be preferred to either direct foreign investment or collaborative ventures in technologies that are not complex, that have strong, well-enforced patents, that are relatively mature, and that do not rely on "user-active" innovation (von Hippel 1976), with its requirements for strong links between marketing and product development. These features are not characteristic of the technology in the steel, automobiles, commercial aircraft, and telecommunications equipment industries, all of which have been active in international collaborative ventures. Licensing has been an important alternative to international collaboration in pharmaceuticals. Licensing and collaboration often complement one another in the microelectronics and robotics industries and in biotechnology.

Direct foreign investment provides the strongest basis for exploiting firm-specific technological capabilities with little danger of leakage or dilution of control. This intrafirm channel for exploitation of such capabilities involves high risks and costs, however, and may not facilitate rapid penetration of foreign markets. The cost penalties of establishing multiple production facilities argue against direct foreign investment in industries in which production technologies exhibit a large minimum efficient scale or strong plant-specific learning and cost-reduction effects; commercial aircraft (primarily airframes) and steel are examples of such industries. The high fixed costs and lengthy delays associated with establishing an offshore production, distribution, and marketing network may preclude direct foreign investment. Licensing, direct foreign investment, and export all force the innovating firm to bear all of the costs

---

[10]In Williamson's terminology, a "small numbers condition" characterizes such markets: "The transactional dilemma that is posed is this: it is in the interest of each party to seek terms most favorable to him, which encourages opportunistic representations and haggling. The interests of the *system*, by contrast, are promoted if the parties can be joined in such a way as to avoid both the bargaining costs and the indirect costs (mainly maladaptation costs) which are generated in the process" (1975, p. 26; emphasis in original).

and risks of research and development, which have grown considerably in many U.S. industries, especially those in the technology-intensive sector.

Uncertainties about economic and political conditions in many foreign markets may further reduce the attractiveness of direct foreign investment. Political barriers have impeded the establishment of wholly owned production facilities in high-technology industries in a number of nations; both the Japanese and the U.S. governments, for example, have opposed some forms of direct foreign investment in their domestic semiconductor industries.[11] International joint ventures have substituted for, and in some cases have complemented, Japanese and U.S. firms' direct foreign investment activities, especially in the integrated circuit, automobile, and pharmaceuticals industries.

What advantages does international collaboration have over licensing or direct foreign investment? Many of the contractual limitations and transactions costs of licensing for the exploitation of technological capabilities can be avoided within a collaborative venture. The problem of determining the value of partners' contributions can be reduced through collaboration. Partner firms make financial commitments to a collaborative venture that back their claims for the value of the assets they contribute; such financial commitments can substitute for the complete revelation of the value and characteristics of the asset that may be necessary to complete a licensing agreement.[12]

---

[11]Steinmueller (1988) discusses the resistance of the Japanese government in the 1960s to the establishment by U.S. firms of wholly owned manufacturing subsidiaries in microelectronics. U.S. government opposition to the sale by Schlumberger of its Fairchild Semiconductor subsidiary to Fujitsu in 1984 is indicative of official U.S. concern over Japanese investment in this industry.

[12]Brodley (1982) summarizes the advantages of joint ventures, defined as separate corporate entities in which all partners hold equity shares, over mergers or market transactions as follows: "By providing for shared profits and managerial control, joint ventures tend to protect the participants from opportunism and information imbalance. The problem of valuing the respective contributions of the participants is mitigated, because they can await an actual market judgment. The temptation to exploit a favored bargaining position by threatening to withhold infusions of capital or other contributions is reduced by the need for continuous cooperation if the joint venture is to be effective. Moreover, a firm supplying capital to the joint venture can closely monitor the use of its contributed capital and thereby reduce its risk of loss. Common ownership also provides a means of spreading the costs of producing valuable information that could otherwise be protected from appropriation only by difficult-to-enforce contractual undertakings. Finally, joint ventures can effect economies of scale in research not achievable through single-firm action. Because of these advantages, joint ventures are especially likely to provide an optimal enterprise form in undertakings involving high risks, technological innovations, or high information costs" (pp. 1528–29).

The noncodified, "inseparable"[13] character of firm-specific assets that may preclude their exploitation through licensing need not prevent the pooling of such assets by several firms within a joint venture, or the effective sale of such assets by one firm to another within a joint venture. Joint ventures enable partner firms to "unbundle" their portfolios of technological assets and transfer components of this portfolio, components that may be worthless in isolation, to a partner. The transfer of technology through a joint venture from a technologically advanced firm to a less advanced enterprise thus may enable the technologically "senior" firm to reap some financial returns to portions of its portfolio of technological capabilities. The difficulties of unbundling the senior firm's technological portfolio for arm's-length transfer mean that these returns cannot be captured through licensing. Technology transfer may also be controlled or regulated more effectively within collaborative ventures than in licensing transactions.[14] Monitoring the behavior of the recipient of technology within a joint venture reduces the risk that the transferor will not benefit from any improvements in transferred technologies made by the recipient. These factors have been influential in collaborative ventures in steel (National Steel and Nippon Kokkan), automobiles (Toyota and General Motors), and commercial aircraft (General Electric and SNECMA; Boeing and the Japan Commercial Transport Development Corporation).

Collaborative ventures offer an alternative to the complete merger of firms as a means of pooling assets. Such ventures may cover only a limited range of products; partner firms often are competitors in other product areas. Collaboration can provide established firms a faster and less costly means than internal development to gain access to new technologies that are not easily licensed. This "technology access" motive for collaboration between established and young firms has been particularly important in industries based on new technologies, such as biotechnology and robotics.

By comparison with direct foreign investment, licensing, or export, collaborative ventures also reduce the financial and political risks of

---

[13] "Inseparability" refers to the fact that much of the firm's noncodified technological know-how may be embedded in the organization. Its transfer therefore will require the transfer of a large number of individuals. Separating and transferring a substantial portion of the parent firm's staff to another enterprise may be infeasible (see Teece 1982 for further discussion).

[14] Kogut (1988) and Porter and Fuller (1986) discuss the potential of joint ventures to reduce incentive conflicts and lower the possibilities for opportunistic behavior that often undercut licensing agreements.

innovation and foreign marketing. The products of a collaborative venture between a U.S. and a foreign firm may well encounter fewer political impediments to access to the domestic market of the foreign firm than would direct exports from the U.S. firm.

Nonetheless, the potential difficulties of collaborative ventures should not be minimized. Management of these undertakings has proven to be extremely difficult. Of the four categories of collaborative ventures described earlier, joint product development ventures appear to be the most complex and costly, occasionally resulting in either failure or significantly higher costs and longer development times than independent development. Even within a jointly financed technology development partnership, the value of partners' contributions may not be easily established, and this difficulty is compounded by uncertainty about technological and market outcomes in such a venture.[15]

Ventures that are further removed from the market may face fewer difficulties in agreeing on the value of partner contributions, if the contributions (e.g., money or technical personnel) are truly homogeneous. Nonetheless, U.S. domestic research consortia like MCC have had serious problems in extracting contributions of well-qualified personnel from participating firms (see Sanger 1984). These problems may be less serious in ventures that combine one firm's product development expertise with the production or marketing skills of another. Such ventures generally are closer to the market, which means that uncertainty about costs, prices, and volumes may be lower, and a value can be more easily assigned to the contributions. Misrepresentation of the value of these contributions nevertheless is possible in this situation as well, and the value of contributions is likely to change over time, forcing renegotiation or the demise of the venture.

Yet another important source of conflict in technology-focused col-

---

[15]Doz (1988) has pointed out the difficulties of measuring partner contributions in product development ventures that lie midway on a continuum between commercial production and fundamental research: "When dealing with basic technology development—usually early in a partnership, much before a competitive stage—parity is maintained between the partners through balance in contribution; and the potential output is still so distant that precise valuation is not an issue. When dealing with well-developed technologies, a precise valuation of the outcome can be made, and the partners are almost at the stage of supply contracts, with precise product or system specifications, costs and prices, and some volume forecasts. Yet a 'danger zone' often separates these stages in the evolution of a partnership, in the transition from precompetitive stages to competitive ones. During that transition one partner, but not the other, may shift from a valuation of the partnership based on contribution to one based on expected results, and show impatience with a divergence from the position of the other partner" (p. 38).

laborative ventures is technology transfer. Especially in ventures that involve firms with different technological capabilities, the senior firm wishes to minimize, and the junior firm to maximize, the amount of technology transfer. Even if conflicts among the participants over the amount of technology transfer can be resolved, actions to control transfer may threaten the viability of the venture.

This problem can be illustrated by the case of International Aero Engines, a consortium founded in 1982 to develop the V2500 jet engine that includes Pratt and Whitney, Rolls Royce, Fiat, Motoren-Turbinen Union, Ishikawajima-Harima Heavy Industries, Kawasaki Heavy Industries, and Mitsubishi Heavy Industries. Pratt and Whitney and Rolls Royce, the "senior partners" within the consortium, attempted to minimize the transfer of engine technology within the consortium by designing the engine in modular form and assigning the development of different modules to different participants. Serious problems in the integration of the engine components, however, led to delays in the delivery of the V2500 engine and to a loss of orders (*Aviation Week and Space Technology*, February 9, 1987; *Aviation Week and Space Technology*, March 16, 1987; Carley 1988). The V2500 project is a good example of the cost and time penalties that may result from international collaboration in product development.

The demands of product development partnerships also may clash with the other, independent activities or products of the participant firms. For example, independently manufactured products may become competitors with the jointly developed product. Such encroachment contributed to the 1977 demise of the collaboration of Pratt and Whitney and Rolls Royce that was intended to develop a high-bypass, high-thrust engine (the JT10D), and contributed to the collapse of a joint venture in airframes between Fokker and McDonnell Douglas in 1982.

Why have international collaborative ventures, which represent a hybrid of interfirm and intrafirm modes for the exchange or sharing of technological and other assets, assumed greater importance recently for U.S. firms, and why do these ventures now incorporate a wider range of activities? The basic answer is simple—changes in the technological and policy environment within which U.S. firms operate have made the potential contributions of foreign firms to collaborative ventures much more attractive to U.S. firms.

3 Causes of Increased Reliance on Joint Ventures

Changes in the technical capabilities of foreign firms and in the nature of product demand have increased U.S. firms' demand for foreign partners in collaborative ventures. The enhanced technological capabilities of many foreign firms mean that their potential contributions to joint ventures with U.S. firms now are more valuable. Foreign firms now are better able to absorb and exploit advanced technologies from U.S. firms in industries in which there remains a substantial technology gap between U.S. and foreign firms. In other industries, foreign firms either are more advanced or are the technological equals of U.S. firms and therefore can contribute managerial or technological expertise to joint ventures with U.S. firms. U.S. firms in the automotive and steel industries and some U.S. firms in the microelectronics industry now collaborate with foreign firms to gain access to superior foreign technologies.

The costs of the research and development necessary to bring a new product or process to market in many high-technology industries have risen considerably in the past 20–30 years—for example, commercial aircraft development costs have grown at an annual rate of nearly twenty percent for decades, despite advances in the application and productivity of the capital equipment used in the R&D process (Mowery and Rosenberg 1982a). Similarly rapid growth in development and marketing costs has characterized the telecommunications equipment, computer, and microelectronics industries. Rising development costs place severe strains on the ability of firms to sustain ambitious R&D programs and increase the importance of penetration of foreign markets to ensure commercial success. Firms in some industries require markets substantially larger than those provided by a single western European nation or even (in some cases) by the huge U.S. domestic economy. Moreover, high development costs raise the risks of new product development, since they increase the fixed costs incurred before introduction of the product.

Another source of cost pressure on R&D programs is a form of technological convergence. Technologies that formerly were peripheral to the commercial and research activities of a firm now have become central to competitive advantage in a number of technology-intensive industries. The increased interdependence of telecommunications and computer technologies is perhaps the best known example of such convergence, but others include the growing importance of biotechnology within pharmaceuticals and food processing, or the greater salience of computer-based machine vision technologies within robotics equipment. Technological convergence means that firms must develop expertise quickly in a broader array of technologies and scientific disciplines, further straining R&D budgets and human resources. This factor has contributed to expansion in domestic interfirm research collaboration and in research collaboration between industry and universities in the

United States and elsewhere (see Mowery and Rosenberg 1989b). Advanced communications and computer-assisted design and production technologies also facilitate the operation of international collaborative ventures.

A reduction in the duration of product cycles in many high-technology industries has increased the urgency of rapid penetration of global markets with new products. Rapid foreign market penetration is also more important because of the declines in the share of global demand accounted for by the U.S. market in many high-technology industries, and by the growing homogenization of demand characteristics across the industrial economies. The "product cycle" model of direct foreign investment hypothesized that differences in local economic conditions gave rise to different firm-specific attributes that eventually were exploited in foreign markets through direct foreign investment. Increasingly, however, economic development and more rapid international technology transfer mean that the characteristics of domestic markets within the industrial world differ less, and the firm-specific assets and products that develop to serve these markets now are less "country-bound."[16] Simultaneous introduction of a product in multiple industrial economies now may be essential to commercial success. Such rapid penetration may require joint production or collaboration with a firm with an established marketing network.

The importance of technical standards also makes rapid access to many markets with a new product particularly important in the microelectronics, computer, and telecommunications equipment industries. The establishment of a product as a *de facto* standard or dominant design may provide a profitable platform for the introduction of related products and subsequent generations of the dominant design.[17] In the global telecommunications equipment market, this dominant design

---

[16]See Vernon 1979 and Dunning 1988 for similar arguments.

[17]The strong incentives to establish a product standard contributed to the practice of "second-sourcing" in the semiconductor industry, which produced a complex network of technology exchange and cross-licensing agreements in the 1960s and 1970s. Farrell and Gallini (1988) and Swann (1987) analyze the monopolist's incentives to establish multiple sources for a new product.

Second-sourcing was also directly encouraged by the Defense Department during the early years of the semiconductor industry. The incentives to establish one's product as a dominant design through encouraging duplication diminish somewhat the economic returns to strong intellectual property protection in these products. The shift in IBM's attitude toward imitators of the PS/2 from hostility and threats of patent infringement suits to liberalized licensing terms, illustrates the trade-offs between intellectual property protection and strategies to establish a dominant design.

motive is supplemented by the recognition that a presence in many markets can contribute to a firm's influence within international standards negotiations. Collaborative ventures and technology exchange agreements now are pursued by some manufacturers of workstations and telecommunications equipment and services in order to achieve a dominant design position.[18]

Still another factor underpinning the recent growth in both domestic and international collaboration involving U.S. firms is the pivotal role of relatively small startup firms in the commercialization of new technologies within the United States (and therefore the world) economy. The successive waves of new product technologies that have swept through the postwar U.S. economy, including semiconductors, computers, and biotechnology, have been commercialized in large part through the efforts of new firms.[19] Small firms appear to have been more important sources of new commercial technologies in the United States than in Japan and Western Europe, where established firms have played a more significant role in new electronics or biotechnology products. In the microelectronics and computer industries, the important role of small firms resulted in part from U.S. government procurement demand, which created a substantial market with comparatively low marketing and distribution barriers to entry. The benefits of the military market were enhanced further by the substantial possibilities for technological "spillovers" from military to civilian applications.

The U.S. military market no longer plays such a strategic role in the computer and semiconductor industries, and the possibilities for military-civilian technology spillovers appear to have declined in many areas of these technologies (see Mowery and Rosenberg 1989b, and Office of Technology Assessment 1988, for further discussion of military-civilian

---

[18]See *Business Week*, July 24, 1989, which describes the licensing strategy of Sun Microsystems in workstations: "Almost anyone can license Sun's basic software and Sparc, the superfast microprocessor that is the brain of its flagship workstation—a $9,000-and-up desktop machine that packs the power of a minicomputer. If enough manufacturers build Sun clones, the software companies will have to take notice. In the end, everyone will prosper. And Sun's Sparc workstation—it makes six other models—will become a desktop standard..." (p. 75).

[19]This is not to deny the major role played by such large firms as IBM in computers and AT&T in microelectronics. In other instances, large firms have acquired smaller enterprises and applied their production or marketing expertise to expand markets for a new product technology. Nonetheless, it seems apparent that startup firms have been far more active in commercializing new technologies in the United States than in other industrial economies. Malerba's analysis of the evolution of the microelectronics industry in Western Europe and in the United States (1985) emphasizes the greater importance of startup firms in the United States.

technology spillovers). Biotechnology firms also face markets that are heavily regulated in the United States and other nations. As a result, the costs of new product introduction and the marketing-related entry barriers faced by startup firms in microelectronics, computers, and biotechnology now are higher. For this and other reasons, including the greater interest by foreign firms in the technological assets of U.S. startup firms, collaborative ventures involving startup and established U.S. and foreign firms have grown considerably in recent years. These ventures often focus on technology exchange (combined in many instances with the acquisition by an established firm of a substantial portion of the equity of the new firm) and/or marketing (including navigating domestic and foreign product regulations), rather than joint development of new products.

The growth of nontariff trade barriers[20] has also increased the incentives for U.S. firms to seek foreign partners. Tariff barriers tend to favor direct foreign investment or joint production ventures as strategies for market penetration, since they affect only the relative prices of domestically produced and imported goods. Nontariff barriers, however, especially government procurement policies or technology transfer requirements, favor the use of collaborative ventures that incorporate product research, development, and marketing, as well as manufacture. Public procurement policies form significant nontariff barriers in export markets for such goods as commercial aircraft and telecommunications equipment, where public ownership or control of major purchasers is common. These procurement decisions can be influenced by the production (or development and production) of components for the purchased product by domestic firms in the purchaser nation. Foreign governments also frequently provide development funding and risk capital to domestic firms as part of industrial development policies. Combined with high product development costs, the availability of capital from public sources for foreign firms has enhanced their attractiveness as partners in product development ventures with U.S. firms in microelectronics, commercial aircraft, telecommunications equipment, and robotics.

Just as foreign government trade and industrial policies have created incentives for U.S. firms to collaborate with foreign firms in export

---

[20]See the estimate by Tyson (1988) that 35 percent of U.S. imports in 1983 were subject to nontariff restrictions, an increase from an estimated twenty percent in 1980. Olechowski (1987) estimated that seventeen to nineteen percent of the imports of developed nations (by value) were covered by nontariff barriers, and concluded that the use of nontariff barriers increased significantly during 1981–85 (p. 125).

markets, nontariff restrictions on foreign access to U.S. markets have led to increased collaboration between U.S. firms and foreign firms wishing to sell in the United States. In several protected U.S. industries, a foreign production presence has been achieved through a joint venture. Examples include the Toyota-General Motors and Nippon Kokkan-National Steel ventures. In the wake of the U.S./Japan Semiconductor Agreement and Fujitsu's failure to acquire Fairchild Semiconductor, joint ventures have become a more important means for Japanese semiconductor producers to gain access to the U.S. market, and vice versa.[21]

Although the postwar growth of multinational firms and direct foreign investment raised the prospect in some assessments of "global firms" to whom national boundaries would mean little or nothing, much of the current wave of international joint venture activity reflects the opposite phenomenon. National governments are able to influence not only production but, increasingly, the product development and technology transfer decisions of firms through the use of trade and other policies. Paradoxically, however, the pursuit by the United States and other industrialized nations of nationalistic or technologically mercantilistic policies of support for domestic industries has encouraged the development of consortia spanning national boundaries. The efforts of these governments to restrict international transfer of technology from publicly funded domestic research consortia that are closed to foreign firms often create strong incentives for the exchange of the fruits of these programs through international collaborative ventures (Mowery and Rosenberg 1989a). This conjunction of nationalistic domestic research policies and expanding international technology transfer through collaborative ventures has been particularly pronounced in the global microelectronics industry.[22]

4 Influences on the Structure of International Collaborative Ventures

One of the most important motives for international collaboration is access to markets, whether in the United States or in a foreign nation. The

[21] A similar argument is made in Borrus 1988.

[22] See Steinmueller 1988 and Chesnais 1988. Chesnais has noted that an interesting complementary relationship may be developing between closed domestic research programs in the EC and the United States, such as JESSI and Sematech, and international product development and technology exchange agreements in microelectronics: "... one finds a combination between *domestic* alliances in *pre-competitive* R&D (with all of the provisos attached to this notion), and a wide range of technology exchange and cross-licensing agreements among oligopolist rivals at the international level" (1988, p. 95; emphasis in original).

asset provided in exchange for market access frequently is technology. The form in which the technology is provided, which is determined by both the motives of the participants and the characteristics of the technology, plays a central role in structuring the collaboration. Thus, where the technology can be provided in essentially "embodied" or codified form—that is, in a finished product or a license—collaboration either is unimportant or focuses largely on marketing, as in the pharmaceuticals industry.

In more mature high-technology industries, such as telecommunications equipment, microelectronics, and commercial aircraft, the high costs of new product development, demanding requirements for systems integration, and the nature of political barriers to market access all mean that many recent collaborative ventures have focused on product development. Because of the strong complementarities between Japanese firms' expertise in microelectronics process (CMOS) technology and U.S. firms' expertise in product design for microprocessors, several of the U.S./Japanese collaborative ventures in microelectronics each span more than one product.[23] Demanding requirements for systems integration in robotics products that involve a widening array of technologies have also led to a number of domestic and international collaborative ventures among suppliers of robotics and factory automation equipment that deal primarily with product development. This recent wave of collaborative activity in robotics follows an earlier generation of user-supplier collaborations between robotics equipment suppliers and users in the automobile industry (see Klepper 1988). In commercial aircraft, telecommunications equipment, and segments of the microelectronics industry, the desire of U.S. firms for risk-sharing partners and for access to foreign capital or technology provides additional motives to focus collaboration on research and product development.

As was noted earlier, much of the pervasive domestic and international collaboration in biotechnology focuses on the marketing and distribution by established firms of the technologies developed by new entrants. The market access motive for collaboration in this industry applies equally to domestic and international collaborative ventures, and domestic collaborative ventures therefore are more important relative to international ones.[24]

---

[23]Examples include agreements between Toshiba and Motorola and between Texas Instruments and Hitachi. See Steinmueller 1988 and *Business Week*, January 16, 1989.

[24]Hagedoorn and Schakenraad (1988) concluded from their analysis of publicly

Much of the domestic and international collaboration in this industry is intended to support entry into new domestic or foreign markets and does not always incorporate joint product development. The major international collaborations in the automotive and steel industries center on the exchange of foreign process technology, managerial expertise, and production systems for access to U.S. markets. These joint ventures accordingly deal with production rather than with product development.

5 Managing International Collaborative Ventures

Large-scale databases on international collaborative ventures' duration, success and failure do not yet allow empirical testing of detailed hypotheses about the factors that contribute to the success and failure of these ventures. Nonetheless, the available empirical studies and a growing body of case study and anecdotal evidence suggest a number of factors that should be addressed in the organization and management of international collaborative ventures that are technology-focused.

Management of these undertakings should be premised first of all on a recognition of their dynamic character. Firms' motives for collaboration often center on knowledge or technology acquisition—once these processes are sufficiently advanced or completed, partners may no longer wish to remain in a joint venture. These motives themselves often change in response to changes in the environment or within the participating firms. The value of the assets contributed by participating firms to the collaborative venture may also shift in response to changes in markets or technology. Virtually all of these statements apply as well to a wholly owned corporate subsidiary or foreign investment, but the presence of other corporate actors in a joint venture increases the potential for rapid and dramatic change.

A second important precondition for management of joint ventures is a realistic assessment of their costs and benefits—in recent years, the latter may have received more attention than the former. The costs, risks, and benefits of alternatives should be compared with those of

announced collaborative agreements that international collaborative ventures were substantially more important relative to domestic (including intra-EC) ventures in information technology than was true of biotechnology: "Over 45 percent of all agreements in biotechnology are intra-U.S. Only twelve percent of all biotechnology agreements are intra-European agreements and almost seventeen percent are between European and U.S. companies. In information technology the number of European-U.S. agreements comes very close to the number of intra-U.S. agreements. Intra-European agreements reach seventeen percent of all such agreements" (p. 15).

a collaborative venture, and comparisons should be made among different structures for a collaborative venture. As was noted earlier, technology-focused collaborative ventures that center on joint development of a product appear to be more difficult to organize and manage than other forms of international technology-centered collaboration. Alternative structures for interfirm collaboration, including technology exchange agreements, collaboration that is restricted to research, and marketing or production agreements, deserve careful scrutiny and comparison.

Since technology transfer is at the center of many joint ventures, the management by partner firms of both technological development and technology transfer is critical to the success or failure of international collaboration. The interests of technological leaders, reluctant to allow the transfer of key technological capabilities, differ from those of technological followers, whose participation often hinges on the amount of technology transfer. The success of a project may be undercut by the attempts of "junior" partners, motivated by their desire to maximize technology transfer and learning, to participate in all aspects of the project, rather than specializing in a particular area or activity. Alternatively, as was noted earlier, efforts by the senior partners to restrict technology transfer create two risks—alienation of the junior partner firm(s) and/or failure of the project to reach its technical goals.

Entry into a collaborative venture therefore must be predicated on an expectation that technology will be transferred to one's partners. But how much and which technologies? Participation in a collaborative venture requires that the firm assess its own technological capabilities and distinguish between those that are critical to its competitive performance and should not be transferred and those components of the corporate technology portfolio that can be transferred profitably with less risk to the competitive performance of the firm. Two other assessments are also necessary. The firm contemplating entry into a collaborative venture must also examine the nontechnological, complementary assets that are needed to realize the commercial returns to its technological capabilities. As Teece (1986) and others have shown, access to these nontechnological assets can determine innovative success or failure. Finally, the firm contemplating a collaborative venture must examine the technological and nontechnological capabilities of its prospective partners in a collaborative venture before the inception of a venture, because of the central importance of choice of partner(s). In all of these assessments, one must recognize that the quality and abundance of the key technological and nontechnological assets may well change over time.

Technology transfer must be managed carefully over the life of a collaborative venture.[25] Participant firms should create internal mechanisms for absorbing technology transferred from other partners. These may involve the regular rotation of personnel through a collaborative project. The importance of personnel flows reflects the tendency for technology transfer to operate more effectively through the movement of people, rather than through a flow of reports and paper. It is not enough, however, to simply capture knowledge (both codified and tacit) or skills from other firms in the head of an engineer or scientist—that individual must be given opportunities to communicate such knowledge to others within the parent firm. Parallel research or engineering activities within the parent firm are used by many firms in university-industry research collaborations to provide an informed audience for the technological and other knowledge gleaned from a collaborative venture. They may be equally useful in a collaborative venture that focuses on commercial technology development. Monitoring the flow of technology transfer also may be difficult in large firms engaged in a collaborative venture that spans several business units. In these circumstances, a central office or point of contact can monitor the requests for technological or other information and data from partner firms to all divisions or subsidiaries of the corporation, enabling senior managers to track the "balance of trade" within the venture more effectively.

Despite the complexities that it creates for management, technology transfer often acts as a source of cohesion within product development ventures, especially those involving a dominant and a subordinate firm. These ventures often are more durable and successful than those among technological equals (Killing 1983). Evidence from the commercial aircraft industry, in which international collaboration in development projects has been widespread for more than a decade, strongly supports this hypothesis. Product development ventures of technological equals, such as those between Rolls Royce and Pratt and Whitney in the JT10D jet engine project, Fokker and McDonnell Douglas in the MDF100 commercial aircraft project, and Saab and Fairchild in the SF340 commuter aircraft project repeatedly have failed to bring a product to market or have been unable to achieve commercial success with a product after its introduction. Product development ventures between technologically dominant and subordinate firms, however, such as the CFM International venture between General Electric and SNECMA of France and the collaborative ventures involving Boeing and the Japan

---

[25]Several of the arguments in this paragraph are also made in Hamel et al. 1989.

Commercial Transport Development Corporation appear to be more manageable.

Technology transfer has also acted as an adhesive rather than a solvent in collaborative ventures that span several products in the microelectronics industry. Monitoring of technology transfer and the clear establishment of benchmarks for reciprocity within a venture may be easier when one firm's proprietary technology in one product is being traded for a partner's expertise in another product. This form of reciprocity appears to have aided the Motorola-Toshiba venture, in which Toshiba's CMOS process expertise is being exchanged for Motorola's microprocessor design capabilities in an enterprise that produces both microprocessors and DRAMs in a Japanese facility.[26]

The dynamic character of collaborative ventures is reinforced by a tendency for the assets contributed by each partner firm to gradually lose their value to the other participants. As technology is transferred through a collaborative venture, learning by the other participants will reduce the value of the technological capabilities that originally were unique to one or another participant. Depreciation is likely to be even more rapid in ventures in which one firm contributes its marketing knowledge and network or other "country-specific" expertise—as the other participants improve their knowledge of the markets in which this partner has specialized, they may well choose to continue without it.[27]

Depreciation in the value of U.S. firms' contributions has played a role in the breakup of a number of collaborative ventures formed between Japanese and U.S. producers of auto parts. As the Japanese partners in these enterprises gain knowledge about local markets and (particularly when selling to Japanese transplant operations in the United States) local production conditions, they frequently withdraw from the joint venture to continue independently, as Phillips (1989) has noted. Although technology-based assets are likely to depreciate more slowly, especially if technology transfer is closely managed, Hamel et al. (1989) suggest that process technologies are less transparent to other participant firms and therefore may depreciate more slowly than product technologies, which venture partners may learn more easily.[28]

---

[26]See Hamel et al. 1989 and Armstrong 1988.

[27]Porter and Fuller have observed that collaborative ventures centered on marketing "may be particularly unstable, however, because they frequently are formed because of the access motive on one or both sides. For example, one partner needs market access while the other needs access to product. As the foreign partner's market knowledge increases, there is less and less need for a local partner" (1986, p. 334)

[28]"The type of skill a company contributes is an important factor in how easily

Depreciation in the value of assets within a joint venture is no less inevitable than depreciation of physical capital assets within a factory. In both cases, participants must take steps to reduce erosion in the value of their contribution. Intrafirm technology development must underpin the technologies contributed to the joint venture; where a firm is providing a "static" asset like market access, the collaborative venture may function most effectively as a means for exit from the industry or as a channel for learning process and product technologies.

The organizational structure of international collaborative ventures, especially those involving joint research and product development, raises additional challenges. There is no optimal management structure for a collaborative venture—its design will depend, among other things, on the character and magnitude of the contributions of the participants. In collaborations of technological equals, an autonomous management structure in charge of a wide range of design, marketing, production, and product support may be preferable. Such a management structure often is costly, since it duplicates some or all of the management structure of the member firms. Nevertheless, the experience of recent collaborative ventures clearly indicates the importance of strong links between the product development and design teams and the organization responsible for marketing and product support. The organization managing the collaborative venture, be it the dominant firm or an independent hybrid of the parent firms, must retain control of a number of downstream activities. On the other hand, in collaborations involving a senior and a junior firm, financial and organizational structure appears to be less important, so long as the technologically more advanced firm retains overall control of technology and management decisions.

Finally, the case of the Anglo-French Concorde partnership, where

---

its partner can internalize the skills. The potential for transfer is greatest when a partner's contribution is easily transported (in engineering drawings, on computer tapes, or in the heads of a few technical experts); easily interpreted (it can be reduced to commonly understood equations or symbols); and easily absorbed (the skill or competence is independent of any particular cultural context).

"Western companies facé an inherent disadvantage because their skills are generally more vulnerable to transfer. The magnet that attracts so many companies to alliances with Asian competitors is their manufacturing excellence—a competence that is less transferable than most" (Hamel et al. 1989, p. 136). Of course, the reverse is also true—a central technological asset contributed by Boeing to its collaborative ventures with Japanese firms is the U.S. firm's expertise in production technology and in the management of fluctuations in production volume for commercial airframes (Mowery 1987).

total project costs rose from $450 million in 1962 to $4 billion by 1978, illustrates the need for building cost controls into the structure of a collaborative venture. This issue is important because of the tendency for participants to be less concerned about minimizing shared costs. In the Airbus and other joint ventures with less disastrous financial consequences than Concorde, fixed-price contracts between a central management organization and the partner firms have preserved incentives for partners to minimize costs. Profit-sharing, rather than cost-sharing, is crucial.

This discussion provides additional reasons to be skeptical about the prospects for collaborative ventures in product development. Within this class of collaborative ventures, those involving partnerships of technological equals appear to be the most difficult. In any consideration of collaborative ventures, then, the product development collaboration of equals should receive the most severe scrutiny and critical analysis. Another issue that spans many of the factors discussed above concerns the relative stability of partnerships of firms with similar assets and the durability of ventures involving firms with complementary capabilities or assets. Despite assertions to the contrary in much of the literature, this analysis suggests that within technology-centered collaborative ventures, complementary capabilities are a greater source of strength and stability than are strong similarities in the technological and other assets contributed by partner firms to a collaborative venture.[29] As in other endeavors, opposites may attract in international collaborative ventures, and the resulting relationship may be more durable than one based on similar endowments.

6 Public Policy Implications of International Collaborative Ventures

Although the full impact of international collaboration on the competitiveness of U.S. firms will not be apparent for some time, a comparative analysis of international collaboration in U.S. manufacturing does not support the critical view of international joint ventures presented by Reich and Mankin (1986), nor does a specific public policy governing such collaborative ventures appear to be advisable. Technology transfer within these ventures is more modest in scope and less uniformly "outbound" than some assessments assume. Just as U.S. industries vary in

---

[29]Both Porter and Fuller 1986 and Hennart 1989 argue that partnerships based on similar assets are more durable, although Hennart applies his observation to a broader class of collaborative ventures than those centered on technology.

their trade balances in goods, the net inflow or outflow of technology through international collaborations varies across industries. Requiring balance in technology transfer on an industry-by-industry basis makes no more sense than a requirement for such balance in goods trade.

Restrictions or controls on international collaborative ventures involving U.S. firms do not appear to be an effective means to improve U.S. international competitiveness and in fact might impair competitiveness. The complexity of international collaborative ventures, the fact that the pattern and impact of these ventures vary considerably across industries, and the historical evidence that restrictions on technology transfer are either ineffective or perverse in their impacts (Harris 1986, COSEPUP Panel on the Impact of National Controls on International Technology Transfer 1987) all argue against controls on collaborations involving nondefense technologies.

Efforts to restrict international collaboration also overlook the fact that in a number of industries, including steel, automobiles, and portions of microelectronics, international collaboration can improve the international competitiveness of the U.S. participants. In other industries, such as robotics, the competitiveness of U.S. systems engineering and software firms and the ability of large U.S. firms to offer a "full line" of factory automation hardware and software depend on access to foreign hardware through joint ventures and licensing.

Concern over the effects on U.S. competitiveness of collaborative ventures between established foreign firms and small startup firms in the United States has grown in recent years. Are these small firms exporting critical technological assets that will strengthen foreign competitors? In many instances, startup firms pursue international collaborative ventures because of their need for capital. Policies to reduce this supposed outflow of U.S. technology must address the availability of capital and/or the willingness of managers in established U.S. firms to support small startup enterprises (e.g., overcoming resistance to technologies "not invented here"), rather than simply attempting to restrict collaboration. Smaller firms' "export" of technology through international collaborative ventures rarely means that opportunities for the exploitation of this technology are lost to U.S. firms—in most cases, the U.S. partner in successful collaborative ventures does not change its management or its location, and protection of its intellectual property is not airtight. The critical agents for the diffusion of these technologies (employees of the small firm) remain in the United States, where they move to other firms, present the results of their research to domestic audiences, and otherwise act to disseminate much of the technology domestically. Location may be as important as ownership in this area.

The basis for the critical view of international collaboration by startup U.S. firms also is open to criticism. Much of the recent concern about U.S. "giveaways" of critical technologies through international collaborative ventures is based on a view of recent history that exaggerates the importance of previous transfers of U.S. technologies in the development of foreign competition. Although such transfers assuredly occurred, a more restrictive approach by U.S. managers to licensing foreign firms might have had little effect on subsequent developments. In most cases, the technologies in question were available from other sources, and in virtually all cases, foreign (primarily Japanese) firms applied considerable engineering ingenuity in both process technology and product design to modify these technologies for sale in domestic and foreign markets.[30]

Although a policy concerned specifically with international collaboration in civilian technology development appears ill-advised, the incentives for U.S. firms to enter into international collaborative ventures may be influenced by the antitrust, trade, and research policies of the U.S. government. Should these policies be revised in order to reduce or modify the incentives for collaboration between U.S. and foreign firms?

There is little evidence to support the argument that U.S. antitrust policy is a central factor in the decisions of American firms to collaborate with foreign enterprises.[31] International collaborative ventures generally are not substitutes for the collaboration among U.S. firms that might develop in the absence of antitrust restrictions. In the semi-

---

[30]See Office of Technology Assessment 1987. The home video cassette recorder, frequently cited as an example of U.S. invention and Japanese commercialization, in fact depended on significant technological advances by the Japanese firms that later commercialized the technology, advances that rested on indigenous R&D efforts that were active for nearly twenty years before the introduction of this product (See Rosenbloom and Cusumano 1987). As the title of Lardner's 1987 article ("The Terrible Truth about Japan: They Didn't Steal Our VCR Technology—They Invented It") indicates, the effects of any U.S. "giveaway" of VCR technology may have been modest.

[31]Nelson (1984) argues that "Although I am less easy about joint design and production ventures than I am about generic research cooperation, it does not seem right that such an international venture would receive totally different treatment from that received by two or more U.S. firms in a design and production venture for which the market is clearly international. . . It seems odd that we would discriminate against a national partnership if each partner judged this venture more promising economically than an international consortium" (p. 84) See Jorde and Teece 1988 for a similar view.

conductor industry, for example, the firms that have developed domestic research collaborations and consortia are among the most active in international collaborative ventures. If U.S. firms were forced to choose between international and domestic collaborative ventures because of antitrust restrictions (see Langlois et al. 1988 for a discussion of the semiconductor case), the opposite relationship would hold, in which firms active in international collaboration did not pursue domestic collaboration.

The primary motive for many collaborative ventures—access to foreign markets—is not affected by restrictions on collaboration between U.S. firms. In other cases, such as the collaborative ventures in the automobile and steel industries, U.S. firms collaborate with foreign firms in order to gain access to technological and other assets that are not available from other U.S. firms. Antitrust policy also has little effect on precommercial research collaboration among U.S. firms in the wake of the National Cooperative Research Act of 1984, and antitrust rarely applies to vertical collaborations between firms and their suppliers or to technology exchange and commercialization agreements between established and startup firms.

Antitrust policy has played an indirect role, however, in international collaborative ventures in several industries. The 1983 Modified Final Judgment in the case of *U.S. v. AT&T* opened the domestic U.S. market to foreign telecommunications equipment suppliers and unleashed AT&T to compete in markets outside of telecommunications and in foreign markets for telecommunications equipment. As a result, AT&T entered into international collaborative ventures to penetrate foreign markets for equipment and developed alliances with domestic (Sun Microsystems) and foreign (Olivetti) producers of computers to aid its entry into businesses outside of telecommunications. Joint ventures between the Bell operating companies and foreign equipment producers also could develop as a means of aiding foreign access to the U.S. telephone equipment market if current judicial restrictions on the activities of the U.S. operating companies were lifted.

Antitrust policy has affected international collaboration in other industries as well. The limited duration of the NUMMI joint venture between General Motors and Toyota is a direct result of the settlement reached in the antitrust suit against NUMMI filed by the Chrysler Corporation. There is some evidence in the commercial aircraft engine industry that the U.S. Department of Justice has allowed international collaboration between competing firms that would not have been allowed between the major U.S. firms, based on an inappropriate distinction between the domestic and international markets for these products (see Mowery 1987). The evidence is not conclusive in this case, however, and the example appears to be an isolated one.

Restrictions on U.S. imports also have resulted in collaboration between U.S. and foreign firms that was prohibited among U.S. firms for antitrust reasons. The conjunction of antitrust and trade restrictions influenced the joint venture between National Steel and Nippon Kokan. That collaboration was established after the Justice Department quashed the sale of a number of National Steel plants to the U.S. Steel Corporation, based on the Department's judgment that the domestic U.S. market was protected from imports.[32] The consequences of this collaboration, however, do not appear to be harmful to the long-term competitiveness of the U.S. firms involved. Indeed, in the absence of such a collaborative venture, National Intergroup (parent firm of National Steel) might have withdrawn entirely from the steel industry. The tendency of trade restrictions to create such incentives for international collaboration is another reason to avoid protectionist policies, rather than an argument for the relaxation of antitrust restrictions in protected industries.

Research and product development subsidies are another potential inducement to engage in international collaboration. But with the exception of spectacular cases of subsidized product development in the commercial aircraft industry (Airbus Industrie, which by some estimates has received billions of dollars in public subsidies since 1968; see Krugman 1984), public financial support for recent Japanese and European technology development programs is modest, both absolutely or by comparison with U.S. programs. Moreover, current foreign programs typically focus on precommercial research, rather than on the development of specific products. Where the results of these publicly supported research programs are published in the open scientific literature, the programs can in fact benefit nonparticipant U.S. firms, which may gain access to many of the results of the basic research without having to bear the full costs of such research. This possibility, however, requires a greater investment by U.S. firms in monitoring foreign scientific and engineering research and, possibly, a greater willingness to utilize scientific and

---

[32]Hufbauer and Rosen (1986) have noted this tendency, arguing that "...a pattern has arisen in which transnational associations raise fewer antitrust concerns than combinations involving two domestic producers...The numerous equity purchases plus one joint venture in the auto industry, as well as several joint ventures in steel, exemplify this phenomenon. While this is the logical consequence of import restraints, it was probably an unintended effect." (p. 62)

technological advances developed outside of the firm, as Japanese firms have done with considerable success.[33]

Two of the three major areas of trade policy that affect collaboration—public subsidies and government procurement—are now covered by multilateral agreements ("Codes") under the General Agreement on Tariffs and Trade that were negotiated during the Tokyo Round of trade talks. In addition, the commercial aircraft industry is covered by an Agreement on Trade in Civil Aircraft that specifies acceptable practices for procurement and development funding. To the extent that international collaboration poses a long-term threat to U.S. international competitiveness and is motivated by the trade-distorting practices of foreign governments, both the Codes and the Agreement on Trade in Civil Aircraft must be strengthened.

The Uruguay Round of trade talks that began in 1986 is discussing the treatment of foreign investment, including the "right of establishment" by firms of production and other operations in foreign markets. Since government restrictions on foreign investment provide a significant motive for international collaborative ventures, the negotiation of multilateral rules for the treatment of foreign investment could alter the incentives for international collaboration in some industries. The development and enforcement of stronger forms of intellectual property protection, another subject of the Uruguay Round negotiations, may also facilitate the licensing of process and product technologies to foreign firms, increasing the attractiveness to U.S. firms of an alternative channel for the exploitation of technical capabilities in such industries as biotechnology. The limited importance of patent, copyright, and trade secret protection for the establishment or preservation of a technological advantage, however (see Levin et al. 1987), means that stronger intellectual property protection probably will have a modest impact on international collaboration in most U.S. manufacturing industries.

The growing role of "managed trade" and bilateral negotiations in U.S. trade policy, especially negotiations over access to foreign (frequently Japanese) markets, may lead to additional collaborative ventures between U.S. and foreign firms. Recent bilateral negotiations between the U.S. and Japanese governments, have produced understandings that stipulate target market shares for U.S. firms in the foreign market (see Bhagwati 1987, Prestowitz 1987, World Bank 1987). Achievement of this outcome in many markets, especially the domestic Japanese

---

[33]See Mansfield 1988, Rosenberg and Steinmueller 1988.

market, requires either direct foreign investment by U.S. firms, which is slow, risky, and costly, or the formation of long-term collaborations between U.S. and foreign firms.

The bilateral talks that led to the U.S./Japan Semiconductor Agreement of 1986 have been followed by negotiations over construction services and may lead to talks covering other industries. If these talks and the associated government monitoring of trade outcomes expand, so too will international collaborative ventures.[34] Trade policy influences international collaboration between U.S. and foreign firms, but the results of such collaboration increasingly may affect trade policy as well. As the national origins of the components in a central office telephone switch, jet engine, or robot become more blurred and complex, national trade policies that are predicated on the protection or promotion of goods with a high domestic content will be unworkable (See Mowery and Rosenberg 1989a, 1989b).

Which U.S. industries are likely to rely on international collaboration as a permanent component of their global competitive strategies? Industries in which collaboration in product development is motivated by a desire to reduce risk, to gain access to technology, or to deal with the effects of a significant government role in procurement or in the resolution and regulation of international trade disputes, will continue to rely on international collaboration. These industries include telecommunications equipment, commercial aircraft, and (potentially) semiconductors. The reluctance of Japanese firms to assume ownership of additional production capacity, as well as restrictions on foreign access to the U.S. market mean that international collaboration in the steel industry is likely to continue. The "cascading" effect mentioned earlier also suggests that smaller firms in the U.S. automobile, aerospace, and electronics industries will increasingly have to deal with international collaboration.

In other U.S. industries, international collaboration may be more transitory. The wave of collaborations between major Japanese and U.S. automobile firms that expanded U.S. production of Japanese auto designs, for example, may have peaked. The major Japanese producers now have established production facilities in the United States, bypassing the barriers to their access to this market. International collaboration that is motivated by market access concerns in industries in which

---

[34]According to one account, the March 1988 agreement between the U.S. and Japanese governments easing access by U.S. firms to Japanese public construction projects has resulted in an extensive set of joint ventures between U.S. and large Japanese construction firms (Rubinfien 1988).

governments are not major customers may also decline in importance in the long run, as small startup firms in these industries either mature and expand their nontechnological capabilities or are acquired by larger firms. These motives figure prominently in biotechnology, segments of the robotics industry, and pharmaceuticals. Commercial development of new technologies (e.g., high-temperature superconductivity applications) by relatively small U.S. firms in the future nevertheless is likely to involve a greater degree of domestic and international collaboration.

7 Conclusion

International collaborative ventures are a relatively new channel for U.S. firms' technology development and commercialization efforts. Their very novelty makes it difficult to render a definitive verdict on their implications for national competitiveness, but current evidence does not support the view that they will do great damage. Any summary evaluation of collaborative ventures must recognize the great differences among industries in the focus and structure of international collaboration, as well as the differences in the international flows of technology that occur within these undertakings. In most cases, international collaborative ventures are a result, rather than a cause, of intensified international competition with the products or technologies of U.S. firms. They reflect changes in the international competitive environment (e.g., the reduction of U.S. technological and economic hegemony as a result of economic reconstruction in Western Europe and Japan, combined with economic development in East Asian economies) that long were the objectives of U.S. foreign and economic policies.

Equally difficult to summarize from the limited experience with these ventures are the optimal approaches to organizing and managing these ventures. Careful consideration of collaborative ventures must recognize their complexity and evaluate alternatives to collaboration. Many of the policies suggested earlier, such as the assessment of the technological and nontechnological assets that are essential to a firm's competitive performance and those of one's competitors, are also necessary for the management of technology development and commercialization within the firm. Recognition of the dynamic nature of collaborative ventures and of the need to invest internally in the strengthening of the assets contributed to the joint venture and in the transfer and exploitation of learning within the joint venture, also is important. International collaboration may be a less costly and faster means of commercializing new technologies, but it is by no means costless.

Substantial additional research is necessary in order to formulate a richer set of prescriptions for public policy makers and private managers. In particular, better data on the structure and performance of collaborative ventures are essential. These data will not be easily collected from the trade and financial press, the source of many of the data sets that have been used thus far in empirical work on international collaborative ventures. It is also important to broaden the data to include firms that are not among the ranks of the global multinationals—international collaborative ventures increasingly will affect U.S. firms that historically have not been involved in international operations. Finally, more research is needed on the microeconomic causes of increased domestic and international collaboration in research and technology development. We know relatively little, for example, about the causes (or even the rate) of escalation in the costs of technology development and commercialization, although this escalation surely contributes to growing collaboration in the international and domestic spheres. The phenomenon of technological convergence is also widely remarked but rather less widely measured, analyzed, or understood. Better understanding of the causes and implications of international collaborative ventures will require analysis of these and other conditioning factors.

References

Armstrong, L. 1988. A Chipmaking Venture the Gods Smiled On. *Business Week*, July 4, p. 109.

Arrow, K. J. 1962. Economic Welfare and the Allocation of Resources for Invention. In Universities-National Bureau Committee for Economic Research, *The Rate and Direction of Inventive Activity*. Princeton, N.J.: Princeton University Press.

Arrow, K. J. 1969. Classificatory Notes on the Production and Transmission of Technical Knowledge. *American Economic Review*, 59, pp. 29–35.

Aviation Week and Space Technology. 1987a. U.S., Europeans Clash Over Airbus Subsidies. February 9, pp. 18–20.

Aviation Week and Space Technology. 1987b. Pratt and Whitney Expands Role in V2500 Compressor Work. March 16, pp. 32–33.

Bhagwati, J. 1987. Trade in Services and the New Multilateral Trade Negotiations. *The World Bank Economic Review*, 1, pp. 549–69.

Borrus, M. 1988. *Competing for Control*. Cambridge: Ballinger Publishers.

Bozeman, B., A. Link, and A. Zardkoohi. 1986. An Economic Analysis of R&D Joint Ventures. *Managerial and Decision Economics*, 7, pp. 263–66.

Brodley, J. 1982. Joint Ventures and Antitrust Policy. *Harvard Law Review*, 95, pp. 1523–90.

Brodsky, N. H., H. Kaufman, and J. Tooker. 1980. *University/Industry Cooperation*. New York: Center for Science and Technology Policy.

Business Week. 1986. Special Report: The Hollow Corporation. March 3, pp. 57–85.

Business Week. 1988. AT&T: The Making of a Comeback. January 18, pp. 56–62.

Business Week. 1989. What's Behind the Texas Instruments-Hitachi Deal. January 16, pp. 93–94.

Business Week. 1989. Is the U.S. Selling Its High-Tech Soul to Japan? June 26, pp. 117–18.

Business Week. 1989. Clonemakers Don't Scare Sun–It's Sending Them Engraved Invitations. July 24, p. 75.

Carley, W. M. 1988. Cancelled Jet Order is a Setback for United Technologies Unit. *Wall Street Journal*, February 9, 1988, p. 6.

Caves, Richard E. 1982. *Multinational Enterprise and Economic Analysis*. Cambridge: Cambridge University Press.

Chesnais, F. 1988. Technical Co-Operation Agreements Between Firms. *STI Review*, pp. 51–119.

COSEPUP (Committee on Science, Engineering, and Public Policy) Panel on the Impact of National Security Controls on International Technology Transfer. 1987. *Balancing the National Interest: U.S. National Security Export Controls and Global Economic Competition*. Washington, D.C.: National Academy Press.

Dertouzos, M., R. Lester, and R. M. Solow. 1989. *Made in America* (Cambridge: M.I.T. Press).

Donne, M. 1987. Dunlop Joins U.S. Group in Airbus Contract Bid. *Financial Times*, September 17, p. 6.

Doz, Y. L. 1988. Technology Partnerships Between Larger and Smaller Firms: Some Critical Issues. *International Studies of Management & Organization*, 17, pp. 31–57.

Dunning, J. H. 1988. *Multinationals, Technology, and Competitiveness*. London: Unwin Hyman.

Farrell, J., and N. Gallini. 1988. Second-Sourcing as a Commitment: Monopoly Incentives to Attract Competition. *Quarterly Journal of Economics*, pp. 673–94.

Ferguson, C. 1988. From the People Who Brought You Voodoo Economics. *Harvard Business Review*.

Florida, R. L., and M. Kenney. 1988. Venture Capital-Financed Innovation and Technological Change in the USA. *Research Policy*, 17, pp. 119–37.

Ghemawat, P., M. E. Porter, and R. A. Rawlinson. 1986. Patterns of International Coalition Activty. In M. E. Porter, ed., *Competition in Global Industries*. Boston, MA.: Harvard Business School Press.

Gomes-Casseres, B. 1988. Joint Ventures in Global Competition. Harvard Business School Working Paper 89–032.

Gomes-Casseres, B. 1989. Ownership Structures of Foreign Subsidiaries: Theory and Evidence. *Journal of Economic Behavior and Organization*.

Gullander, S. 1976. Joint Ventures in Europe: Determinants of Entry. Working paper, Columbia University Graduate School of Business.

Hagedoorn, J., and J. Schakenraad. 1988. Strategic Partnering and Technological Co-operation. Presented at the EARIE conference, Rotterdam, August 31–September 2.

Hamel, G., Y. Doz, and C. K. Prahalad. 1989. Collaborate with Your Competitors—and Win. *Harvard Business Review*, January/February, pp. 133–39.

Harrigan, Kathryn R. 1984. Joint Ventures and Competitive Strategy. Working paper. New York: Columbia University Graduate School of Business.

Harrigan, Kathryn R. 1985. *Strategies for Joint Ventures*. Lexington, MA.: D. C. Heath.

Harris, J. 1986. Spies Who Sparked the Industrial Revolution. *New Scientist*, May 22, pp. 42–47.

Hennart, J.-F. 1988. A Transactions Cost Theory of Joint Ventures. *Strategic Management Journal*, pp. 361–74.

Hippel, Eric von. 1976. The Dominant Role of Users in the Scientific Instrument Innovation Process. *Research Policy*, 5, pp. 221–39.

Hirsch, S. 1976. An International Trade and Investment Theory of the Firm. *Oxford Economic Papers*, 28, pp. 258–69.

Hladik, Karen. 1985. *International Joint Ventures*. Lexington, MA.: D. C. Heath.

Hof, R., and N. Gross. 1989. Silicon Valley is Watching Its Worst Nightmare Unfold. *Business Week*, September 4, 1989, pp. 63–67.

Hufbauer, G. C., and H. F. Rosen. 1986. *Trade Policy for Troubled Industries*. Washington, D.C.: Institute for International Economics.

Jorde, T., and D. J. Teece. 1988. Innovation, Strategic Alliances, and Antitrust. Presented at the Conference on the Centennial Celebration of the Sherman Act, University of California, Berkeley, October 7–8.

Killing, J. P. 1983. *Strategies for Joint Venture Success*. New York: Praeger.

Klein, R., R. G. Crawford, and A. A. Alchian. 1978. Vertical Integration, Appropriable Rents, and the Competitive Contracting Process. *Journal of Law and Economics*, 21, pp. 297–326.

Klepper, S. 1988. Collaborations in Robotics. In D. C. Mowery, ed., *International Collaborative Ventures in U.S. Manufacturing*. Cambridge: Ballinger Publishers.

Kogut, B. 1988. Joint Ventures: Theoretical and Empirical Perspectives. *Strategic Management Journal*, pp. 319–32.

Krugman, P. W. 1984. Import Protection as Export Promotion: International Competition in the Presence of Oligopoly and Economies of Scale. In H. Kierzkowski, ed., *Monopolistic Competition and International Trade*. Oxford: Oxford University Press.

Lane, H. W., R. G. Beddows, and P. R. Lawrence. 1981. *Managing Large*

Research and Development Programs. Albany: State University of New York Press.

Langlois, R. N., T. A. Pugel, C. S. Haklisch, R. R. Nelson, and W. G. Egelhoff. 1988. *Microelectronics: An Industry in Transition*. Boston: Unwin Hyman.

Lardner, J. 1987. The Terrible Truth About Japan. *Washington Post*, June 21, p. B19.

Leiberman, M., and D. Montgomery. 1988. First-Mover Advantages. *Strategic Management Journal*.

Levin, R. C., A. K. Klevorick, R. R. Nelson, and S. G. Winter. 1987. Appropriating the Returns from Industrial Research and Development. *Brookings Papers on Economic Activity*, pp. 783–820.

Lineback, J. R. 1987a. Can MCC Survive the Latest Defections? *Electronics*, January 22.

Lineback, J. R. 1987b. It's Time for MCC to Fish or Cut Bait. *Electronics*, June 25.

Malerba, F. 1985. *The Semiconductor Business: The Economics of Rapid Growth and Decline*. Madison, WI: University of Wisconsin Press.

Mansfield, E. 1988. Industrial Innovation in Japan and the United States. *Science*, September 30, pp. 1769–74.

McCulloch, R. 1984. International Competition in High-Technology Industries: The Consequences of Alternative Trade Regimes for Aircraft. Presented at the National Science Foundation Workshop on the Economic Implications of Restrictions to Trade in High-Technology Goods, Washington, D.C., October 3.

Mowery, D. C. 1987. *Alliance Politics and Economics: Multinational Joint Ventures in Commercial Aircraft*. Cambridge, MA.: Ballinger Publishers.

Mowery, D. C., ed. 1988. *International Collaborative Ventures in U.S. Manufacturing*. Cambridge, MA: Ballinger Publishers.

Mowery, D. C. 1989. Collaborative Ventures Between U.S. and Foreign Manufacturing Firms. *Research Policy*, 18, pp. 19–32.

Mowery, D. C., and N. Rosenberg. 1979. The Influence of Market Demand Upon Innovation: A Critical Review of Some Recent Empirical Studies. *Research Policy*, 8, pp. 102–53.

Mowery, D. C., and N. Rosenberg. 1982a. Government Policy and Innovation in the Commercial Aircraft Industry, 1925–75. In R. R. Nelson, ed., *Government and Technical Change: A Cross-Industry Analysis*. New York: Pergamon Press.

Mowery, D. C., and N. Rosenberg. 1982b. Commercial Aircraft: Cooperation and Competition Between the U.S. and Japan. *California Management Review*.

Mowery, D. C., and N. Rosenberg. 1989a. New Developments in U.S. Technology Policy: Implications for Competitiveness and Trade Policy. *California Management Review*, 32, pp. 107–24.

Mowery, D. C., and N. Rosenberg. 1989b. *Technology and the Pursuit of Economic Growth*. New York: Cambridge University Press.

Mytelka, L. K., and M. Delapierre. 1987. The Alliance Strategies of European Firms in the Information Technology Industry and the Role of ESPRIT. *Journal of Common Market Studies*, pp. 231–53.

National Academy of Engineering. 1985. *The Competitive Status of the U.S. Civil Aircraft Manufacturing Industry*. Washington, D.C.: National Academy Press.

Nelson, R. R. 1984. *High-Technology Policies: A Five-Nation Comparison*. Washington, D.C.: American Enterprise Institute.

Office of Technology Assessment. 1987. *International Trade In Services*. Washington, D.C.: U.S. Government Printing Office.

Office of Technology Assessment. 1988. *Commercializing High-Temperature Superconductivity*. Washington, D.C.: U.S. Government Printing Office.

Okimoto, D. 1989. *Between MITI and the Market: Japanese Industrial Policy for High Technology*. Stanford: Stanford University Press.

Olechowski, A. 1987. Nontariff Barriers to Trade. In J. M. Finger and A. Olechowski, eds., *The Uruguay Round: A Handbook for Negotiators*. Washington, D.C.: World Bank.

Perlmutter, H. V., and D. A. Hennan. 1986. Cooperate to Compete Globally. *Harvard Business Review*, pp. 136–52.

Phillips, S. 1989. When U.S. Joint Ventures with Japan Go Sour. *Business Week*, July 24, pp. 30–31.

Pisano, G. P., W. Shan, and D. J. Teece. 1988. Joint Ventures and Collaboration in the Biotechnology Industry. In D. C. Mowery, ed., *International Collaborative Ventures in U.S. Manufacturing*. Cambridge, MA: Ballinger Publishing Co.

Porter, M. E., and M. B. Fuller. 1986. Coalitions and Global Strategy. In M. E. Porter, ed., *Competition in Global Industries*. Boston: Harvard Business School Press.

Prestowitz, C. 1987. *Trading Places*. New York: Basic Books.

Reich, R. B., and E. D. Mankin. 1986. Joint Ventures with Japan Give Away Our Future. *Harvard Business Review*, March/April, 79–86.

Rosenberg, N., and W. E. Steinmueller. 1988. Why Are Americans Such Poor Imitators? *American Economic Review*, pp. 229–34.

Rosenbloom, R. S., and M. A. Cusumano. 1987. Technological Pioneering and Competitive Advantage: The Birth of the VCR Industry. *California Management Review*, 29, pp. 51–79.

Rubinfien, E. 1988. U.S. Contractors Forge Alliances in Japan. *Wall Street Journal*, June 28, p. 28.

Sanger, D. 1984. Computer Consortium Lags. *New York Times*, September 5, p. D1.

Sigurdson, J. 1986. *Industry and State Partnership in Japan–The Very Large*

Scale Integrated Circuits (VLSI) Project. Lund, Sweden: Research Policy Institute, University of Lund.

Steinmueller, W. E. 1988. Integrated Circuits. In D. C. Mowery, ed., *International Collaborative Ventures in U.S. Manufacturing*. Cambridge: Ballinger Publishers.

Stuckey, J. S. 1983. *Vertical Integration and Joint Ventures in the Aluminum Industry*. Cambridge, MA: Harvard University Press.

Swann, G. M. P. 1987. Industry Standard Microprocessors and the Strategy of Second-Source Production. In H. L. Gabel, ed., *Product Standardization and Competitive Strategy*. Amsterdam: North-Holland.

Teece, D. J. 1980. Economies of Scope and the Scope of the Enterprise. *Journal of Economic Behavior and Organization*, 1, pp. 223–47.

Teece, D. J. 1982. Towards an Economic Theory of the Multiproduct Firm. *Journal of Economic Behavior and Organization*, 3, pp. 39–63.

Teece, D. J. 1986. Profiting from Technological Innovation: Implications for Integration, Collaboration, Licensing, and Public Policy. *Research Policy*, 15, pp. 285–305.

Tyson, L. 1988. Making Policy for National Competitiveness in a Changing World. In A. Furino, ed., *Cooperation and Competitiveness in the Global Economy*. Cambridge: Ballinger Publishers.

Vernon, R. S. 1979. The Product Cycle Hypothesis in a New International Environment. *Oxford Bulletin of Economics and Statistics*, 41, pp. 255–67.

Williamson, O. E. 1975. *Markets and Hierarchies*. New York: The Free Press.

Williamson, O. E. 1979. Transaction-Cost Economics: The Governance of Contractual Relations. *Journal of Law and Economics*, 22, pp. 233–62.

Williamson, O. E. 1981. The Economics of Organization: The Transaction Cost Approach. *American Journal of Sociology*, 87, pp. 548–77.

Williamson, O. E. 1985. *The Economic Institutions of Capitalism*. New York: The Free Press.

World Bank. 1987. *World Development Report*. New York: Oxford University Press.

Part V

Managing the Commercialization of Technology

14

The Technology-Product Relationship: Early and Late Stages

RALPH GOMORY

This paper will examine, and distinguish between, two kinds of relationship between technology and product: One which is characteristic of the early stages of an industry and one which is characteristic of later stages. The first one is more or less familiar; and it is the one that has shaped most people's thinking on the subject of science, technology, and product. The second is less familiar. My terminology for these, which you will find in other articles of mine, are the "ladder" and the "cycle."

First of all, let me discuss the more familiar relation of science, technology and product. This is the one that relates to why people came to believe that scientific supremacy should mean supremacy in product markets. This first type of relationship is exemplified by the transistor. The transistor was the product of decades of fundamental scientific research which eventually reached a point of practicality and then, through a series of rapid developments, resulted in the first semiconductor chips and went on to be the start of an enormous industry. And it is that paradigm which is the more familiar one to most people who discuss this subject. This I call the ladder paradigm because the new thing descends from the realm of science—step by step—into practice and becomes the genesis of an industry. Molecular biology is, today, in that state, having emerged from a period of tremendous scientific progress, and is starting to generate a whole new industry.

The belief that this kind of scientific dominance should translate into product dominance is probably, in many cases, the residue of the Second

World War. Those of us who can remember that with any clarity remember the enormous impression made by the science-led, science-developed process of the atomic bomb. And, after the war, we emerged with a picture that scientific dominance does translate into economic dominance. The processes of which that is an example—the atom bomb, the transistor, molecular biology—are scientist-led. Scientists play the dominant role both in the basic research and in the early phases of the industry because they are the only people who understand what's going on in enough detail. So, in the early stages of a new industry, the ladder paradigm predominates. Everything revolves around the new technology. There are no old plants to accommodate; there are new people, new ideas, and new facilities—you are writing on a blank slate. That this kind of activity should produce industrial dominance is an attractive idea. But we should have known better, because the United States, before the Second World War, was already the dominant industrial power of the world, and had been for several decades. And it was not this kind of breakthrough on which that dominance was built, but much more on the manufacturing skills and so forth which are characteristic of the second type of innovation, which I will describe.

Secondly, we should realize that our problems today are not caused by the lack of this "ladder" type of innovation because those industries which cause the negative balance of payments—semiconductor memories, consumer electronics such as TVs and VCRs, and the automobile industry—were not industries that someone else started by stunning innovations. They were industries which U.S. people started and reduced to practice, led and dominated. And it was only in the later stages of these industries that we lost control of them. So, the familiar paradigm is really a paradigm for getting things started. But it is not a paradigm for winning the longer race.

Let me turn then to a second paradigm, a second relationship of technology and science, which I call "the cyclic process." The cyclic development process is a process of repeated, continuous, incremental improvement. It's a problem of getting out a better semiconductor chip next year, based on what you already have in production; or, you are already in production with automobiles, and you are working on next year's model. It is that process of following up what you have in manufacturing with the next model, which is designed, built and prototyped, tested, redesigned for manufacturing, put into manufacturing, and then you turn around and start on the next generation. This process is characteristic of the later (not the earlier) states of an industry.

The type of industry that I am talking about is discrete manufacturing, of which automobiles and transistors are very good examples. It is this cyclic development process that determines in the long run, then, who will be dominant in this industry. It is not as glamorous as the breakthrough type of thing; but, nevertheless, the progress which it brings about is enormous. By and large, the small, innovative companies tend to be very early phases—not necessarily of an industry, but of an industry sector, or of something new. Whenever that settles down and becomes major, then it turns into the cyclic development process. And I think that one way to characterize the difficulties of the American semiconductor industry, perhaps a superficial way, but with a grain of truth, is that it consists of the companies that started it. And those companies have the characteristics that enabled them to succeed in the ladder phase, but they do not necessarily have the scale or other characteristics that would enable them to succeed in the cyclic development process.

Certainly, firms do make a transition from being small startup firms to becoming large ones. But, it is notorious that many of the firms do not; that they have the wrong management style or whatever. Now, I think to succeed today in the DRAM business requires the ability to make enormous investments in plant and in tooling, and to make those steadily because the competition will make them steadily; which means to make them in bad times as well as in good. Now, how would you develop such a capability? I am not quite sure. I think Texas Instruments, for example, has reached the scale where they seem to be able to do that. So, scale is, clearly, one indicator. The second might be to be a vertically integrated company and be a part, say, of a company that makes things that depend on semiconductors. And that when the semiconductor cycle is down, the rest of the thing may still be up. And that might give you the stability needed to make those continuing investments. But you have got to change gears from the rapid one-pass motion of the early stages, to the ability to invest, sustain, and rapidly turn over the crank. That does happen, but a lot of firms fall by the wayside.

For example, in semiconductor chips, by the process of cyclic development, they have gone from sixteen bits on a chip to one or four million bits on a chip, in some twenty years. All of this by the straightforward process of turning the crank: refinement, manufacturing, refine, manufacture. This process is the one which is not clearly visible in most people's minds, and which I want to describe in a little more detail.

First of all, the length of the cycle itself—the cycle from design to manufacture—is a very important parameter and is often mistaken for technological innovation or technological leadership. But consider, if you have one company that can bring a product from design to manufacture in three and a half years and another company that can do it in two and a half years—and suppose they both came out in 1989—the technical notions embedded in the product with the shorter cycle will be one year more advanced. This is simply because, even if both companies are using the same set of ideas, the product with the shorter cycle will embody a later technology. And it will be technically more advanced, though no invention at all has taken place and both companies are simply using the common stream of ideas.

Thus, it is difficult to overestimate the importance of getting through each turn of the cycle more quickly than a competitor. It requires only a few turns for the company with the shortest cycle time to build up a commanding lead. Even if a company starts out with an inferior product, it is possible to overtake the industry leader if it has the capacity to turn out a new line six to twelve months more quickly. In fact, our Japanese competitors believe that, in the shortest of long runs, quick development triumphs over market research every time. I once made the mistake of asking a Japanese colleague, my counterpart at that time in an electronics company, whether he had undertaken any research on how customers were likely to respond to a particular kind of ink jet for printers. Why, he politely retorted, should he study whether customers are likely to respond positively to this or that jet if his company can get out a wholly redesigned printer in a year to eighteen months? Why not simply adapt to actual buying patterns? (Why, he implied, should I be bothering with such questions in the first place?)

My conclusion is that a company that can establish and maintain a shorter cycle can develop a decisive advantage over its competitors. The speed with which the design is translated into manufacture, turned around and redone, is enormously important in determining commercial success.

I have taken the case in which technology is changing and, therefore, there is something new to be exploited a year or so later. If market is the issue, of course, the short cycle also gives you quicker ability to adjust to market feedback. This kind of cyclic process is very different from the ladder process. It is not scientist-based. It is based on what is already there, the existing product and its restrictions. If you made a printer last year, and you want to put a new head in the next round of cycle, it has to fit into that printer. If you want to put a better metallurgy in the next round, it has to be one that the development team can deal with, and accept, and get done quickly. If there are only eighteen months for development, there is not necessarily time to do something over from the beginning. So this type of development is very much restrained, not by a totally new idea, but rather by what is already there, whether that is the plant, or the tools, or the engineering team, or what they understand. And if new technologies are going to be part of this, they must fit into that very special world.

If the engineers are partway through the cycle, that is, if they started eight months ago to design the next round of printers, and you approach them with a new head design, you will find that they are not interested. A typical reaction to that is to decry the development people as having the not-invented-here syndrome, to conclude that the company engineers are simply impervious to ideas from outside sources. But this is an inappropriate psychological phrase to describe a genuine, objective difficulty. The resistance of designers to new ideas cannot simply be ascribed to a certain mental inertia—an inertia that accounts for the resistance of U.S. car designers to disc brakes, radial tires, and computer-governed electronic fuel-injection systems or that accounts for how long it took consumer electronics companies to replace metal parts and casings with molded plastics. All this reflects a lack of perception of what the real problem is. To revert to the case of the new printer, the real problem is that, if the engineers are going to bring the new printer out on time, and they are already eight months down the pipeline, they cannot accept a new head. Thus, if you want to get new technology into the cyclic development process, you have a genuine problem: the new technology has to arrive very close to the beginning of the next cycle.

Further, new ideas from any source need to be very thoroughly fleshed out. Many things are hard to accept into this process because the process is bound by the necessities of schedule. It is simplistic to ignore the scheduling compulsions of the cyclic development process and to conclude that what is at issue can be reduced to resistance to outside ideas. If people are building something as part of a development and manufacturing team, what they can accept depends a great deal on what they presently have. It depends on what tools they have, whether they accept square shapes or round, whether their engineers are familiar enough with any particular approach being proposed.

A related point is that it is often also quite difficult for ideas originating in the university community to be quickly accepted into this rather closed world of industry. You can have a great idea at the university, and you can then go to the product developers and say: "Look, I can make a better valve." But even if it is a better valve, the odds are overwhelming that they cannot accept it. Putting a better valve into their equipment may mean that they have got to change 43 tools, and they may quite possibly conclude that is not worth that much disruption.

On the other hand, if the people in industry are familiar with ongoing advances in science and technology, they can go out and pick the ones that they can accept—the ones that fit. That is the phenomenon of "pull." By contrast, it is very important not to assign to universities the responsibility for "push"—for transferring ideas into industry. I do not think they can do it (I am distinguishing here between the "ladder" stages, when universities do indeed play a major role, and the later "cyclical," or product improvement stages, which is the context of the present discussion).

In this respect, one of Japan's great advantages is that much of its technical strength is in entities that are tied to industry or actually in industry, whereas America's scientific strength is heavily concentrated in the universities. As a result, even though Japan's universities have a weak research capability compared to those of the United States, and do not help as much in the idea-generation process, Japan's industries function extremely well in the process of appropriating new ideas.

The essential point is that the primary way that ideas enter the product development cycle is not outside push, but inside pull. The engineers, who are themselves the prime movers in this, go out and select from the great world of outside ideas which are visible to them at engineering society meetings or at universities, those few which fit into their constraints. If you are the owner of a new, technical idea and you go to them, the odds are overwhelming that it will not fit. Your notion will call for stamping something round, but their stamps will stamp square things; and both the time and the cost of changeover make your idea unacceptable.

So, in the cyclic development process, which is the main process of a developed industry, not at startup phase, the key pullers of technological process have to be the people there, not the people outside. This is very different from the ladder process, the early-stage process. Again, because of the importance of cyclic development time, the close cooperation between the design and manufacturing is absolutely essential.

Of course, the manufacturing skill, itself, is a tremendous factor, as are the closely related issues of quality, but I don't propose to get into that issue here because they have been exhaustively described by many other people.

I just want to mention, though, the importance in a short cycle of designing for manufacture, on the part of the design team; and I will illustrate that by an example which perhaps you have heard of, but perhaps not in as much detail. And that is the example of the IBM ProPrinter.

The IBM ProPrinter was a matrix printer, exactly the sort of a thing you would put on a table next to your personal computer. It hammers out letters by driving little wires into a ribbon, the choice of wires determining the letter that emerges. It is basically a mechanical device. IBM was not originally competitive in the production of these printers. And when the early PCs, the first PCs, came out, they were all equipped with competitors' printers—notably Epson printers. And when IBM's development team looked into how it might have a competitive machine, one thing that was immediately discovered was that the IBM designs had many more parts than the competitive designs. The number of parts in the IBM design was, I think, roughly 120 or so in the Epson-type designs. For the first time, the development team took very seriously the consequences of designing things that way. They concentrated on designing the next printer as a manufacturable printer. And they took as their goal 60 parts. The number of parts is not a bad surrogate for the complexity of assembly.

There also was a second theme, which was pure manufacturing. They put together a robotic line. They cleaned out an old plant and put in 35 robots, two-armed, highly programmable robots to assemble this thing—and spent a lot of money doing that. I, myself, was very pleased to see the two-arm robots go in there because they were the product of a project which I had started in IBM Research in 1972. So, I thought this would be very spectacular. Here is what happened.

The design goal was almost met. They emerged with 62 parts. They designed the machine to make it easy for the robots to put it together; and that meant that the parts were all put down top-to-bottom, no insertion sideways. There was also a rule—no screws and no springs— because screws and springs are also hard for a robot to assemble. They made it with 62 parts, no screws, and no springs. Now, when they assembled this printer, they found that a human being could assemble this in roughly three to four minutes. It was that easy to put together because it had been designed to be manufacturable.

Another thing that occurred, by the way, was that the development time was cut approximately in half from previous development times. So, they affected the development cycle by making it simple, and they made it so easy to manufacture that, in fact, the gain due to the investment in the robots was almost completely obviated. Because if the human being could put it together in three and a half minutes, a very expensive robot had a real problem in just funding itself.

I have difficulty separating the managerial issues of this process from the structure. Is it structure that makes you turn the cycle fast—or other things—or is it management? I do not know. The point is, one way or another, you do it. Now, in the case of the ProPrinter, someone, I do not know whether it was management or other people, took the competing thing apart, and found a very small number of parts. And someone had the wit to say, "Let's have less." Now that may seem very trivial, but it is a change, or change of view. It is a change to say, "Our job, as engineers, is to make this thing manufacturable." I don't know whether that is management or organization. I am really describing what has to be done. I think, in some cases, it may be a spectacular individual that does that. And, sometimes, it will simply be the sharp spur of necessity, knowing that the other guy is going to bring out the next round two years from today and you'd better have something there. And there may be no spectacular individual at all involved. Perhaps the "Champion" concept matters less in the world of cyclic development. I think the system itself can drive it, if you know where you are going wrong.

I bring all this out to emphasize the importance of the cycle and of design for manufacturing. This printer went on to become the best-selling printer of that type in the United States. This was not due to a scientific breakthrough. It was due to proper design for manufacturing. By the way, that is a learnable thing; and, when people talk, as they often do, about the difficulties of competing with the Japanese because of various and mystical factors, that we must change the entire U.S. culture in order to compete, we have got to redo the educational system, etc.— let me point out that competitiveness was achieved, in this instance, by changing the culture, not of a country, but of 70 people; the 70 people who made up that design team. So, there is a great deal of competitiveness that is concrete, focusing on the manufacturing/development cycle, on finding ways to shorten it, on finding ways to design for manufacturing, and finding ways for the engineering team to pull ideas into the next product.

And let me say once more that, in the taxonomy that I use, this cyclic development process is, in the long run, the decisive one for many manufacturing industries.

Now, if we do look at the competition—at the Japanese competition—we find that they do all the things I am talking about. They do design for manufacturability extremely well. In fact, the design and manufacturing teams are often very much tied together. They do have very short cycle times. If you read the press, you would think that the

Japanese owe a great deal of their success to the very advanced technology programs which MITI has sponsored. Now those programs are useful, but they relate to the early stages of technology, not to the rapid turnover of the cyclic development process. And in the areas which I have seen, and with the people with whom I have discussed these matters, it is the cycle, not the ladder, that has been decisive in the industries in which we are in trouble—automobiles, semiconductors, TV; that sort of thing.

Another striking contrast with the Japanese is relevant here. In most American firms the people who are responsible for development, and those who are responsible for manufacturing, tend to live in different worlds. In U.S. industry, the high prestige is in development. The manufacturing task is considered to be relatively dull, repetitive, and for the uneducated. Therefore, production considerations have not been an early part of the design, say, of a new car or computer or printer. Until very recently, the product was not designed to be manufactured. It was designed to work, to make good print, or to run fast, but it was not designed to be manufactured. When the prototype was ready, you simply handed it to a different group of people—the manufacturing people—and you said, "Okay, I made one, you make 10,000."

In Japan, curiously enough, it is the other way around: Manufacturing calls the shots. Manufacturing people are highly trained, but people also move back and forth a good deal between various functions. The distinction we tend to make between the engineers in development and those who replicate their designs does not exist in Japan.

So, let me sum up what I am saying here. Much of our thinking about innovation comes from the ladder process because it has been so visible and so spectacular, involving the emergence of a new scientific effort and new products emerging from that effort for the first time. The United States has, in fact, historically been very good at that, while it has been steadily losing in the cyclic development process which is the phase characteristic of the mature industry. Now, that does not mean that we should let go of the first process because you can lose at that, too, and the Japanese are making much more of an effort in those early phases than they did in years past. But it does mean that we also have to make the cyclic development process more visible and to understand it better.

I believe strongly that the invisibility of that process does hurt us. Many of the words that we use to describe this subject hide, by their very meaning, the fact that there is a cyclic development process.

I think the wrong mental picture hurts us a great deal. Let me be concrete. Consider the picture brought to mind by the phrase, "commercialization of new technology." If you talk to most people, I think you will find there is a notion that commercialization means to take something new and make it commercial, whereas the essence of the cyclic development process is you are refining something which is already commercial. When you talk about the commercialization of scientific discovery, those words already specialize you to phase one, or the ladder process.

The United States, as a whole, seems to resonate far more to the spectacular events of ladder innovation. Take, for example, superconductivity—high-temperature superconductivity. High-temperature superconductivity, from the point of view of industrial competitiveness, is something that is way, way out there—at least, optimistically, ten years away, if ever. And yet, when that great scientific event occurred, we had meetings in Washington, with the President of the United States in attendance, special legislation in Congress, set up special committees to advise the government, etc. A tremendous focus on a ladder-type event and, in all the discussion about it, the close tie with industrial supremacy which, in fact, does not exist.

It is necessary, therefore, for us to learn the ins and outs of the cyclic process, because, very often, that is the decisive process. And, if you look at that picture, I think you will find that it does turn around what you have to think about.

Take, for example, the notion of investing in R&D. A lot of people, if you are talking to them, will say, "Gee, the Japanese are investing a lot in R&D and that means they are going to jump past us," or something like that. And, again, the picture is, you put the R&D in here and the product pops out. That is very, very different from the one you really encounter if you are a participant in a cyclic development process, because in the cyclic development process it runs much more like this. You have a product. The product is selling. That gives you a certain stream of revenue. You can take that stream of revenue and put some of it into R&D for the next round. Some of it has to be reserved for the manufacturing, some of it for profits. Now, if you are on an upward swing and your product is succeeding, you have a flow back of money to invest into R&D; and if it isn't, you don't. And, in my experience, and the experience of many other people, oddly enough, R&D is determined, more or less, as a percent of sales. It is not an independent variable.

Let me say that once more. R&D is often a fixed percent of sales. Now I exaggerate to make the point. Ten percent is a very reasonable sort of number in a high-tech industry. Now, there may be some nonlinearity in relation with production. But, as you produce more, you spend more on R&D. It is not at all clear that in this phase of the industry, the cyclic development phase, that the R&D decision is an independent decision. As it is often discussed, "We have decided to invest in R&D" is a phrase that fits, again, much better at the front end of an industry. Later, what you can afford is determined by your success in the cyclic process. And it may be that, in the correlation, which has often been remarked on, between R&D spending and industrial success, it is the industrial success which causes the R&D spending, not the other way around.

I think that high-definition TV is just one more step along the continuum of TV products, but it is being represented as something totally new, and a tremendous opening to go in. I think that it would not be very different to decide that you want to get back into TV. I think getting into HD-TV is not that different a decision; and the only way to do it would not be the sort of things that were originally happening—which is, "Let's have DARPA spend some money in R&D." This is a total misconception related to mistaking the ladder for the cycle. You are still in a cycle process, in my opinion. The only way that you can enter HD-TV is to build up a manufacturing capability, on a very large scale, and back it up with a major R&D, and start the cycle churning. It is not that kind of an event where you can come in from the outside with a new technology. It is a cycle event, not a ladder event.

So, I think we need to rework our mental picture if we are to make progress with this matter. And in the subject of cycle versus ladder, and the characteristics of the cycle, there is much to learn. Of course, the ladder process is not always quite as separate from the cyclical process as I depict. A good company is continually investing in both kinds of developments and the problems within both of them. It is how to move things along that is the problem. But if we do not realize that there is a cyclic development process with its own peculiar resistances, then things will not move. Believe me, I had a very expensive and extensive education along those lines. So, I do very much believe in doing the research. But you then have to pay attention to how it will get into product. Much research is done, and it never will get in, because it will violate the rules. It will deal with an object that is not there. It will not understand that there is an issue of timing, that there is an issue of familiarity, and so forth. You have to understand that this cyclic beast is grinding away, and this is the beast you have to affect, and that your invention is only a part, and you have to match the timing and you have to match what is there. And, otherwise, it just does not happen. It is not enough to have done it. To have done it is to put it on a shelf, along with many other inventions or innovations; and then, various companies will come along and pull from that shelf the thing that fits their cycle. Now, which company does that—who knows? I think it is very wise for companies to be assiduous in that process. If you are talking about things that a company itself develops, then it had better pay attention to what is known as technology transfer, a very nontrivial process, because the thing you are transferring to has its own life structure.

Let me close with an important caveat with regard to the transfer of technology. Success or failure in technology transfer depends critically on the characteristics of the receptor. If the receptor knows very little, he can do very little even with a simple idea, because he cannot generate the mass of detail that is typically required to put a new technology into execution. On the other hand, if he knows a great deal and is capable of generating the necessary details, then from just a few sentences or pieces of technology he will be capable of filling in all the rest. That is why it is hard to transfer technology to the Third World and very hard not to transfer it to Japan.

15

Profiting from Innovation

WILLIAM G. HOWARD, JR., AND BRUCE R. GUILE

Industrial competitiveness is influenced by many factors. Some, such as the macroeconomic environment and the legal and regulatory systems, are determined primarily by forces external to companies vying for places in world markets. Other factors are internal. Two internal competences, manufacturing skill and the effectiveness with which individual firms translate technical innovations into marketplace advantage, are complementary and lie at the heart of efficient industrial practice. Since private companies form the front line in the contest for share of world markets, success at practicing manufacturing and utilizing the entire range of innovation available to the company is an important factor in a nation's industrial competitiveness.

This paper is a preliminary report of a National Academy of Engineering study of industrial commercialization: the translation of innovative ideas into marketplace success as profitable products, processes, and services.[1]

The manufacturing successes of the Japanese, our primary competitors for many world markets, have been well publicized. Terms such as *kanban*, just-in-time inventory management, quality circles, total quality management, and Taguchi method have become accepted terms by U.S. company managers. Leading U.S. manufacturers have come to realize that they have serious manufacturing problems, and are beginning to address them. Progress has been evident at companies such as Ford, Motorola, Xerox, USX, and many other manufacturing-based companies, and additional firms are following suit.

---

[1]Since this paper was prepared, the Academy's study has been completed and was published in *Profiting From Innovation*, 1992, New York: The Free Press.

On the commercialization front, we have not been as responsive. In addition to manufacturing skills, the Japanese have cultivated the purchasing dollars of many customers. Unlike the case with manufacturing, U.S. industry has not shown a widespread appreciation for its commercialization shortcomings, and many firms still retreat to the observation that Americans are still the most inventive people on earth. That may be true—but translating that inventiveness into manufacturing processes and marketed products is essential if we are to turn our ideas into industrial strength. Microelectronics, gene splicing, computer communications, and efficient airfoil design are great technologies. It is products based on these technologies, however, that provide competitive edge.

We have been seeking clues to effective management of commercialization that apply across a wide range of industrial circumstances and which are useful for manufactured products, production processes and services—those practices that seem to distinguish companies who are consistently successful at commercializing from those who have not performed as well.

The study was conducted as a series of workshops where senior managers discussed aspects of their commercialization experiences, both good and bad. Several additional interviews were conducted to supplement workshop results. Participant experiences covered about 30 firms, plus the financial community, business schools, research consortia, U.S. government laboratories, and independent research institutes. A limited number of European and Japanese experiences were sought, in addition to U.S. perspectives, for comparison.

This study makes no pretense of completeness; our goal was to identify those practices that characterize successful commercialization in the widest possible spectrum of enterprises. Many successful techniques, which apply to particular companies or industries, did not survive the winnowing process that our methodology employed.

Our activity focused primarily on the role and methods of management in promoting the commercialization of technological innovations—in particular the senior decision-making leadership of a firm—reasoning that these people most directly shape the environment within companies.

As expected, we found no magic bullets. Our findings re-emphasize the importance of the fundamentals of business management—of a leadership preoccupation with producing a better product or service with higher quality at a lower price than anyone else.

The results we will describe are preliminary results, as the report is still in preparation (as of this writing, in 1989; see note 1). Regrettably, there is not enough time to describe all of our conclusions, so we have selected a number of observations that seem to apply to the theme of this book.

1 The Prepared Mind

Louis Pasteur observed that "fortune favors only the mind that is prepared." Nowhere is that more true than in the case of commercialization management. As we shall see, there are several types of commercialization processes, each with its own time frame, indicators of progress, organizational approaches, and risk factors. Understanding when each of these is under way, and the nature of the true challenges faced, is a major factor in consistently successful efforts. Intimate understanding of customer needs and technological potentials is essential if the two are to be mated. Finally, we have observed that successful commercialization leaders are constantly re-examining their business, always seeking the "better way" that must lie out there somewhere. They are prepared to continuously re-evaluate their activities.

The ideal of a "detached manager" capable of overseeing a business at arm's length did not apply to the successful leaders we found.

2 Commercialization Efforts

We have encountered three distinct types of commercialization, distinguished according to their driving forces. Each has unique characteristics, each is measured according to different analytical tools, and each requires its own management techniques for successful application.

Technology-enabled commercialization is driven by a revolutionary scientific or technical discovery. Rarely is technology revolutionary. Developments such as the transistor, dry copying, the industrial robot, nylon, perhaps someday cold fusion or room temperature superconductors, are singular events. They sweep away the old ways and replace them with entirely new products and processes whose expanded capabilities were previously unimaginable. However, such revolutionary new technologies rarely live up to expectations initially. Development of successful products embodying such technological "supernovas" typically takes fifteen to twenty years and entails a great deal of risk, as technological developments, without which the "revolution" is nothing more than a curiosity. Only after this long investment in time and resources is the enormous market potential of such developments realized.

In technology-enabled commercialization, the major risk lies in understanding problems entailed in producing the product in volume and in the market for products, much more than in the technology. DuPont's CORFAM (tm) plastic replacement for shoe leather was a major technological development, and DuPont was able to solve the problems of producing the material in high volume. DuPont reasoned that shoes made of this material that were longer lasting, water resilient, less expensive, and required less polishing were sure to be a market success. They did not correctly appreciate that failure of the shoes to break in to the wearer's feet, the inability of the material to breathe moisture, and the failure of the public to develop an appreciation for foot odor would be problems. These market miscalculations led to the eventual failure of the CORFAM enterprise.

A flash of brilliance, followed by dogged application of "Yankee Ingenuity" to overcome all obstacles is the basic stuff of the American industrial legend. This is where we Americans see that our distinctive competence lies.

However, our discussants felt that, as a nation, our grasp of this process is slipping. In addition to accelerating diffusion of basic science and technology around the world, they felt that pressures for a shorter-term payoff and a general aversion to risk work against our success in technology-enabled commercialization efforts. This was observed to be a problem both in large corporations, where concentration on existing businesses is often the rule, and small, venture-capital-financed concerns where the need on the part of financial backers to "cash out" within five to seven years has led to the dismemberment of otherwise promising enterprises.

Successful practitioners of technology-enabled commercialization appear to understand how these projects progress. They are less dazzled by initial announcements of discoveries than they are dedicated to the long, steady pull required to reduce truly new ideas to practice. In many instances, this is the result of tolerance of stubborn individuals, but a handful of companies (DuPont and 3M, for example) seem able to consistently make a success of technology-enabled innovation through more systematic means. Technology-enabled commercialization, however, is a very risky undertaking. Along with its successes in Kevlar, Rayon, Lycra, and Tyvek, DuPont can point to numerous products (such as CORFAM) that did not make the grade. Clearly, persistence at technologically driven efforts is a valuable trait.

A major industrial nation cannot depend principally on technical revolutions. They occur infrequently, entail a level of risk that makes them haphazard, and they do not provide the sustained push that an economy must have to endure in the market marathon.

Although technology-enabled commercialization occurs rarely, it is a particularly important part of the overall commercialization activity, for "orthogonal" innovations do not follow logical extrapolation of business growth. When successful, technology-enabled products, processes, and services not only produce extremely large and rapid business growth, but also render the competition's products or devices obsolete.

The second type of commercialization effort is production or design driven. Once a concept has been proven in the market, competition shifts to price, performance, and design features with less competitive emphasis being put on fundamental differences in product or service concepts. In short, after one or two initial successes, competitors swarm in quickly, and the focus shifts from making it work to making it work better. Production- or design-driven commercialization is propelled principally by competition—the need to contend with others in the same field.

The dominant pattern is illustrated by medical imaging equipment, overnight package express, and automated bank teller services, which were as recently as 1975 new and different products. These items have now become standard offerings, and early providers and later entrants are now engaged in competition over incremental improvements in the product or service.

Unlike the technology-enabled case, success at production- and design-driven commercialization depends upon a company's ability to push incremental innovation in the product design and production process. This is a very different management challenge from running a revolution. This process must be a continuous series of small triumphs. Changes take place rapidly—in fact, the faster the better. In spite of the need for speed, there is little tolerance for mistakes or slipshod work. Stumbling costs irreplaceable time and customer confidence.

Two fundamental issues underlie product and design commercialization success—rapid product cycle time, and starting early. As has been pointed out by Ralph Gomory, the effect is cumulative. As a company beats its competitors to the market time and time again, its rivals fall ever further behind, less likely ever to be able to catch up.

Continuous product and process improvement doesn't just happen—it must be driven by the expectation that such improvement is the basis of future competitive advantage, and must be eagerly sought if the firm is to succeed.

U.S. industry's track record at continuous improvement is spotty—we became complacent in the halcyon days of success over the past 40 years. The story of our steel industry's failure to continually upgrade its process and plant technology is well known. The Japanese, on the other hand, have rightly earned their reputation for relentlessly honing their products (such as television sets, cameras, and automobiles) until they drown their competition in a tidal wave of improvements.

One illustration of how this process works can be seen in the example of the competition between U.S. and Japanese semiconductor firms for share of the dynamic random access memory (or DRAM) merchant market. The U.S. semiconductor industry, as inventor of the DRAM, led early with 1K, 4K, and 16K DRAM products. The business became fiercely competitive in the late 1970s when Japanese companies began to enter the world market with well-designed high-quality products. Both countries' industries reached the market at about the same time with the 64K product. New generations reached the market about every two and a half years.

Most U.S. merchant market semiconductor firms chose to concentrate their attention on one product generation at a time—either designing the next product or fixing problems with the last one. Their speed in developing new product generations was as fast as any. But need for haste forced many technical shortcuts, which were reflected in customer applications problems, necessitating a series of redesign efforts that dissipated initial product momentum.

Japanese firms that now lead in this business, followed a different strategy. At the time the 64K DRAM was introduced, they also had simultaneous efforts under way with two or three additional generations of product. The 256K DRAM was in advanced design, and the 1 megabit version was actively being researched. Very likely, a group was off thinking about what had to happen to make the 4 megabit product a reality as well. Products, when introduced to the market, were second or third generation designs, and thoroughly debugged.

Thus while U.S. companies turned their swat teams from one generation to the next, they were confronted by a continuing series of well-designed competing new products in rapid succession. Japanese management drove DRAM development with the long haul in mind, and with commitment to development in depth.

Continuous improvement of existing products and processes is where the United States can improve most. The sense of disquietude that prompts the management and work force of an entire company to continually seek better ways requires sustained attention to detail and persistence that comes harder to Americans than to those from some other cultures. Even our "Cash Cow" businesses require continuous investment to keep them current and competitive—we have neglected to appreciate this over the past several decades, and must change.

The risk in production- and design-driven commercialization is in torpor. The failure to keep making improvements, and to keep aware of progress being made in competing products is to ensure eventual problems in currently successful businesses. The "you can have it any color as long as it's black" mentality displays a rigidity of thought that provides entry opportunities for others.

There are a number of U.S.-based firms that have recognized the need to continually update their products and process. Boeing's commercial aircraft technology has undergone almost continual change as it has sought to incorporate the latest innovations into production airframes. Bill Boeing's dictum to "Let no advance in flying pass us by" has been accepted as a basic philosophy of this successful company's business.

The third commercialization category is market driven. In this instance, perception of customer needs in one product sector prompts a search for technology from other applications to open new opportunities. The perceived new market is the driving force in this case as opposed to a search for applications of a given new technology.

In market-driven commercialization, the emphasis must be on detailed understanding of the customer's needs (as opposed to wants) and on the selection of the best technology to fulfill those needs.

The primary risk element is technological—choosing the best technology on which to base a product to fill an identified customer need from the wide range of possible alternatives.

The Bell & Howell Company confronted these tasks when it responded to a request from GM to apply its skill at microfiche publication to simplify the parts counter catalog used in many dealerships and parts stores. The problem was well known—looking up parts numbers and prices in existing paper catalogs was a time-consuming effort that resulted in a great many errors. Further, the effort required to keep a comprehensive parts catalog up to date on a daily basis is substantial. Bell & Howell, as the leading publisher of microfiche documents, sought to use its traditional technology to meet the need, but discovered that the ease of access, document quality, and ability to incorporate illustrations on microfiche were insufficient to provide an attractive alternative. After a five-year search for the best technology fit, and three tries, involving microfiche, computer terminals, and a variety of document storage systems, Bell & Howell settled on a CD-ROM based system which not only solved the parts catalog problem at lower cost and provided a continuing publishing business for the company, but enabled the mechanic to specify and order parts directly without going through the parts counter clerk—a major customer labor saving.

Each of the types of commercialization—technology enabled, product and design driven, and market driven—is a different kind of problem. Each requires its own management approach.

Few medium- and large-size firms can thrive on a single type of commercialization activity. Some firms have all three activities under way at any given time. The simultaneous search for ways to apply new technology, for means to serve new market opportunities, and for improvements in existing products, processes, and services requires leaders who understand the characteristics and who differentiate approaches to each.

3 Management Tools

Traditionally, the primary management measures of the commercialization process are financial. Indeed, financial success (and being able to demonstrate its likelihood at the outset) is crucial to commercial technological innovation. In the language of financial analysis, the value of current and future returns must exceed the investment. The logic is inexorable and sound. The problem arises in making the simple and tight logic of financial analysis work for decision making about technological issues in real companies, about matters of great uncertainty using information provided by less than neutral parties to the decision. In fact, in the absence of other indicators of the promise and progress of commercialization efforts, financial data, with their apparent precision, fill the void to dominate decision making.

Consider, for example, a decision to proceed with development of a technologically new product for an undeveloped market. If one assumes absolute technological certainty, the calculation of return on investment for the development expense still requires estimation of

o the cost of development of a prototype;
o the cost of manufacturing a product for which a prototype does not yet exist; and
o the market for and likely sales of a product that does not yet exist over the next several years (remembering that the product must be priced at a sufficient margin above a manufacturing cost that is not yet known).

Given that few technological developments proceed without many trial-and-error iterations, any decision maker will recognize the biases and negotiating positions that those who provide information are likely to bring to the table. Technological champions may underestimate development costs out of enthusiasm. The production staff may pad the estimates of production costs in an effort to give themselves some slack in scale-up. The marketing staff, really out on a limb measuring markets that do not exist, may bring any number of biases—underestimation if they see their role as brakes on the technological champions or overestimation if they themselves have become champions of the product.

There is yet another set of constraints on the usefulness of financial analysis: business strategy. What is the value of this product as part of the company's portfolio? Does the company need to offer this service to keep competitors out of the company's primary market? Will the company-wide learning-by-doing that comes from a new product in this area of technology be of long-term value even if the product loses money? What is the strategic cost of not proceeding with this development?

Underlying all of these uncertain and often fuzzy judgments that must be translated into numbers for financial analysis are often fundamental questions about the technology and the technological capability of the company. Will the technology work? When somebody turns on the switch, will the product grind, whir, and whiz in ways that meet consumer demand? And, can the people in the company pull it off faster or as fast as the competition? Given the uncertainty of the whole process, it is a wonder that any manager ever considers financial analysis of corporate technological matters—but the fact is that virtually all decision makers do look at financial calculations, probably because there is no other tool so pervasive or so easily understood. Regrettably, this breeds a reliance on financial indicators as a substitute for a real understanding of the business—financial measurements such as discounted cash flow and the profit and loss statement become an end in themselves rather than the short-term indicators of performance that they really are.

An important aspect of our study, therefore, has been the search for useful tools to aid the management process. As a result of our activities, we have assembled a small catalog of commercialization indicators and techniques that assist the leader in assembling business information in a way to make sense of the chaos that pervades the commercialization activity. In addition to traditional financial tools that managers use to measure and plan the health of the business, we have identified methods such as technology road maps, learning curves, time to break-even analysis, and logistic curve considerations that can prompt action by leaders before chronic problems become acute. Many of these tools are not new, but use of them throughout the range of industry we have considered is spotty, and a more widespread understanding of them is useful. Particularly important is the fact that these tools are, in their very nature, more sensitive to the characteristics of the technological challenge being faced than traditional financial analysis.

For technology-enabled instances, the tools of value—fuzzy though they may be—are those that allow a project team to develop the ability to look through current consumers to the market that will exist. In new product or service developments, an internally generated vision of the market success factors must be developed and refined by exposure to critics and customers. Additional tools can help leaders keep track of project progress and to understand the characteristics of each stage of development.

New technologies characteristically must pass a number of critical decision points in the life of the development. At each such point, the investment commitment rises—typically by an order of magnitude. Descriptions of such decision points exist, and are useful particularly to managers engaged in technology-enabled commercialization.

In production- and design-driven commercialization—continuous improvement in existing products and processes—analytical tools are much more developed if not widely used. These tools help build a leader's anticipation of possible improvements in products and processes and where to seek them, and help to compare internal company operations with efforts of other organizations. Experience curves, benchmarking, product life cycle analysis, and product-age portfolio approaches are all part of a growing set of tools that are demonstrated in a growing literature on manufacturing and technology management.

In the final type of commercialization challenge—moving developed or developing technologies into new markets—the approaches are again different. Most applications can be satisfied with a variety of technological approaches. Sometimes the application of an existing technology to a new market is quite dramatic, for example, technology-enabled creation of a totally new product or service drawing mostly on existing knowledge. In other cases, efforts at moving existing products and services lines into new markets are more incremental, and the technological challenge is in packaging and applying an existing technology in such a way that it matches the particular characteristics of the market sought. These are nontrivial technical challenges. As with production- and design-driven developments, there is no substitute for knowing what the customer wants, but the challenge is similar to that in technology enabled developments—envisioning how customers will respond to a product or service that does not yet exist. Tools for this commercialization type emphasize frequent market trials and iterations in the real-life environment.

4 Joint Development Efforts

As our study progressed, it became apparent that important changes are taking place in the technology development structure of several U.S. industries. Increasing research and development costs, limited skill pools, and external competitive pressures are forcing companies to rely upon government laboratories, universities, and consortia for longer range developments. The problems of establishing constructive joint programs, technology transfer, and timely utilization of the results of such disaggregated joint efforts are major new challenges facing industry.

Universities, consortia, and government laboratories that participated in our study agree that most companies do not manage joint activities well. Firms that commit to programs such as the MCC consortium, or university industrial liaison programs generally honor their financial obligations, but the relationship often stops there. Those companies who join such programs, making operational, as well as financial commitments, seem to do much better.

The clue came from observing the way Japanese companies band together in co-development projects. Since each participant fears that another partner will gain a competitive advantage over them as a result of the joint activity, and establishes its own "mirror" effort which closely follows consortium progress, it is not uncommon for researchers to move back and forth between joint and proprietary groups. Thus, as results are produced by the joint program, each participant is immediately on board and ready to use the fruits of the joint investment. Those firms that have made the best use of American joint efforts have followed a similar course, making a commitment to an internal "shadow group" to track joint activities in each university and consortium program at the same time that the financial obligation is assumed.

Companies that see joint research and development efforts strictly in terms of financial obligations are poorly positioned to realize the benefits of the results, since they must create internal uses after the ideas are developed, rather than concurrently. As a consequence, the lead time advantage potential of a pooled relationship is frittered away in internal decision-making and organizational activities. Joint research and development activities are not a substitute for internal efforts; they can only complement them.

Successful participants see joint research and development activities as dual commitments—half external and half internal. Companies that do not understand the need for the internal part of the effort become disillusioned with the idea of joint research and development, and soon drop out. The prosecution of the internal portion of the joint effort is just as important as the joint program itself.

In conclusion, this has been a very fast look at some of the results emerging from our commercialization study. Clearly, commercialization is not a simple process, and consists of at least three distinct kinds of activities, each with its own management challenges. If we were to summarize a single conclusion, it would be that the chances of commercialization success are highest when the leader is personally involved with the process, has a restless desire to do better, and understands the nature of his particular activity. He must be engaged in a continual search for new ways to characterize his business and the competition that go beyond the limitations of strictly financial considerations.

16

Managing the Growth of Technical Information

Sergio Barabaschi

1 Introduction

The subject I wish to treat is the methodology that firms employ to acquire the technical knowledge needed to maintain a dynamic balance between the requirements of the markets they operate in and the attractiveness of the products they manufacture.

In the past such knowledge was acquired mainly through in-house R&D activities, which meant a considerable consumption of both economic resources and time. There are several reasons for believing that in the future there will be a better utilization of external sources of technical knowledge. This will speed up industrial innovation and cut costs.

I intend to analyze these reasons, describe what we mean by technical knowledge and external sources and suggest specific action to improve the effectiveness of industrial innovation by managing the growth of technical information.

First, however, I must explain the background against which I make these observations and state the specific industrial sectors to which they may apply.

My standpoint is not that of an economist but that of a technologist working in an electromechanical firm, Ansaldo, which produces power plants, railway systems and industrial automation systems. According to

I would like to express my thanks to Professor Rosenberg for the stimulating discussions which helped me a great deal, and to Daphne Martin for the invaluable help she has given in offering ideas and in the drafting of this presentation.

TABLE 16.1
Definition of Assembly Industry

1. Designs systems.
2. Defines major and minor sub-systems that will be assembled to form a whole artifact.
3. Defines certain parts of these systems but mostly subcontracts design of sub-systems.
4. Components for sub-systems are specified by subcontractors and made and supplied to subcontractors by specialist companies.
5. Subcontractors assemble major and minor subsystems which are supplied to assembly company.
6. Assembly company integrates subsystems into final artifact.

Examples of Assembly Industries

Aircraft construction	Motor vehicle construction
Locomotive construction	Shipbuilding
White goods	Consumer durables
Retail stores	Building construction
Computer manufacture	Photocopier manufacture

Dominant Principle
"design and integrate"

the classification suggested by Working Group 43 of EIRMA (European Industrial Research Management Association) Ansaldo is an "assembly industry" (A.I.—see Table 16.1), with products structured in many different subsystems, components and parts. It does not, therefore belong to the other important class, the "whole product industry" (W.P.I.), which manufactures major products very different from Ansaldo's because they are not assembled from various parts; such as chemicals, glass, food products, and pharmaceuticals.

It is worth noting though that whole product production plants are often themselves the products of an assembly industry. Some similarities may therefore be found between innovations in A.I. products and innovations in W.P.I. production plant, since the latter are also structured in many subsystems, components, and parts.

I have noted this so as to emphasize that some of the points that I shall be making, which are specific to products made up of different components (i.e., typical of the first class of industries) are also applicable to the second class of industries, but only, in this case, with respect to the innovation affecting their production process.

Now that my industrial background is clear, we can move on to the innovation model most used at Ansaldo. It is the "cyclic innovation process," to use R. Gomory's term, that is, a process which improves a

TABLE 16.2
Four-Step Approach to Innovative Projects

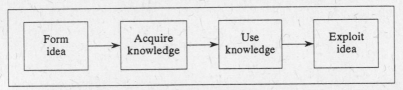

product by means of periodic upgradings of its characteristics or of its manufacturing procedures.

A third characteristic of Ansaldo products is that they are expensive, durable products that often cannot be tested in a laboratory. For example, an innovative steam turbine or a new electric powered locomotive can only be built and tested if you find a customer willing to share the risk and order a prototype after merely examining the blueprints. The cyclic process in Ansaldo therefore needs, in such a case, not only the in-house capability to come up with innovative designs, but also an innovation oriented customer willing to participate.

In the Ansaldo environment, which is typical of the electromechanical industry as a whole, the single innovative step in the cyclic process is bound to be a small one in terms of technological risk. It must, however, offer the customer appreciable benefits in order to secure the placing of the first order. Such is the industrial framework to which my comments will refer, a framework in which cyclic development may be divided into four phases (see Table 16.2). For the sake of simplicity we may consider these four phases sequential in time, being connected by forward links only and not with feedback links and loops. A more realistic model of a typical Ansaldo cyclic process is illustrated in Table 16.3.

In our simplified four-step approach we have the following phases:

Idea Formation: Starting with an idea, which may come from a customer, a supplier or a perceived new technology, the innovation process develops a conceptual design of the improved product and defines the "knowledge package" needed to implement the idea. This is then broken down into "elementary knowledge components," thus defining precisely what technical knowledge must be acquired. (For more details see Table 16.4.)

Acquisition of Knowledge: Having defined what technical knowledge is required, the innovation process must select the most suitable channels to obtain the various components of this knowledge "mosaic."

TABLE 16.3
Innovation Model

TABLE 16.4
Forming Ideas

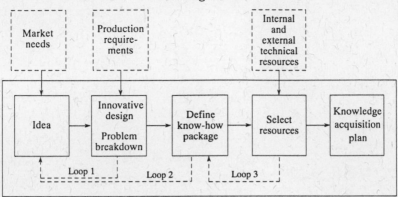

The result is a set of technical documents (drawings, design guides, technical specifications, design codes) that we can call, for simplicity's sake, the "blueprints" of the new product (and of its production process).

Utilization of Knowledge: The innovation process now uses these blueprints to build a prototype that can be tested or simulated, in full or reduced scale, in the field or in a simulator.

Exploitation of Knowledge: Now that it has a "qualified" new product, the innovation process must recoup development costs through the classic industrial activities of production and sales.

Let's take a closer look at the second phase, that of knowledge acquisition, which is central to my thesis. (and still widespread) method of acquiring knowledge is to rely too much on in-house R&D, which is becoming more and more difficult, costly, and time-consuming. This is mainly due to the increasing complexity of each product upgrade, which often requires many new technologies (lateral technologies) outside the classic "core technologies" typical of previous electromechanical innovations. These lateral technologies are gaining greater importance in each new upgrade and are gradually assuming the role of key technologies in each cyclic process within our company.

In addition to this, while core technologies are present in most periodic product upgradings, the key lateral technologies may change from one product upgrading to the next. For example, an improvement to a locomotive may require, as a key lateral technology, information on the cooling of a stack of power-semiconductors. The next improvement might involve a totally different lateral technology, such as the

TABLE 16.5

Some Typical Key Lateral Technologies Used
in Recent Ansaldo Product Upgrades

---

o Freon Heat Transfer Coefficients
o Expert System For Locomotive Diagnostics
o Image Processing For Flame Sensor
o Viscosity/Temperature Law For Coal-Water Mixture
o Hard Ceramic For Burner Head
o Surface Coating By Ion Sputtering
o Ceramic/Metal Brasing For Fusion Reactors
o High Reliability Communication Protocol
o Neutron Irradiation For Semiconductor Doping
o Liquid Helium
o Computer Code For Boiler Modeling
o Signature Analysis For Turbine Bearings

---

know-how needed to develop an expert diagnostics system for monitoring the "health" status of a mechanical trolley.

Table 16.5 provides some examples of lateral technologies that have been employed in recent product developments at Ansaldo. This partial list makes it clear that specific knowledge necessary for certain recent improvements to classic Ansaldo products covers a very wide field of technologies and that the latter also change during the progressive development (in small steps) of the product.

We are speaking about technologies quite remote from our core technologies, which concentrate on the burning of various types of fuel, on the production of steam, on its utilization in steam turbines to produce mechanical power, and on the final generation, transportation and use of electrical energy. These key lateral technologies, which come new to Ansaldo, are essential, indeed more and more so, to each product improvement that we plan. The "knowledge package" that we must acquire in order to develop each new idea contains large portions of elementary knowledge components belonging to these lateral technologies and for which internal resources are poorly equipped, if at all. In fact, our in-house R&D skills are limited to the core technologies, where we possess a considerable knowledge base consisting of both existing know-how and R&D tools and expertise for the development of new know-how when needed. But to extend these skills to in-house production of information on lateral technologies would mean spending time and money to establish knowledge producing structures geared to one innovative project and which would then remain idle and become a li-

TABLE 16.6
Steps in a Typical R&D Activity
(From Lab Point of View)

1. Analyze knowledge demand.
2. Design and build experimental tools (new lab or test rig).
3. Hire and train research personnel and form team.
4. Design and build specific test section.
5. Calibrate specific experiments and run test.
6. Interpret results and prepare test report.
7. Express knowledge in required transferable form.

ability instead of an asset. In practice these idle knowledge producing structures will tend to protect themselves and insist on producing unessential knowledge. They will also look for problems to solve instead of following the opposite approach (technical push versus market pull).

A far more rational way to acquire these new technologies is to invest in the perception of available outside knowledge and of outside knowledge producing structures. Outside opportunities can then be integrated into the company's knowledge acquisition process where appropriate.

To get a better understanding of the merits of the in-house and outside knowledge acquisition options, we should first analyze the sequence of actions involved in developing a new knowledge component in an R&D lab, either in-house or outside. Table 16.6 shows the steps in a typical R&D project. This table has been prepared from the laboratory point of view. The deliverable is what we previously called the knowledge component. This may come in a wide variety of forms and content, which can also change considerably from case to case. It might be a simple number or a correlation of two variables in a plot, a technical recommendation, a process flow-chart, or a mathematical code derived from multiple test results. Table 16.7 provides examples of deliverables from electromechanical industry R&D labs.

All these knowledge components have been obtained from outside the company, sometimes in the form of already existing data available in the literature, sometimes through consulting experts or commissioning research from a laboratory that has the proper tools and experienced experimentation personnel.

If we compare this investigating of outside sources of knowledge with the traditional method, which gives priority to in-house production of knowledge (the seven steps in Table 16.6), we can get some idea of the savings (both in time and money) to be made by creating a structure to

TABLE 16.7

Some Examples of Deliverables from an R&D Laboratory

1. Erosion rate of coal powder on Inconel.
2. Thermal shock resistance of copper/graphite joints.
3. Nil ductility temperature of copper.
4. Hardness of laser induced surface alloying.
5. Magnetic properties of Fe-Bo permanent magnets.
6. Precision of electro-ceramic sensors.
7. Critical current in a high temp. superconductor.
8. Conversion efficiency of a solid cell.
9. Shock resistance of an electronic circuit.
10. Performance of a plasma torch on chromium compounds.
11. Electrical properties of an insulator at low temperatures.

manage external knowledge channels instead of limiting yourself exclusively to in-house R&D.

Cost and time benefits that industry can obtain by using external technologies depend on two things: the efficiency of those industries' "Perception Systems" and the temporal and technical matching between what industry wants at a given moment and what is currently available in the relevant technological environments. The first requirement is easily understood: to choose between external possibilities and gather information (or use information sources) one must know where and how to search. The search process must obviously be as quick and cheap as possible. This is what we mean by the efficiency of an industrial perception system. As we will see, such systems depend on industry's level of investment in the necessary tools and training for analysis and assessment of technological information and laboratory tools available outside.

The second requirement is more complex and varies substantially according to circumstances. Let's start with a "fortuitous" case. To develop a superconductive magnet it was essential to know the mechanical properties of a new insulating plastic at 1.8°K (above absolute zero). It was the exceptionally low temperature that made the search for the mechanical properties so difficult, both in the technical literature and in the documentation of the company that manufactures the plastic. It therefore seemed essential to plan ad hoc laboratory experimentation activities to obtain the required data. This in-house development would have cost hundreds of thousands of dollars and set the project back nearly a year. Our labs, not being equipped for measuring mechanical properties at such low temperatures, would require all seven steps in Table 16.6 to have been necessary.

In this event, we were able to save time and money thanks to the determination of a researcher who had the idea that the data we needed might also be relevant to the aerospace industry. A careful search through the European Space Agency data bank led us to a German laboratory where we were able to obtain the required data in a matter of days. This was, as I said, a fortuitous case. What happens rather more often is that the available information is not exactly what we want. It can, however, be an excellent point of departure for achieving our aims.

Another example will serve to illustrate a more complex mechanism that has to be employed in these situations. For the cooling of power superconductors, which are used to control the speed of the motors in underground trains, we had decided to use freon heat exchangers, an indirect system, instead of direct air. To develop this innovative project we had to obtain graphs of the heat transfer coefficients against freon velocity and temperature. As in the previous example, our labs were not equipped for this, so "in-house" development would have meant building a new experimental capability and training a couple of research workers in a highly specialized field—researchers for whom there would be no other tasks after the one in hand. This would have meant building an experimental structure destined to inactivity. On the other hand, the data would have been difficult to find in the required form because the freon we were interested in had only just come on the market. It was, in fact, a new organic fluid, with improved safety and environmental characteristics, developed to substitute traditional freons. Searching directly in the data banks rapidly proved useless. The data banks did, however, reveal the address and telephone numbers of laboratories that had done research into the heat transfer coefficients of other freon based cooling fluids. At this point enquiry by telephone led us to a French laboratory which was conducting research on exactly the freon we were interested in. We visited the lab, saw the high quality of its equipment and staff and reached an agreement with them that enabled us to obtain our data much quicker and more cheaply than we could have done through in-house R&D activities.

In the first example—which I called a "fortuitous case"— the external resource was actually the "knowledge component" we needed and we were able to acquire and use it immediately. In the second example we were dealing with a structure equipped to produce the required "knowledge component." Matching Ansaldo R&D project requirements with what is available in the outside technological environment is usually *even more complicated* than either of these examples.

An innovative project involving above all the technologies that we earlier called "lateral" usually entails a mix of situations in which a wide variety of knowledge acquisition mechanisms are used. To give an idea of this variety of situations I will cite a third case, which will illustrate just some of the possible complexities.

One of our basic products is the oil-fired boiler, which burns oil to produce steam to operate steam turbines which generate electricity. Development of these boilers over recent years has been influenced by environmental protection considerations and has therefore involved using all the most advanced technologies both for optimizing combustion and for reducing the environmental impact of flue gases.

Automation technologies have become so important to competing boiler manufacturers that a whole series of sophisticated improvements now depend almost exclusively on our capacity to acquire technological knowledge across a wide range of "lateral technologies" (of which our experience is limited) and to integrate them in our "core technologies." In this way we satisfy our customers' demands and face the continuing challenge from our formidable competitors, especially the Japanese.

In a recent Product Upgrading project the idea on which the innovation was based required the development of a sensor which could "see" the burner's flames so that the preset optimization criteria could be checked and, if necessary, corrected.

Initial analysis identified three technologies relevant to the project and showed that the use of outside information was indispensable. The required knowledge regarded the maximum working temperatures of the optical fibers and the digital video camera and also the software to process signals from the video camera. It goes almost without saying that these technologies are far removed from the fields in which boiler designers and manufacturers work. We were thus dealing with technologies that were typically "lateral" for Ansaldo but which were essential to the innovative project and to our survival on the world boiler market.

Once again it didn't seem sensible to create specific laboratories and research capabilities within Ansaldo, given the extremely low work load such structures would have after completion of the Project. So we decided to give priority, at least in the initial phase, to searching outside for the necessary information. The analyses of outside possibilities relevant to the three knowledge components in question differed considerably and deserve brief illustration.

The first, regarding optical fiber temperatures, had seemed the simplest, but turned out to be the most complicated. No information could be found in the conventional literature, neither of optical fiber manufacturers (at least the European ones) nor of the major customers. No one in fact seemed interested in studying what happens to optical fibers at over 200°C. One data bank, however, did give us the address of a Norwegian laboratory that had studied the instrumentation of certain geothermal wells in New Zealand. In the laboratory we found an expert who became our consultant and quickly solved the problem.

The second knowledge component—concerning the coupling between the optical fibers and the video camera—was also found through a data bank, this time in a small United Kingdom company that produces endoscopes for visual analysis of the stomach. The idea of searching in this direction came from one of our researchers, who is related to a doctor of medicine.

The third knowledge component—that of the image-processing software—was where we had the least experience. Various programs were needed, above all to produce a stable image (a flame, as we know, is in intense movement) and to use the stabilized image of the flame to obtain temperature maps for comparison with the reference temperature profiles.

We immediately thought of searching the data banks of European laboratories that use video cameras as sensorial elements in industrial robots. This search, however, proved fruitless. Success came from a lab which used video cameras to read addresses in an automated mail handling system. Part of the software was thus found in a postal automation lab. Another program highly specific to our project had to be developed by a local software house, which also tailored the rest of the software we had found to the project requirements. These examples show how using external resources can bring cost and time benefits and also avoid the creation of in-house labs which no one knows how to use afterwards.

We will now take a closer look at what it means to implement a knowledge methodology (at least on the basis of our experience at Ansaldo) instead of relying exclusively on in-house R&D. It means, among other things, using the growing and increasingly available quantity of technical information outside the company. We will also see the numerous barriers that tend to slow down the development of this important modification to the process of knowledge acquisition in industry—a modification that is increasingly essential for keeping our products competitive.

One thing does emerge clearly from the few examples I've given: that greater use of technical information available outside *also requires a high degree of in-house creativity*. Knowing what to look for entails considerable inventiveness and up-to-date technological culture. As we have seen, the knowledge useful to our innovative product may have been developed in a lab working in a sector very different from Ansaldo's and on an application without any apparent connection with our innovative project.

We must understand these correlations and stimulate the curiosity of research workers and of industry at large in what is happening outside, even in remote sectors. To search in data banks we have to be able to formulate the right questions to ask the computer. Often the answer we get has to be integrated by consultants and experts and joint research activities and, finally, may also need to be completed by in-house research laboratory activities.

2   From In-House R&D to Multi-Channel Knowledge Acquisition

In assessing the various alternatives for improving the marketing of innovations produced by R&D, people at Ansaldo are aware that conventional industrial knowledge acquisition (based on in-house R&D) needs to be replaced by a more effective methodology. We are still relying too much on the intramural production of knowledge and are not paying enough attention to outside sources of information, which can cut the costs of innovation projects and help get the product on the market quicker. Very different is the approach of our Japanese competitors, who have been applying *multi-channel knowledge acquisition* (MKA) for years.

The multi-channel approach deserves some explanation. The process of acquiring required industrial technical knowledge was once performed mainly through in-house R&D activities. Recently this principal channel was backed up by a second one, the purchasing of technology in licensing agreements, know-how supply, acquisition of High Tech companies, and contracting out of research.

Joint development programs with various outside organizations (customers, suppliers, adjacent industries, and research establishments) are becoming a third channel for an in-flow of technical knowledge.

With the growth of available information produced in universities and other publicly funded research establishments, we have yet another useful channel for gathering knowledge (see Table 16.8).

In essence what was once managed by in-house R&D alone, is now obtained through a much more complicated, multi-channel knowledge acquisition system. In fact, in order to reduce costs and accelerate development we must optimize all these channels and in particular the distribution of knowledge in-flow.

To achieve this we must appreciate the need to have both a *Modern Perception System* to monitor outside technical opportunities and

TABLE 16.8
Knowledge Acquisition Channels

Make	In-house R&D Activities Innovative Design
Buy	Through Contracting: o K-H Acquisition (Full Rights) o License Agreements (Right of Use) o Acquisition of High Tech Companies o Outside Production of K-H
Cooperate	Through Joint Development Projects with: o Customers o Suppliers o Complementary Industries o Competitors o Research Institutions
Gather	Free Technological Information from: o University and Public Labs o Technical Literature o Data Banks o Operating Products o Inside the Industrial Group

one manager in charge of all knowledge acquisition channels. We should move from the R&D manager concentrated on in-house skills to a manager open to outside opportunities.

This is no small change, as it runs against the widespread conviction that proprietary information must be developed in-house. Commercial successes have been scored using outside information that was available to our competitors as well. Experience also shows that sufficiently competitive proprietary skills can be developed even when a large portion of the basic technical knowledge is acquired outside.

Putting off this evolution gives an advantage to our competitors, who have understood its importance.

The main reasons for adopting the new approach may be summarized as follows:

1. There is a lot of technical knowledge available, just waiting to be picked up and used. For instance, the available material relevant to an electrics/electronics company like Ansaldo is estimated at twenty million pages (see Table 16.9).
2. As we have already seen, in each new product or process there is a growth of so-called "lateral technologies." These require knowledge, expertise, and R&D tooling quite different from those

TABLE 16.9
Volume of Published Technical Information
Relevant for Ansaldo

1988	
o Computer and Control	65,000 Entries/Yr.
o Electric and Electronic	79,000 Entries/Yr.
o Physics (Solid. Liquid, Gas, Plasma)	145,000 Entries/Yr.
o Outside Production of K-H	
TOTAL	289,000 Entries/Yr.

EQUIVALENT TO 2 MILLION PAGES OF FULL TEXT EACH YEAR

(From Dialog Data Banks)

employed in the core technologies and for which in-house resources
are therefore inadequate.

3. New information technologies (data banks, on-line information services, and technical journals on CD-ROM) make it profitable to
   scan a wide spectrum of the available outside technologies and
   select useful data in the required form.

The need to adopt MKA urgently is implicit in the three points above.
We can restate them, however, in a form that speaks more directly to
the user company:

o Point one means there is a high probability of finding what you
  need outside.

o Point two indicates that our search should concentrate on our requirements in the "lateral technologies," where the ratio between
  outside opportunities and internal skills is maximum.

o Point three emphasizes that the search will get easier and easier
  and that investment in outside perception will therefore pay high
  dividends.

A change in the mentality of top management and R&D people would,
of course, only be a first step, though far from easy. The actual implementation of a Multi-Channel System will present several critical problems, including the creation of a good Perception System and the breakdown of know-how packages into elementary questions understandable
by electronic data banks. Another critical point is the attitude (very
common in Europe) of putting too much emphasis on the production of
knowledge by in-house R&D.

By placing R&D at the center of industrial innovation we were mistaking ways and means and oversimplifying the complexity of the inno-

vation process. The means of industrial innovation is the *application of a new information package* in order to put profitable products on the market. The required know-how, therefore, is just a tool and as such subject to the usual make or buy choice, or to be more precise, make, buy, cooperate, or gather. A lot of information is in fact available in university and national research laboratories. To bring home the required know-how, part of which may be available outside, we must invest in perception. This outside information costs much less than in-house R&D production. *Nevertheless, it is not free.*

A new information package for a new product can be *assembled* using pieces of information from various sources. Some may be available inside, some outside. Others may be commissioned from a university or bought. Certain activities may be carried out by in-house R&D to complete the package and to assemble the pieces acquired into a harmonious system. The Japanese success in the development and commercialization of videotapes is a good example of this more effective methodology.

My opinion is that *in-house R&D, even if properly managed, is not enough.* It must be considered as just one channel in a more complex industrial knowledge acquisition system: the integration of many parallel channels. Naturally, the point of all this is to open up the development structure, and the mentality behind it, to the technological world outside—which means investing in the appropriate perception systems.

To do this we had to redirect the creativity and ideas of R&D people towards the formation of know-how packages, optimal interaction of the various information channels and the development of the global innovation plan. In fact, *considerable delays and unforeseen costs* often derive from improper forecasting of the activities that follow up knowledge acquisition. Our former emphasis on in-house R&D made us miss this point. The assumption (wrong) was that once you had good ideas and a good R&D structure, success was within easy reach. Most failures, on the other hand, are due to *bad assessment of internal production tooling* and of the market's real needs. Lack of perception has imposed the activation of feedback loops and repetition of work that could easily have been avoided. More creative resources had to be initially devoted to the perception of key information and to proper overall plans.

3 An Industrial Perception System

I think it is clear at this point what we have been aiming at at Ansaldo in our efforts to cope with the growth of lateral technologies in our products. We have been trying to balance our limited internal skills in the

TABLE 16.10
Sources of External Knowledge

1. University Laboratories
2. National Research Centers
3. Science Parks
4. High Tech Companies
5. User Laboratories
6. Supplier Laboratories
7. Complementary Industries
8. Users and Operating Products
9. Technical Journals
10. Patent Publications
11. Databank Networks

new technical disciplines with a more active search into the opportunities offered by the outside world. An acceptable solution to the problem is still a long way off. This is because we still have to overcome barriers in the form of the attitudes of R&D people (not easy to change) and to master search techniques.

On the other hand, the pressure to translate a new idea as fast as possible into a product upgrade to sell to the customer imposes such short times for knowledge acquisition that the option of in-house development often has to be discarded. In fact, there are so many useful, external sources of technical information that answer the company's needs, that one is naturally induced to exploit this vast heritage of knowledge. Table 16.10 lists the main external sources of technological knowledge. The usefulness of this external knowledge is so crucial to the innovative process in Ansaldo that MKA methodology is now being applied and developed, in spite of the various barriers to overcome and the problems to solve.

However, to implement MKA a vital requirement must first be met: an industrial perception system. This means first persuading top management to allocate resources and wait for the results, which will not be immediately forthcoming. In fact, rapid dynamics in lateral technologies and in the market are justified and in this connection those responsible for industrial innovation often express concern. Ask them to invest in a perception system, however, and they will ask you why.

For me, since my technical and research background is in Control Engineering, the answer is quite simple. If you accept the idea that the Management of an Innovation Project can be considered the Control System of a Production Process, you will agree that the system must have all the necessary sensors, as well as a fast operating structure.

Investment in industrial perception systems is therefore the prime requisite for a fast reaction. In the words of P. A. Bode, "Without fast sensors, even the most advanced and rapid control system will never be able to follow a fluctuating variable." To this we might add that "without good sensing capabilities, even the most skilled and quick-witted human manager will never be able to follow a fluctuating market or technology."

Unfortunately our top management will often ask our innovation project leaders to perform fast, without providing them the means to build the necessary perception systems. In my personal contacts with many Japanese companies I've noted their surprise when I talk about this problem, as if I'd said you must keep your eyes open when you walk over uneven ground. They smile with a courteous bow. The Japanese are masters of industrial perception systems and moving information around because they have been investing for years in all aspects of this problem. In Europe we are supposed to have a lot of creativity and intuition, but good ideas do not provide a substitute for information. As Oscar Wilde said, even "intuition thrives upon information." Naturally, there are many sources of information that need to be monitored, selected and transferred to various people and which relate to many technical decisions.

We all know that most good ideas come from the customers and from the performance of operating products, that many repetitive development activities come from a lack of perception of production tools. We also know that many failures derive from mistakes in the selection of suppliers and that most of the specific knowledge that we need exists somewhere outside the company. Anyhow, no matter what source you start from, your Perception System will cause problems at the beginning because there will be little immediate return on investment. At some subsequent point the system will look a complete mess. Eventually it will be properly focused and you will start "to see."

With regard to the primary objective, "Outside Scientific and Technical Opportunities," for Ansaldo these opportunities could be the following:

o available information, for example, a new correlation of heat transfer, material properties, a mathematical model or a new laboratory process, or information available on software which can be transferred via an electronic information network;

o new hardware, such as a new sensor, a sample of a new material or chemical compound, or a new type of high-temperature super-conductor. It could also be something available elsewhere which can be sent by post;

o a person or group of people skilled in a specific field, either offering phone-in advisory services or research work under contract; or

o a research tool needed in a specific phase of industrial research, for example, a sintering furnace, an ion implanter, a heat transfer rig, measuring or testing equipment. Something already existing and for which skilled operators are usually available.

The whole industrial innovation procedure would start with an initial assessment, followed by the design phase, when the problem is broken down into fundamental areas. The know-how required is then divided into two parts, *what the company already possesses* (an old design or existing project, a computer code, previous research results) and *what must be acquired*. At this point, the latter is defined in more detail and decisions made on *how to acquire it*.

In-house R&D should always be compared with other solutions, such as buying in the technology from outside. This can be done in various ways: you can sign a license agreement with a company owning a similar product and operating in a different market; you can buy a company owning a particular technology; or you can pay to enter an R&D program already started up by another company. There are therefore two basic ways of obtaining technological know-how: *make it or buy it*.

The problem then remains of who decides and on the basis of what information. Technological know-how is not the easiest product to eval-uate, and must often be integrated with other knowledge to adapt it to specific requirements. Common sense indicates that the decision should be taken by someone who knows what is *available outside* and is open-minded about both solutions. This is not always the case at Ansaldo. How can the overall knowledge acquisition process be optimized when two different people are in charge of the two different methods? All that will happen is that two positions will be taken up: "Not invented here" and "Useless research," both unproductive and leading nowhere except to a waste of everybody's time.

It seems timely to note at this point that our Japanese competitors have solved this problem very effectively, using simple consensus. I re-cently attended a meeting concerning a joint venture with a Japanese firm, at which the make or buy problem was dealt with. This proved to be a most interesting experience. The decision-making process itself was slow, but followed a very precise pattern and, most important of all, was greatly enhanced by an extremely efficient information system. At one point in the proceedings, I suggested a partial solution requiring experiments in a heat transfer facility which we have available. The Japanese went on to discuss this without checking on the screen. At the end, I asked them if the facility had been recorded in their data bank, and was shown that it had been, together with its source: an Ansaldo publication from 1986. A note from a visit last June reported that the facility was still in operation and that skilled personnel were available.

Returning to the problem of knowledge acquisition, I believe there are not two, but four, ways: *making*, *buying*, *cooperating*, and *gathering*. The fourth, gathering, is becoming more and more important, since a large amount of public funds are now being allocated to finance research activities producing information which is open to everybody, and which generates expertise and low-cost research tools.

An analysis of the European GERD in the five years 1987–91 shows that 184 billion ECU's are allocated to the building of R&D tools and the generation of information open to all. Industrial enterprises taking up these opportunities in their innovation projects are going to have a clear advantage over those relying solely on in-house R&D. However, to be able to benefit from this information, you must know what is available and how to get hold of it, and for this you must invest in specific perception activity concerning both existing knowledge and knowledge producing tools.

This approach is perhaps best summed up in the words of Soichiro Honda: "Before you begin any development, you must go out and see what is available, you must touch, you must feel, because your sensors are vital." "*How much you see and how much you touch*" should be measured in the future and should become an index to correlate with the many others used by economists to evaluate the various attitudes towards innovation.

Even electromechanical industries, which still tend to rely on in-house R&D, often require extensive bibliographical search to be presented before starting a new research program. Most research organizations subscribe to the major technical journals, and once a year top scientists participate in an international meeting where they may learn of new ideas or discuss their own with other scientists. Industrial scientists spend their careers building up their own personal data banks on laboratories similar to their own and on where to find the principal skills and expertise. But when the number of journals reaches 1,500, as in the case of Siemens, and the number of new papers being published annually is around 250,000, *this system becomes impractical*. Useful information for a typical electromechanical company could be contained in over ten to twenty million pages (see Table 16.9).

Fortunately, this exponential growth in the availability of technical know-how has been accompanied by a corresponding one in *Information Technology*. This parallel growth means that a perception system which takes full account of the developments in information technology is conceivable. The main elements of this system are: an internal data bank; a communications network; I/O terminals and a network of inputs.

At Ansaldo, we don't feel there is much point in making large investments to gather external information unless a system for storing it and distributing it to all potential Ansaldo users is firmly in place. We had to progress from a researchers' data bank to a company-oriented one. Fortunately, at Ansaldo we had an internal information system which, with a few improvements, made it possible to put a Company Technology Perception System into operation.

It should not, however, be forgotten that there are some specific requirements. To begin with, the users are the research and development people, and a search is normally performed, on average, only a few times a year. To cite some examples: Hitachi claim that they average twenty searches per year per scientist, while Bayer (most of the chemical and pharmaceutical companies are among the European leaders in this) do fifteen a year and Siemens eight (Siemens also gave the rate of increase in on-line searches, which is 25 percent p.a.). Occasional users must therefore be provided with *intelligent* terminals and a user-friendly interface to the data bank, and eventually with intelligent selective access and *proprietary* software with automatic searching.

Our early experiences in this at Ansaldo were hindered by the complexity of the search process. We assumed that this difficulty could be overcome with training, but this proved not to be the case, and only the introduction of an easy access system solved the problem. An essential improvement is that of keeping not only the author's name and abstract of the technical paper in the data bank, but the entire text as well. This requires a *full page text and graphics communication protocol*. The famous Siemens library is still operational and sends a page of text by fax a few hours after receiving the request for it.

4 Network of Inputs

Once an internal data bank exists, a network of inputs can be activated. The most common inputs are:

o Commercial information data banks, for example: EEC Information Market, Dialog, Inspec, Pergamon Info Line, Chemical Abstract, GE Information Services, which already contain a large amount of data accessible by phone. In the last EEC study (*Information Market*, no. 56), it is reported that: "three out of four users are now logging on automatically to their host computer." Over 84 percent of European users of these Technology Information Services are expanding their internal networks and an average increase in searches of thirteen percent p.a. is expected.

o Full texts of scientific and technology journals and proceedings. Unfortunately, at present only a very small percentage is sold in electronic form. However, the recent decision by IEEE to publish all its journals in CD-ROM may lead other important publishers to do the same. Philips, who still have to search for texts in their library, photocopy, and fax them to users, consider this to be an *important breakthrough*. At present, only 50 of the 1,500 technical journals considered important by Siemens are available in the Dialog database.

o Information obtained from university professors, with whom it can be very useful to build up a good network. In Europe Alfa Laval relies almost entirely on this system, and over the years the company has established a network of 30 professors in Europe, the United States, and Japan, each of whom has been given a specific monitoring field on which to make a monthly report. New ad hoc consulting contracts can be added to the standard ones, and names of experts can easily be obtained from the data bank abstracts.

o At the beginning of the exploration phase special monitoring of new technologies by institutions, such as the U.K. Innovation Group, Battelle in Geneva, Futurtech, or the European Institute of Technology in Paris. Demand for high level advisory services is increasing, since consultants are considered to have their finger on the pulse with regard to technical publications.

Other sources, too numerous to mention here, also exist, such as a patent index, the various venture capital organizations (messy but effective), EEC programs and results, public funding of industrial R&D contracts.

All these channels of information are useful for an initial screening of the situation. They also serve to direct further detailed searching in order to acquire specific information (if it exists) or to find the tools to produce it. In an efficient industrial perception system, the first stage

TABLE 16.11

Elements in an Industrial Perception System

---
o Internal databank for text and graphics.

o Distribution of *friendly* I/O terminals.

o Subscription to external databanks and information update systems.

o Automatic search procedures.

o Widespread training.

---

may cost $100 per search and last only a day, whereas the second may cost up to $10,000 and take up to a month. After these initial phases, the cost and time taken will depend on a variety of factors difficult to generalize about. The investment always carries a high risk, but it can also bring high dividends. Experience and internal skills are clearly a deciding factor in all this. Information may be available in many forms (papers, drawings, detailed reports) of varying degrees of reliability. The decision on where to focus, after the initial screening phase, is often dictated by considerations which do not appear in public data banks.

At this point, I would like to summarize the Ansaldo experience in the formation of an industrial perception system by underlining two important factors: (1) the specific investment of time, money and skilled personnel required to organize the system hardware (including the elements mentioned in Table 16.11); (2) the specific investment required to master use of this hardware, in order to integrate external opportunities with the internal innovation structure; the important phases for this are listed in Table 16.12.

Overall, our experience has been interesting and has provided evidence of reduced time and development costs in certain cases. It is still too early to state with absolute certainty that the multi-channel knowledge acquisition system offers net advantages over the single in-house R&D channel, but evidence is building up.

What is undeniable is that this wider approach, so efficiently used by our Japanese friends, makes it possible to expand both the internal know-how base and the internal know-how production structure to include a part of the much more important external factors. The cost/benefit ratio depends very much on how this is done and for which specific projects.

An evaluation of these developments might be of particular interest to an economist, particularly in view of the current development of full text technical papers in electronic form and subsequently their accessibility via data banks.

TABLE 16.12
Mastering Search Techniques

o Breakdown of product know-how package.
o Define elementary knowledge components.
o Direct automatic searching according to past experience.
o Use multiple techniques (databanks consulting network, information services, etc.).
o Develop experience with proprietary software.

The hardest barrier to overcome in this was changing the mentality of those directly concerned, particularly when it came to convincing them that they could rely on information also available to our competitors.

There is a parallel with problems the industry had to solve years ago in switching from products with in-house manufactured mechanical components to new products containing components acquired externally.

Looking through Ansaldo's archives, I discovered a letter sent by the Chief Engineer to the Company Chairman in 1906, complaining about the decision to use valves bought from a supplier (i.e., not manufactured in-house) on a new locomotive. To quote his words: "This decision will have very negative effects, because the same valves will immediately be used by our competitors." Fortunately, his doubts proved unfounded, since Ansaldo has continued to be competitive in its production.

In 1906, the components on Ansaldo's steam locomotives were manufactured almost entirely in-house (even the bolts and steel plates). Now, over 90 percent of the components on one of our most up-to-date electronic locomotives are acquired externally, that is, they can also be acquired by our competitors. This does not in any way represent a problem for new and innovative products.

The same principle applies to knowledge components, but, as was the case 84 years ago, the hardest task of all is to change the attitude of those who only trust what is produced in-house. It becomes even harder when you require unified management of acquisition channels that used to be beyond the individual responsibility of R&D managers (e.g., license agreements and the acquisition of small high-tech companies).

Japanese companies usually solve this problem by using a steering committee which operates on consensus, but their real strong point is their industrial perception system. They cannot understand why people as inquisitive as the Europeans have so far failed to create sufficiently inquisitive industrial structures.

5 Public Funding Policy

Before concluding this presentation, I would like to say a few words on a matter which seems to go against common sense, that is, *the low public investment in spreading the results of publicly funded* R&D to industrial users. This lack of public investment in such an important area, like the lack of industrial investment in perception, comes as somewhat of a surprise.

Mother Nature herself understood how important this is: strawberries are red to make them easily distinguishable against the green background of their leaves. Yet of the 184 billion ECU's allocated to the generation of new know-how and research tools, only 90 million that is, five in a thousand, is to be spent on the spreading of this information. This figure is clearly not high enough. At least one percent of this will be needed to improve existing data banks, to include full texts, to classify information by technology and application, and to include viable ideas which have not received funding, or funded research that has failed. Fortunately, there are signs that the situation may improve over the next few years, in particular at EEC level.

EIRMA has offered European governments several suggestions on what is needed to improve the commercialization of R&D through higher visibility of its results and applications. IRDAC (Industrial R&D Advisory Committee) has also recently suggested to the European Commission that more attention must be paid to making existing technical know-how available, particularly to smaller companies.

6 Conclusions

Commercialization of R&D results is not one of Ansaldo's strong points. This is partly due to the way in which development project objectives are selected and the methods used to achieve them:

o *Objectives* often give priority to novelty value, rather than to meeting production requirements and real market needs;

o *Methods* chosen to achieve the objective often give priority to *in-house production*, instead of acquiring know-how and concentrating on its *in-house application*.

A closer examination of industrial innovation demonstrates that the reasons for poor commercialization often take root in the initial project phases, when objectives and plans of action are established. In order to make improvements to the final phase, a careful examination must first be made of the initial phase.

Costs in this phase are low, and choices are endless, whereas the final stage is expensive, and the constraints are numerous and *for the most part determined at the start*. For example, costs involved in perception of the *required decision elements* are low in proportion to the entire project cost, yet these early decisions can be fundamental in significantly reducing times and costs and in setting objectives which will aid commercialization. With just a little extra investment in the first phase, it would be possible to (1) avoid carrying out expensive R&D which may be aimed at gathering know-how already available at low cost; and (2) reduce reworking when the product reaches the manufacturing stage or the customer.

Higher investment in the initial phases can therefore help to determine commercialization potential and is sure to pay high dividends. This is especially true if investment is aimed at setting-up an Industrial Perception System which provides facts essential to decision-making early on, that is, information on customer requirements or production tools and the *available technical know-how*.

Lack of adequate perception can prove to be particularly costly when the parameters involved are subject to rapid variation, which is precisely the case with the following:

○ Production tooling is in a state of continuous evolution, due to the introduction of new electronic technologies such as CAD, CAE, CIM, and FMS.

○ Market forces are constantly changing.

○ *Available knowledge* is produced at an annual cost of more than $100 billion in OECD countries.

There are valid reasons for closely analyzing the importance of monitoring all three types of information, and I have chosen to concentrate on the third simply because it is the one which up to now has received the least attention, despite the fact that a growth in the amount of available technical know-how is an important asset to industrial innovation.

Perception of available external know-how opens up a *multi-channel process of industrial knowledge acquisition* which compares highly favorably with the traditional method of acquiring know-how through in-house R&D activities. Besides reducing the time and costs needed for industrial developments, it can offer more general benefits which merit close attention.

It is frequently stated that Western societies are at an advantage in this due to the *enormous asset* of free availability of scientific and technical know-how developed with public funding in universities and national laboratories. It is also frequently emphasized that (1) our scientific know-how must *also* support our economic growth; and (2) industrial innovation in all its forms, that is, development, design, manufacturing, etc., must draw on available research. These are merely statements, with which most of us are in full agreement. Unfortunately they are rarely followed by the necessary action, either by industry or the public sector.

What precisely do the words "enormous asset" actually mean? Each year, over ten million pages of journals, conference proceedings, technical books, and reports are printed. This information on technical data, expertise sources, and research tools is naturally of great importance and interest to industry.

So it is an "enormous asset" in the sense that it offers a very high probability of finding what is needed in terms of know-how acquisition. However, the search is not easy. For it to be an asset, a company must therefore, in economic terms, make considerable investments in perception, which will then produce important economic returns.

The same reasoning can be applied on the public side, since scientific know-how can play an important role in promoting economic growth. However, investment in research by itself is not enough—additional investment in *improving the visibility of scientific results* for the benefit of industrial users must also be made.

In this new trend, priorities should be established on both sides with an aim to promoting scientific disciplines of wider industrial interest, such as the study of sensors, electronics, automation, mathematical models, advanced materials, surface technologies, etc.

As has already been mentioned, besides encouraging industrial innovation, perception of outside know-how also goes a long way towards changing the attitude of the R&D people to the outside world, particularly with regard to (1) perception of real market needs, which can provide useful information on competitors and opportunities offered by suppliers; and (2) perception of in-house production needs, which should be taken into consideration right from the very beginning, in order to avoid costly and time-consuming redevelopment activities.

To sum up, investments a company makes in perception of outside technologies (and on other relevant information) may become an important index in the evaluation of a company's innovative capability. Similarly, public investment in improving visibility of scientific results could become an index for evaluating genuine public commitment towards the use of science in promoting economic growth.

I have tried to give you a general idea of innovation at Ansaldo and the actions I consider necessary if commercialization of R&D results is to be improved. Basically, most ideas stem from market needs, and the *internal application of new know-how* is essential to the survival of any industrial enterprise. However, *internal production of know-how* is not essential; a company must simply find the most appropriate method of obtaining the technological information required. *For industry, know-how is a means to an end*, so to maintain profitability, we must compete in the most innovative use of technological opportunities. Their origin, whether "in-house" or external, is not relevant.

Science must remain free to flow to wherever there is someone who can make good use of it for the benefit of society as a whole. But scientific results do not appear by themselves—they must be searched for and gathered.

I cannot at this point provide the statistics which support what I have written. I do hope, however, that what I have written will encourage some economists or others to make a professional study of this in the not too distant future. Whatever happens, I believe that simply monitoring R&D expenditures and in-house R&D management capabilities is not sufficient for a correct evaluation of industrial innovative potential. Other investments, above all in the perception of outside technological opportunities, must be given serious consideration in the future.

While the process of acquiring technical know-how at Ansaldo in the past was done almost exclusively via in-house R&D activities, this main acquisition channel is now flanked by a second, that of buying technology in its various forms: license agreements, know-how, acquisition of high-tech companies, and the contracting of research activities outside.

Cooperation programs constitute a third possible in-flow of technical know-how. Now, however, a fourth channel has been opened up by the growth in information available from universities and other research organizations and laboratories, funded mainly by public investment, combined with a corresponding development in information technology.

What, in essence, was once carried out via the single channel of in-house R&D is now obtained via a more complicated *Integrated Knowledge Acquisition System*, which has not been easy to master. On the other hand, optimization of all these channels, and the distribution of know-how flow among them, which will reduce costs and speed up development, can only be guaranteed if there is one management body in charge of all of them. The creation of a new post in know-how in-flow management would seem to be an ideal solution, and could be the logical development of the Industrial R&D Manager.

The fields of industrial innovation in which this new mechanism could be put to the best effect are not yet fully defined, and further research is obviously required. I have only been able to evaluate it with regard to industrial innovation projects in the electromechanical sector. The findings leave me firmly convinced that there is still ample room for growth in investments in outside technology research and that this offers excellent dividends. We must strive more than ever to change the attitude of our Research Managers towards an open system of knowledge acquisition. Greater efforts must also be made to transfer the emphasis from *in-house production of know-how* to *in-house commercialization of products*.

Name Index

Subject Index

Library of Congress Cataloging-in-Publication Data

Technology and the Wealth of Nations / edited by Nathan Rosenberg, Ralph
  Landau, and David Mowery.
      p.   cm.
  Includes bibliographical references and index.
  ISBN 0-8047-2082-7
  ISBN 0-8047-2083-5 (pbk)
      1. Technology—Economic aspects. 2. Technological
  innovations—Economic aspects. 3. Research, Industrial—Economic
  aspects. I. Rosenberg, Nathan, 1927– . II. Landau, Ralph. III. Mowery,
  David C.
  HC79.T4T4395 1992
  338′.064—dc20
                                                      92-24181
                                                         CIP